Welcome to the 72nd International Symposium on Molecular Spectroscopy
June 19-23, 2017
Urbana-Champaign, IL

I0500354

On behalf of the Executive Committee, I extend a heartfelt welcome to all the attendees of the 72nd Symposium and welcome you to the University of Illinois at Urbana-Champaign.

The Symposium presents research in fundamental molecular spectroscopy and a wide variety of related fields and applications. The continued vitality and significance of spectroscopy is annually re-affirmed by the number of talks, their variety, and the fact that many are given by students. These presentations are the heart of the meeting and are documented by this Abstract Book. Equally important is the information flowing from informal exchanges and discussions. As organizers, we strive to provide an environment that facilitates both kinds of interactions.

The essence of the meeting lies in the scientific discussions and your personal experiences this week independent of the number of times that you have attended this meeting. It is our sincere hope that you will find this meeting informative and enjoyable both scientifically and personally, whether it is your first or 50th meeting. If we can help to enhance your experience, please do not hesitate to ask the Symposium staff or the Executive Committee.

Ben McCall
Symposium Chair

SCHEDULE OF TALKS

ABSTRACTS

VENUE AND SPONSOR INFORMATION FOLLOWS AUTHOR INDEX

72nd INTERNATIONAL SYMPOSIUM ON MOLECULAR SPECTROSCOPY

International Advisory Committee
Tucker Carrington, Queen's University
Emilio Cocinero, Universidad del País Vasco
Stephen Cooke, Purchase College, SUNY
Gary Douberly, University of Georgia
Shuiming Hu, University of Science and Technology of China
Kaori Kobayashi, University of Toyama
Laurent Margules, Laboratoire de Physique des Lasers,
Atomes et Molécules
Terry Miller, Ohio State University
Takamasa Momose, University of British Columbia
Hiroyuki Sasada, Keio University
Trevor Sears, Brookhaven National Laboratory, Past Chair
Stephan Schlemmer, University of Cologne
Tim Steimle, Arizona State University
Nick Walker, Newcastle University
Mathias Weber, University of Colorado and JILA
Susanna Widicus Weaver, Emory University
Yunjie Xu, University of Alberta, Chair
Shanshan Yu, Jet Propulsion Laboratory

Executive Committee
Ben McCall, Chair
Brian DeMarco
Dana Dlott
Gary Eden
Nick Glumac
Martin Gruebele
So Hirata
Leslie Looney
Josh Vura-Weis
Dave Woon

Please send correspondence to
Ben McCall
International Symposium on Molecular Spectroscopy
Department of Chemistry
600 S. Mathews Avenue
Urbana IL 61801 USA
e-mail: chair@isms.illinois.edu
http://isms.illinois.edu

Mini-Symposia

ALMA'S MOLECULAR VIEW

Organized by **Brett McGuire** (NRAO) and **Anthony Remijan** (NRAO). This mini-symposium will consist of discussions of laboratory, computational, and observational work at any wavelength/facility which support, complement, or directly involve spectroscopic observations with ALMA. Invited Speakers: **Susanne Aalto** (Chalmers University of Technology), **Arnaud Belloche** (Max-Planck-Institut für Radioastronomie), **Michael McCarthy** (Harvard-Smithsonian Center for Astrophysics), **Chunhua "Charlie" Qi** (Harvard-Smithsonian Center for Astrophysics), **Naoki Watanabe** (Hokkaido University)

CHIRALITY-SENSITIVE SPECTROSCOPY

Organized by **Yunjie Xu** (University of Alberta) and **Melanie Schnell** (Max-Planck Institute, Hamburg). This mini-symposium will explore spectroscopic methods that can provide detailed stereochemical information of chiral molecules, such as enantiomeric excess, absolute configuration, and enantiomeric recognition. Invited Speakers: **Rina Dukor** (BioTools, Inc.), **Laurent Nahon** (Synchrotron SOLEIL), **Patrick Vaccaro** (Yale)

MULTIPLE POTENTIAL ENERGY SURFACES

Organized by **Jinjun Liu** (University of Louisville) and **Chris Elles** (University of Kansas). This mini-symposium will focus on spectroscopy and dynamics on multiple potential energy surfaces covering different experimental and computational aspects such as vibronic interactions, conical intersection dynamics, photoisomerization, and photodissociation. Invited Speakers: **Robert W. Field** (MIT), **Roseanne J. Sension** (University of Michigan), **John F. Stanton** (University of Florida)

Picnic (please note day of week!)

The Symposium picnic will be held on **Tuesday evening** at Ikenberry Commons. The cost of the picnic is included in your registration (at below cost to students), so that all may attend the event. The **Coblentz Society** is the host for refreshments for one hour. Please see your packet for additional details.

Sponsorship

We are pleased to acknowledge the many organizations that support the 72nd Symposium. Principal funding comes from the **Army Research Office** (ARO) and the **National Radio Astronomy Observatory** (NRAO). We are most grateful to ARO for their long-standing support. We also acknowledge the many efforts and contributions of **The University of Illinois** in hosting the meeting, including financial contributions from the Departments of Chemistry, Electrical and Computer Engineering, Astronomy, and Physics.

Our Corporate Sponsors are **BrightSpec, Bristol Instruments, Elsevier/JMS, Ideal Vacuum Products, Journal of Physical Chemistry/ACS,** and **Quantel**. Please see the back of this book for their advertisements.

We are also pleased to acknowledge **Bruker, JASCO, M Squared Lasers,** and **Menlo Systems** as Contributing Sponsors.

IOS Press has a special insert in your conferee packet. Our sponsors will have exhibits at the Symposium and we encourage you to visit their displays.

Rao Prize

The three Rao Prizes for the most outstanding student talks at the 2016 meeting will be presented. The winners are **David Grimes**, MIT; **Kasper Mackeprang**, University of Copenhagen; **Josey E. Topolski**, Indiana University Bloomington. The Rao Prize was created by a group of spectroscopists who, as graduate students, benefited from the emphasis on graduate student participation, which has been a unique characteristic of the Symposium. This year three more Rao Prize winners will be selected.

The award is administered by a Prize Committee chaired by Gary Douberly, University of Georgia, and comprised of David Anderson (University of Wyoming); Brooks Pate (University of Virginia); Rebecca Peebles (Eastern Illinois University); Jennifer van Wijngaarden (University of Manitoba); and Tim Zwier, (Purdue University). Any questions or suggestions about the Prize should be addressed to the Committee. Anyone (especially post-docs) willing to serve on a panel of judges should contact Gary Douberly (douberly@uga.edu).

Miller Prize

The Miller Prize was created in honor of Professor Terry A. Miller, who served as chair of the International Symposium on Molecular Spectroscopy from 1992 to 2013. The Miller Prize for the best presentation given by a recent PhD at the 2016 meeting will be presented. The winner, **Marie-Aline Martin-Drumel** (Harvard-Smithsonian Center for Astrophysics), will give a lecture on Thursday.

The Miller Prize winner and his or her co-authors will be invited to submit an article to the Journal of Molecular Spectroscopy based on the research in the prize-winning talk. After passing the normal review process, the article will appear in the Journal with a caption identifying the paper with the talk that received the Miller Prize.

The award is administered by a Prize Committee chaired by Mike Heaven, Emory University and comprised of Lan Cheng (Johns Hopkins University); Richard Dawes (Missouri University of Science and Technology); Helen Leung (Amherst College); Trevor Sears (Stony Brook University); Steve Shipman (New College of Florida); John Stanton (University of Texas at Austin); Tim Steimle (Arizona State University); Susanna Widicus Weaver (Emory University); Shanshan Yu (Jet Propulsion Laboratory). Any questions or suggestions about the Prize should be addressed to the Committee. Anyone willing to serve on a panel of judges should contact Mike Heaven (mheaven@emory.edu).

Information

ACCOMMODATIONS

The check-in for dormitory accommodations is located in Bousfield Hall, 1214 South First Street, opens at noon on Sunday, June 18th, and remains open 24 hours a day through the Symposium. Hotel information is listed on the ISMS website. The desk at Wassaja Hall will not be staffed during the meeting.

PARKING

Parking permits are for lot E14 (see map at end of book). Please purchase parking as part of your check-in process at the dorm. If you need to purchase meter hang-tags for parking near the meeting rooms, you can do so at the registration desk.

REGISTRATION (NEW LOCATION THIS YEAR!)

The registration desk is located in the Chemistry Library in Noyes Lab, and is open on Sunday from 4:00-6:00 PM, and Monday through Friday from 8:00 AM-4:30 PM.   Refreshments will be available from 8:00 AM-4:30 PM.

CHEMISTRY LIBRARY

The Chemistry Library will be the new home for our Registration desk and exhibitor space (plus coffee and donuts) this year. The library has a few small conference rooms, and comfy chairs (and books!).

READY ROOM/STATION

We will have 2 desks in the Library with computers that you can use to test your powerpoint presentation. If you have any problems, the staff at the "Ready Station" (right next to registration) can assist you.

COMPUTER LAB (VizLab)

Noyes Lab 151 is a small computer lab with Apple computers that is available for your use during the meeting. Please look in your packet for an access code to enter the room.   You can also use the PCs in the Chemistry Library.

INTERNET ACCESS/Wi-Fi

Each attendee will receive a login and password to access campus WiFi (SSID: IllinoisNet) as a guest. This access should work in most locations through campus. Please read the Internet Acceptable Use Policy below.

AUDIO/VIDEO INFORMATION

Each session room is equipped with a laptop computer, onto which presentation files will be pre-loaded by Symposium staff. To submit your presentation file, you must go to the **Manage Presentations** link on our web site and follow the instructions. All files must be submitted by **11:59 PM CDT THE DAY BEFORE** your presentation session. All submitted files will be loaded onto the presentation computer one half-hour prior to the beginning of the session.

ACKNOWLEDGMENTS

The Symposium Chair wishes to acknowledge the hard work of numerous people who made this meeting possible. First and foremost is the Symposium Coordinator Birgit McCall, who has smoothly and single-handedly taken care of almost all of the electronic and logistical aspects of the meeting. Second are our symposium assistants, Scott Dubowsky and Charlie Markus, who have handled innumerable important details to ensure the sessions and exhibitions go well. The other students in my group also play vital roles in other aspects of the meeting. I wish to acknowledge the hospitality of the Chemistry Department and the School of Chemical Sciences (as well as the School of Molecular and Cell Biology) in tolerating our takeover of their buildings.

LIABILITY

The Symposium fees DO NOT include provisions for the insurance of participants against personal injuries, sickness, theft, or property damage. Participants and companions are advised to obtain whatever insurance they consider necessary. The Symposium organizing committee, its sponsors, and individual committee members DO NOT assume any responsibility for loss, injury, sickness, or damages to persons or belongings, however caused. The statements and opinions stated during oral presentations or in written abstracts are solely the author's responsibilities and do not necessarily reflect the opinions of the organizers.

DISCLAIMER

The views, opinions, and/or findings contained in this report are those of the authors and should not be construed as an official Department of the Army position, policy, or decision, unless so designated by other documentation.

INTERNET ACCEPTABLE USE POLICY

Each attendee will receive a login and password to access campus WiFi (SSID: IllinoisNet) as a guest. Guest accounts are intended to support a broad range of communications. Professional and appropriate etiquette is required. Anonymous access and posting through guest accounts is forbidden. All users must accept that their identity may be associated with any content they provide while using the service. By accessing the campus WiFi network, you expressly acknowledge and agree to the following:

Use of the guest account service is at your sole risk and the entire risk as to satisfactory quality and performance is with you. You agree not to use the guest account intentionally or unintentionally to violate any applicable local, state, national or international law, including, but not limited to, any regulations having the force of law. To the extent not prohibited by law, in no event shall the university be liable for personal injury, or any incidental, special, indirect or consequential damages whatsoever, including, without limitation, damages for loss of profits, loss of data, business interruption or any other commercial damages or losses, arising out of or related to your use or inability to use the guest account, however caused, regardless of the theory of liability (contract, tort or otherwise) and even if the university has been advised of the possibility of such damages. The use of the guest account is subject, but not limited to, all University policies and regulations detailed at the Campus Administrative Manual (http://www.cam.illinois.edu). See the University's Web Privacy Notice (http://www.vpaa.uillinois.edu/policies/web_privacy.cfm) for all applicable laws and policies.

MA. Plenary
Monday, June 19, 2017 – 8:30 AM
Room: Foellinger Auditorium

Chair: Martin Gruebele, University of Illinois at Urbana-Champaign, Urbana, IL, USA

Welcome   8:30
Peter Schiffer, Vice Chancellor for Research
University of Illinois at Urbana-Champaign

MA01   8:40 – 9:20
MOLECULES FROM CLOUDS TO PLANETS: SWEET RESULTS FROM ALMA, Ewine van Dishoeck

MA02   9:25 – 10:05
INFRARED SPECTROSCOPY OF NEW MOLECULES AND CLUSTERS, Mingfei Zhou

Intermission

MA03   10:40 – 11:20
ELECTRONIC WAVE PACKET INTERFEROMETRY OF GAS PHASE SAMPLES: HIGH RESOLUTION SPECTRA AND COLLECTIVE EFFECTS, Frank Stienkemeier

FLYGARE AWARDS   11:25
Introduction by Trevor Sears, Brookhaven National Laboratory

MA04   11:30 – 11:45
DUAL CRYOGENIC ION TRAP SPECTROMETER FOR SPECTROSCOPY OF COLD ION-MOLECULES COMPLEXES, Etienne Garand

MA05   11:50 – 12:05
HIGH-RESOLUTION LASER SPECTROSCOPY OF FREE RADICALS IN NEARLY DEGENERATE ELECTRONIC STATES, Jinjun Liu

MF. Mini-symposium: ALMA's Molecular View
Monday, June 19, 2017 – 1:45 PM
Room: 274 Medical Sciences Building

Chair: Susanna L. Widicus Weaver, Emory University, Atlanta, GA, USA

MF01   1:45 – 2:00
THE ATACAMA LARGE MILLIMETER/SUBMILLIMETER ARRAY - FROM EARLY SCIENCE TO FULL OPERATIONS., Anthony Remijan

MF02   *INVITED TALK*   2:02 – 2:32
EXPLORING MOLECULAR COMPLEXITY IN THE INTERSTELLAR MEDIUM WITH ALMA , Arnaud Belloche

MF03   2:36 – 2:51
ROTATIONAL SPECTROSCOPY OF THE LOW ENERGY CONFORMER OF 2-METHYLBUTYRONITRILE AND SEARCH FOR IT TOWARD SAGITTARIUS B2(N2), Holger S. P. Müller, Nadine Wehres, Oliver Zingsheim, Frank Lewen, Stephan Schlemmer, Jens-Uwe Grabow, Robin T. Garrod, Arnaud Belloche, Karl M. Menten

MF04   2:53 – 3:08
SUBMILLIMETER WAVE SPECTRUM OF THIOACETALDEHYDE AND ITS SEARCH IN SgrB2, L. Margulès, R. A. Motiyenko, J.-C. Guillemin, Y. Ellinger, Brett A. McGuire, Anthony Remijan

MF05   3:10 – 3:25
THE SURPRISING COMPLEXITY OF DIFFUSE AND TRANSLUCENT CLOUDS TOWARD SGR B2: DIATOMICS AND COMs FROM 4 GHz TO 1.2 THz, Brett A. McGuire, Joanna F. Corby, Marie-Aline Martin-Drumel, P. Schilke, Michael C McCarthy, Anthony Remijan

MF06   3:27 – 3:42
SIO EMISSION FROM THE INNER PC OF SGR A*, Farhad Yusef-Zadeh

Intermission

MF07   4:01 – 4:16
ASTROCHEMICALLY RELEVANT MOLECULES IN THE W-BAND REGION, Benjamin E Arenas, Amanda Steber, Sébastien Gruet, Melanie Schnell

MF08   4:18 – 4:33
ASTRONOMICAL TRIPLETS: ALMA OBSERVATIONS OF $C_2H_4O_2$ ISOMERS IN SGR B2 (N), Ci Xue, Anthony Remijan, Andrew M Burkhardt, Eric Herbst

MF09   4:35 – 4:50
DIMETHYL ETHER BETWEEN 214.6 AND 265.3 GHZ: THE COMPLETE, TEMPERATURE RESOLVED SPECTRUM, James P. McMillan, Christopher F. Neese, Frank C. De Lucia

MF10   4:52 – 5:07
H_2S: A NEW PROBE OF HIDDEN LUMINOSITY IN ORION KL, Maria Kleshcheva, Nathan Crockett, Geoffrey Blake

MF11   5:09 – 5:24
TRACING THE ORIGINS OF NITROGEN BEARING ORGANICS TOWARD ORION KL WITH ALMA, Brandon Carroll, Nathan Crockett, Olivia H. Wilkins, Edwin Bergin, Geoffrey Blake

MG. Mini-symposium: Multiple Potential Energy Surfaces
Monday, June 19, 2017 – 1:45 PM
Room: 116 Roger Adams Lab

Chair: Christopher G. Elles, University of Kansas, Lawrence, KS, USA

MG01   *INVITED TALK*   1:45 – 2:15
THE ELECTRONIC GROUND STATE OF THE NITRATE RADICAL: A DECADE OF CONTROVERSY, John F. Stanton

MG02   2:19 – 2:34
A ZERO-ORDER PICTURE OF THE INFRARED SPECTRA OF CH_3O AND CD_3O: ASSIGNMENT OF STATES, Britta Johnson, Edwin Sibert

MG03   2:36 – 2:51
DVR3DUV: A SUITE FOR HIGH ACCURACY CALCULATIONS OF RO-VIBRONIC SPECTRA OF TRIATOMIC MOLECUlES, Emil J Zak, Jonathan Tennyson

MG04   2:53 – 3:08
THE ROVIBRONIC SPECTRA OF THE CYCLOPENTADIENYL RADICAL, Ketan Sharma, Terry A. Miller, John F. Stanton, David Nesbitt

MG05   3:10 – 3:25
ROTATIONAL PARAMETERS FROM VIBRONIC EIGENFUNCTIONS OF JAHN-TELLER ACTIVE MOLECULES, Scott M. Garner, Terry A. Miller

Intermission

MG06   3:44 – 3:59
RYDBERG STATES OF ALKALI METAL ATOMS ON SUPERFLUID HELIUM DROPLETS -
THEORETICAL CONSIDERATIONS, Johann V. Pototschnig, Florian Lackner, Andreas W. Hauser, Wolfgang E. Ernst

MG07   4:01 – 4:16
OBSERVATION OF HEAVY RYDBERG STATES IN H_2 AND HD, Maximilian Beyer, Frederic Merkt

MG08   4:18 – 4:33
THE ROLE OF PERTURBATIONS IN THE B-X UV SPECTRUM OF S_2 IN A TEMPERATURE-DEPENDENT MECHANISM FOR SULFUR MASS INDEPENDENT FRACTIONATION, Alexander W. Hull, Robert W Field, Shuhei Ono

MG09   4:35 – 4:50
THE ORGIN OF UNEQUAL BOND LENGTHS IN THE $\tilde{C}^1B_2$ STATE OF SO_2: SIGNATURES OF HIGH-LYING POTENTIAL ENERGY SURFACE CROSSINGS IN THE LOW-LYING VIBRATIONAL STRUCTURE, Barratt Park, Jun Jiang, Robert W Field

MG10   4:52 – 5:07
COMPUTATIONAL MODELING OF ELECTRONIC SPECTROSCOPY OF 3-PHENYL-2-PROPYNENITRILE, Claudia I Viquez Rojas, Khadija M. Jawad, Timothy S. Zwier, Lyudmila V Slipchenko

MG11   5:09 – 5:24
INFLUENCE OF THE RENNER-TELLER COUPLING IN CO+H COLLISION DYNAMICS, Steve Alexandre Ndengue, Richard Dawes

MH. Small molecules
Monday, June 19, 2017 – 1:45 PM
Room: 1024 Chemistry Annex

Chair: Wei Lin, University of Texas Rio Grande Valley, Brownsville, TX, USA

MH01   1:45 – 2:00
A SEMI-CLASSICAL APPROACH TO THE CALCULATION OF HIGHLY EXCITED ROTATIONAL ENERGIES FOR ASYMMETRIC-TOP MOLECULES, Hanno Schmiedt, Stephan Schlemmer, Sergei N. Yurchenko, Andrey Yachmenev, Per Jensen

MH02   2:02 – 2:17
NEW WAYS OF TREATING DATA FOR DIATOMIC MOLECULE 'SHELF' AND DOUBLE-MINIMUM STATES, Robert J. Le Roy, Jason Tao, Shirin Khanna, Asen Pashov, Joel Tellinghuisen

MH03   2:19 – 2:34
HIGH-RESOLUTION INFRARED SPECTROSCOPY AND ANALYSIS OF THE ν_2/ν_4 BENDING DYAD AND ν_3 STRETCHING FUNDAMENTAL OF RUTHENIUM TETROXIDE, Mbaye Faye, Sébastien Reymond-Laruinaz, Jean Vander Auwera, Vincent Boudon, Denis Doizi, Laurent Manceron

MH04   2:36 – 2:51
A MOLECULAR FOUNTAIN, Cunfeng Cheng, Aernout P.P. van der Poel, Wim Ubachs, Hendrick Bethlem

MH05   2:53 – 3:08
COMB-ASSISTED CAVITY RING DOWN SPECTROSCOPY OF ^{17}O ENRICHED WATER BETWEEN 7443 AND 7921 CM^{-1}, Didier Mondelain, Semen Mikhailenko, Ekaterina Karlovets, Magdalena Konefal, Serge Béguier, Samir Kassi, Alain Campargue

MH06   3:10 – 3:25
MILLIMETER-WAVE SPECTROSCOPY OF MgI ($\tilde{X}^2\Sigma^+$), Mark Burton, K. M. Kilchenstein, Lucy M. Ziurys

Intermission

MH07   3:44 – 3:59
QUANTIFICATION OF FLUORESCENCE FROM THE LYMAN-ALPHA PHOTOLYSIS OF WATER FOR SPACECRAFT PLUME CHARACTERIZATION., Justin W. Young, Christopher Annesley, Jaime A. Stearns

MH08   4:01 – 4:16
LIF SPECTROSCOPY OF ThF AND THE PREPARATION OF ThF$^+$ FOR THE JILA eEDM EXPERIMENT, Kia Boon Ng, Yan Zhou, Dan Gresh, William Cairncross, Tanya Roussy, Yuval Shagam, Lan Cheng, Jun Ye, Eric Cornell

MH09   4:18 – 4:33
LASER-INDUCED FLUORESCENCE SPECTROSCOPY OF TWO RUTHENIUM-BEARING MOLECULES: RuF AND RuCl , Hanif Zarringhalam, Allan G. Adam, Colan Linton, Dennis W. Tokaryk

MH10   4:35 – 4:50
OBSERVATIONS OF LOW-LYING ELECTRONIC STATES OF NiD, AND MULTI-ISOTOPE ANALYSIS, Mahdi Abbasi, Alireza Shayesteh, Patrick Crozet, Amanda J. Ross

MH11   4:52 – 5:07
$B\,^1\Pi \to A\,^1\Sigma^+$ ELECTRONIC TRANSFER IN NaK, Amanda J. Ross, Heather Harker, Maxime Giraud, Ella Wyllie

MH12   5:09 – 5:24
QUANTUM CONTROLLED NUCLEAR FUSION, Martin Gruebele

MI. Comparing theory and experiment
Monday, June 19, 2017 – 1:45 PM
Room: B102 Chemical and Life Sciences

Chair: Luis A. Rivera-Rivera, Ferris State University , Big Rapids , MI, USA

MI01   1:45 – 2:00
EXPERIMENTAL AND COMPUTATIONAL INVESTIGATIONS OF THE THRESHOLD PHOTOELECTRON SPECTRUM OF THE HCCN RADICAL, B. Gans, Cyril Falvo, L. H. Coudert, Gustavo A. Garcia, J. Küger, J.-C. Loison

MI02   2:02 – 2:17
EXPERIMENTAL AND NUMERICAL CHARACTERIZATION OF A PULSED SUPERSONIC UNIFORM FLOW FOR KINETICS AND SPECTROSCOPY, Nicolas Suas-David, Shameemah Thawoos, Bernadette M. Broderick, Arthur Suits

MI03   2:19 – 2:34
DETERMINATION OF THE OSCILLATOR STRENGTHS FOR THE THIRD AND FOURTH VIBRATIONAL OVERTONE TRANSITIONS IN SIMPLE ALCOHOLS, Jens Wallberg, Henrik G. Kjaergaard

MI04   2:36 – 2:51
AUTOMATED SPECTROSCOPIC ANALYSIS USING THE PARTICLE SWARM OPTIMIZATION ALGORITHM: IMPLEMENTING A GUIDED SEARCH ALGORITHM TO AUTOFIT, Katherine Ervin, Steven Shipman

MI05   2:53 – 3:08
SUBSTITUTION STRUCTURES OF LARGE MOLECULES AND MEDIUM RANGE CORRELATIONS IN QUANTUM CHEMISTRY CALCULATIONS, Luca Evangelisti, Brooks Pate

Intermission

MI06   3:27 – 3:42
NEW VARIATIONAL METHODS FOR COMPUTING VIBRATIONAL SPECTRA OF MOLECULES WITH UP TO 11 ATOMS, James Brown, Phillip Thomas, Tucker Carrington

MI07   3:44 – 3:59
A NUMERICALLY EXACT FULL-DIMENSIONAL CALCULATION OF RO-VIBRATIONAL LEVELS OF WATER DIMER, Xiao-Gang Wang, Tucker Carrington

MI08   4:01 – 4:16
COMPUTER SPECTROMETERS, Nikesh S. Dattani

MI09   4:18 – 4:33
JET-COOLED INFRARED LASER SPECTROSCOPY OF DIMETHYL SULFIDE: HIGH RESOLUTION ANALYSIS OF THE ν_{14} CH_3-BENDING MODE, Atef Jabri, Isabelle Kleiner, Pierre Asselin

MI10   4:35 – 4:50
WEAK HYDROGEN BONDS FROM ALIPHATIC AND FLUORINATED ALOCOHOLS TO MOLECULAR NITROGEN DETECTED BY SUPERSONIC JET FTIR SPECTROSCOPY, Soenke Oswald, Martin A. Suhm

MI11   4:52 – 5:07
PHOTOELECTRON VELOCITY MAP IMAGING SPECTROSCOPY OF BeS^-, Amanda Reed Dermer, Mallory Theis, Kyle Mascaritolo, Michael Heaven

MI12   5:09 – 5:24
EXTENDING THE LOCAL MODE HAMILTONIAN INTO THE CONDENSED PHASE: USING VIBRATIONAL SUM FREQUENCY GENERATION TO STUDY THE BENZENE-AIR INTERFACE, Britta Johnson, Edwin Sibert

MJ. Atmospheric science
Monday, June 19, 2017 – 1:45 PM
Room: 161 Noyes Laboratory

Chair: Jacob Stewart, Connecticut College, New London, CT, USA

MJ01　　　　　　　　　　　　　　　　　　　　　　　　　　　　　　　1:45 – 2:00
SPECTROSCOPIC CHARACTERIZATION OF THE REACTION PRODUCTS BETWEEN HCl AND THE SIMPLEST CRIEGEE INTERMEDIATE CH_2OO, Carlos Cabezas, Yasuki Endo

MJ02　　　　　　　　　　　　　　　　　　　　　　　　　　　　　　　2:02 – 2:17
PROBING THE CONFORMATIONAL BEHAVIOR OF THE C_3 ALKYL-SUBSTITUTED CRIEGEE INTERMEDIATES BY FTMW SPECTROSCOPY, Carlos Cabezas, J.-C. Guillemin, Yasuki Endo

MJ03　　　　　　　　　　　　　　　　　　　　　　　　　　　　　　　2:19 – 2:34
MICROWAVE CHARACTERIZATION OF PROPIOLIC SULFURIC ANHYDRIDE AND TWO CONFORMERS OF ACRYLIC SULFURIC ANHYDRIDE, CJ Smith, Anna Huff, Becca Mackenzie, Ken Leopold

MJ04　　　　　　　　　　　　　　　　　　　　　　　　　　　　　　　2:36 – 2:51
FACILE FORMATION OF ACETIC SULFURIC ANHYDRIDE IN A SUPERSONIC JET: CHARACTERIZATION BY MICROWAVE SPECTROSCOPY AND COMPUTATIONAL CHEMISTRY, Anna Huff, CJ Smith, Becca Mackenzie, Ken Leopold

MJ05　　　　　　　　　　　　　　　　　　　　　　　　　　　　　　　2:53 – 3:08
LINE SHAPE PARAMETERS OF WATER VAPOR TRANSITIONS IN THE 3645-3975 cm^{-1} REGION, V. Malathy Devi, D. Chris Benner, Robert R. Gamache, Bastien Vispoel, Candice L. Renaud, Mary Ann H. Smith, Robert L. Sams, Thomas A. Blake

Intermission

MJ06　　　　　　　　　　　　　　　　　　　　　　　　　　　　　　　3:27 – 3:42
SPECTROSCOPIC STUDY OF AIR-BROADENED NITROUS OXIDE IN THE ν_3 BAND, Robab Hashemi, Hossein Naseri, Adriana Predoi-Cross, Mary Ann H. Smith, V. Malathy Devi

MJ07　　　　　　　　　　　　　　　　　　　　　　　　　　　　　　　3:44 – 3:59
ATMOSPHERIC ISOTOPOLOGUES OBSERVED WITH ACE-FTS AND MODELED WITH WACCM, Eric M. Buzan, Christopher A. Beale, Mahdi Yousefi, Chris Boone, Peter F. Bernath

MJ08　　　　　　　　　　　　　　　　　　　　　　　　　　　　　　　4:01 – 4:16
THE ATMOSPHERIC CHEMISTRY EXPERIMENT (ACE): LATEST RESULTS, Peter F. Bernath

MJ09　　　　　　　　　　　　　　　　　　　　　　　　　　　　　　　4:18 – 4:33
ACCURATE LASER MEASUREMENTS OF THE WATER VAPOR SELF-CONTINUUM ABSORPTION IN FOUR NEAR INFRARED ATMOSPHERIC WINDOWS. A TEST OF THE MT_CKD MODEL. , Alain Campargue, Samir Kassi, Didier Mondelain, Daniele Romanini, Loïc Lechevallier, Semyon Vasilchenko

MJ10　　　　　　　　　　　　　　　　　　　　　　　　　　　　　　　4:35 – 4:50
PHOTOACOUSTIC SPECTROSCOPY OF PRESSURE- AND TEMPERATURE- DEPENDENCE IN THE O_2 A-BAND, Matthew J. Cich, Elizabeth M Lunny, Gautam Stroscio, Thinh Quoc Bui, Priyanka Rupasinghe, Daniel Hogan, Caitlin Bray, David A. Long, Joseph T. Hodges, Timothy J. Crawford, Charles Miller, Brian Drouin, Mitchio Okumura

MJ11　　　　　　　　　　　　　　　　　　　　　　　　　　　　　　　4:52 – 5:07
ULTRAVIOLET STUDY OF THE GAS PHASE HYDRATION OF METHYLGLYOXAL TO FORM THE GEMDIOL , Jay A Kroll, Anne S. Hansen, Kristian H. Møller, Jessica L Axson, Henrik G. Kjaergaard, Veronica Vaida

MJ12　　　　　　　　　　　　　　　　　　　　　　　　　　　　　　　5:09 – 5:24
DEVELOPMENT OF A QCL-BASED SPECTROMETER FOR SPECTROSCOPIC ANALYSIS OF BIOGENIC VOLATILE ORGANIC COMPOUNDS, Michael Cyrus Iranpour, Minh Nhat Tran, Jacob Stewart

MK. Instrument/Technique Demonstration
Monday, June 19, 2017 – 1:45 PM
Room: 140 Burrill Hall

Chair: Thomas Giesen, University Kassel, Kassel, Germany

MK01　　　　　　　　　　　　　　　　　　　　　　　　　　　1:45 – 2:00
DUAL-COMB SPECTROSCOPY OF THE $\nu_1 + \nu_3$ BAND OF ACETYLENE: INTENSITY AND TRANSITION DIPOLE MOMENT, Kana Iwakuni, Sho Okubo, Koichi MT Yamada, Hajime Inaba, Atsushi Onae, Feng-Lei Hong, Hiroyuki Sasada

MK02　　　　　　　　　　　　　　　　　　　　　　　　　　　2:02 – 2:17
HIGH-RESOLUTION DUAL-COMB SPECTROSCOPY WITH ULTRA-LOW NOISE FREQUENCY COMBS, Wolfgang Hänsel, Michele Giunta, Katja Beha, Adam J. Perry, R. Holzwarth

MK03　　　　　　　　　　　　　　　　　　　　　　　　　　　2:19 – 2:34
PROGRESS TOWARD INNOVATIONS IN CRYOGENIC ION CLUSTER SPECTROMETERS, Casey J Howdieshell, Etienne Garand

MK04　　　　　　　　　　　　　　　　　　　　　　　　　　　2:36 – 2:51
CAVITY-ENHANCED SPECTROSCOPY OF MOLECULAR IONS IN THE MID-INFRARED WITH UP-CONVERSION DETECTION AND BREWSTER-PLATE SPOILERS, Charles R. Markus, Jefferson E. McCollum, James Neil Hodges, Adam J. Perry, Benjamin J. McCall

Intermission

MK05　　　　　　　　　　　　　　　　　　　　　　　　　　　3:10 – 3:25
CONTINUOUS-WAVE CAVITY RING-DOWN SPECTROSCOPY IN A PULSED UNIFORM SUPERSONIC FLOW, Shameemah Thawoos, Nicolas Suas-David, Arthur Suits

MK06　　　　　　　　　　　　　　　　　　　　　　　　　　　3:27 – 3:42
MIR AND FIR ANALYSIS OF INORGANIC SPECIES IN A SINGLE DATA ACQUISITION, Peng Wang, Sergey Shilov

MK07　　　　　　　　　　　　　　　　　　　　　　　　　　　3:44 – 3:59
HIGH PRECISION 2.0 μm PHOTOACOUSTIC SPECTROMETER FOR DETERMINATION OF THE $^{13}CO_2/^{12}CO_2$ ISOTOPE RATIO, Zachary Reed, Joseph T. Hodges

MK08　　　　　　　　　　　　　　　　　　　　　　　　　　　4:01 – 4:16
OPTICAL DETECTION AND QUANTIFICATION OF RADIOCARBON DIOXIDE ($^{14}CO_2$) AT AND BELOW AMBIENT LEVELS, David A. Long, Adam J. Fleisher, Qingnan Liu, Joseph T. Hodges

MK09　　　　　　　　　　　　　　　　　　　　　　　　　　　4:18 – 4:33
USING WIDE SPECTRAL RANGE INFRARED SPECTROSCOPY TO OBTAIN BOTH SURFACE SPECIES AND CHANGES OF CATALYST ITSELF UNDER THE REACTION CONDITIONS, Xuefei Weng, Ding Ding, Huan Li, Yanping Zheng, Mingshu Chen

MK10　　　　　　　　　*Post-Deadline Abstract*　　　　　　　4:35 – 4:50
MEASUREMENTS OF ELECTRIC FIELD IN A NANOSECOND PULSE DISCHARGE BY 4-WAVE MIXING , Edmond Baratte, Igor V. Adamovich, Marien Simeni Simeni, Kraig Frederickson

MK11　　　　　　　　　*Post-Deadline Abstract*　　　　　　　4:52 – 5:07
N2 VIBRATIONAL TEMPERATURES AND OH NUMBER DENSITY MEASUREMENTS IN A NS PULSE DISCHARGE HYDROGEN-AIR PLASMAS, Yichen Hung, Caroline Winters, Elijah R Jans, Kraig Frederickson, Igor V. Adamovich

TA. Astronomy
Tuesday, June 20, 2017 – 8:30 AM
Room: 274 Medical Sciences Building

Chair: Leslie Looney, University of Illinois, Urbana, IL, USA

TA01   8:30 – 8:45
DISCOVERY OF ^{13}CCC in SgrB2(M), Thomas Giesen, Bhaswati Mookerjea, Jürgen Stutzki, Alexander A. Breier, Thomas Buechling, Guido W Fuchs

TA02   8:47 – 9:02
A SEARCH FOR THE HOCO RADICAL IN THE MASSIVE STAR-FORMING REGION Sgr B2(M), Takahiro Oyama, Mitsunori Araki, Shuro Takano, Nobuhiko Kuze, Yoshihiro Sumiyoshi, Koichi Tsukiyama, Yasuki Endo

TA03   9:04 – 9:19
PRECISE DETERMINATION OF THE ISOTOPIC RATIOS OF HC_3N IN THE MASSIVE STAR-FORMING REGION Sgr B2(M) , Takahiro Oyama, Mitsunori Araki, Shuro Takano, Nobuhiko Kuze, Yoshihiro Sumiyoshi, Koichi Tsukiyama, Yasuki Endo

TA04   9:21 – 9:36
THE ^{12}C/^{13}C RATIO IN THE GALACTIC CENTER: IMPLICATIONS FOR GALACTIC CHEMICAL EVOLUTION AND ISOTOPE CHEMISTRY, DeWayne T Halfen, Lucy M. Ziurys

Intermission

TA05   9:55 – 10:10
A STUDY OF THE c-C_3HD/c-C_3H_2 RATIO IN LOW-MASS STAR FORMING REGIONS., Johanna Chantzos, Silvia Spezzano, Paola Caselli, Ana Chacon-Tanarro

TA06   10:12 – 10:27
DETECTIONS OF LONG CARBON CHAINS CH_3CCCCH, C_6H, LINEAR-C_6H_2 AND C_7H IN THE LOW-MASS STAR FORMING REGION L1527, Mitsunori Araki, Shuro Takano, Nami Sakai, Satoshi Yamamoto, Takahiro Oyama, Nobuhiko Kuze, Koichi Tsukiyama

TA07   10:29 – 10:44
POTENTIAL LINE STRUCTURE VARIABILITY IN DIB FEATURES OBSERVED IN PATHFINDER TRES SURVEY, Charles Law, Dan Milisavljevic, Kyle N. Crabtree, Sommer Lynn Johansen

TA08   10:46 – 11:01
MODIFICATIONS OF THE RELATION BETWEEN COSMIC RAY IONIZATION RATE ζ AND H_3^+ COLUMN DENSITY IN THE CENTRAL MOLECULAR ZONE OF THE GALACTIC CENTER, Takeshi Oka

TA09   11:03 – 11:18
VIBRATIONAL SPECTROSCOPY OF He–O_2H^+ AND O_2H^+, Hiroshi Kohguchi, Koichi MT Yamada, Pavol Jusko, Stephan Schlemmer, Oskar Asvany

TB. Mini-symposium: Multiple Potential Energy Surfaces
Tuesday, June 20, 2017 – 8:30 AM
Room: 116 Roger Adams Lab

Chair: Zhou Lin, Massachusetts Institute of Technology, Cambridge, MA, USA

TB01   *INVITED TALK*   8:30 – 9:00
MORE SPECTRA! A LOT MORE! BETTER TOO! NOW WHAT?, Robert W Field

TB02   9:04 – 9:19
TIME-RESOLVED MEASUREMENT OF THE C_2 $^1A\Pi u$ STATE POPULATION FOLLOWING PHOTODISSOCIATION OF THE S_1 STATE OF ACETYLENE USING FREQUENCY-MODULATION SPECTROSCOPY, Zhenhui Du, Jun Jiang, Robert W Field

TB03   9:21 – 9:36
PRECISION MEASUREMENT OF THE ROVIBRATIONAL ENERGY-LEVEL STRUCTURE OF $^4He_2^+$, Luca Semeria, Paul Jansen, Josef A. Agner, Hansjürg Schmutz, Frederic Merkt

TB04   9:38 – 9:53
FORMATION OF H_2^+ AND ITS ISOTOPOMERS BY RADIATIVE ASSOCIATION: THE ROLE OF SHAPE AND FESHBACH RESONANCES, Maximilian Beyer, Frederic Merkt

TB05   9:55 – 10:10
HOT BAND ANALYSIS AND KINETICS MEASUREMENTS FOR ETHYNYL RADICAL, C_2H, IN THE 1.49 μm REGION, Anh T. Le, Gregory Hall, Trevor Sears

Intermission

TB06   10:29 – 10:44
PROBING THE STRUCTURES OF NEUTRAL B_{11} AND B_{12} USING HIGH-RESOLUTION PHOTOELECTRON IMAGING OF B_{11}^- AND B_{12}^-, Joseph Czekner, Ling Fung Cheung, Lai-Sheng Wang

TB07   10:46 – 11:01
VARIABLE MIXED ORBITAL CHARACTER IN THE PHOTOELECTRON ANGULAR DISTRIBUTION OF NO_2, Benjamin A Laws, Steven J Cavanagh, Brenton R Lewis, Stephen T Gibson

TB08   11:03 – 11:18
VIBRONIC COUPLING IN THE GROUND STATE OF VINYLIDENE $\tilde{X}^1A_1$ H_2CC, Stephen T Gibson, Benjamin A Laws, Hua Guo, Daniel Neumark, Carl Lineberger, Robert W Field

TB09   11:20 – 11:35
CONFORMATIONAL STUDY OF DIBENZYL ETHER, Alicia O. Hernandez-Castillo, Chamara Abeysekera, Daniel M. Hewett, Timothy S. Zwier

TB10   11:37 – 11:52
THE EXOTIC EXCITED STATE BEHAVIOR OF 3-PHENYL-2-PROPYNENITRILE, Khadija M. Jawad, Claudia I Viquez Rojas, Lyudmila V Slipchenko, Timothy S. Zwier

TB11   11:54 – 12:09
VIBRONIC EMISSION SPECTROSCOPY OF JET-COOLED BENZYL-TYPE RADICALS FROM CORONA DISCHARGE OF CHLORO-SUBSTITUTED O-XYLENE MOLECULES , Young Yoon, Sang Lee

TC. Structure determination
Tuesday, June 20, 2017 – 8:30 AM
Room: 1024 Chemistry Annex

Chair: Wolfgang Jäger, University of Alberta, Edmonton, AB, Canada

TC01   8:30–8:45
MICROWAVE SPECTRA OF THE TWO CONFORMERS OF PROPENE-3-d_1 AND A SEMIEXPERIMENTAL EQUILIB-
RIUM STRUCTURE OF PROPENE, Norman C. Craig, J. Demaison, Heinz Dieter Rudolph, Ranil M. Gurusinghe, Michael
Tubergen, L. H. Coudert, Peter Szalay, Attila Császár

TC02   8:47–9:02
CONFORMATIONAL STUDIES OF 1-OCTYNE FROM ROTATIONAL SPECTROSCOPY, Mark P. Maturo, Daniel A.
Obenchain, Robert Melchreit, S. A. Cooke, Stewart E. Novick

TC03   9:04–9:19
HIGH-RESOLUTION ROTATIONAL SPECTROSCOPY OF A MOLECULAR ROTARY MOTOR, Sergio R Domingos,
Arjen Cnossen, Cristobal Perez, Wybren Jan Buma, Wesley R Browne, Ben L Feringa, Melanie Schnell

TC04   9:21–9:36
THE MOLECULAR STRUCTURE OF MONOFLUOROBENZALDEHYDES, Issiah Byen Lozada, Wenhao Sun, Jennifer
van Wijngaarden

TC05   9:38–9:53
CHIRPED-PULSE FOURIER TRANSFORM MICROWAVE SPECTROSCOPY OF THE 2,3-DIFLUOROPYRIDINE-
CARBON DIOXIDE COMPLEX, Sydney A Gaster, Cameron M Funderburk, Gordon G Brown

Intermission

TC06   10:12–10:27
THE IMPORTANCE OF A GOOD FIT: THE MICROWAVE SPECTRA AND MOLECULAR STRUCTURES OF *TRANS*-
1,2-DIFLUOROETHYLENE-HYDROGEN CHLORIDE AND *CIS*-1,2-DIFLUOROETHYLENE-HYDROGEN CHLO-
RIDE, Helen O. Leung, Mark D. Marshall, Leonard H. Yoon

TC07   10:29–10:44
TAKING THE NEXT STEP WITH HALOGENATED OLEFINS: MICROWAVE SPECTROSCOPY AND MOLECULAR
STRUCTURES OF TETRAFLUORO- AND CHLORO-TRIFLUORO PROPENES AND THEIR COMPLEXES WITH THE
ARGON ATOM, Mark D. Marshall, Helen O. Leung, Miles A. Wronkovich, Megan E Tracy, Laboni Hoque, Allison M
Randy-Cofie, Alina K Dao

TC08   10:46–11:01
GERMANIUM DICARBIDE: EVIDENCE FOR A T–SHAPED GROUND STATE STRUCTURE, Oliver Zingsheim,
Marie-Aline Martin-Drumel, Sven Thorwirth, Stephan Schlemmer, Carl A Gottlieb, Jürgen Gauss, Michael C McCarthy

TC09   11:03–11:18
CARBON CHAINS CONTAINING GROUP IV ELEMENTS: ROTATIONAL DETECTION OF GeC_4 AND GeC_5,
Michael C McCarthy, Marie-Aline Martin-Drumel, Sven Thorwirth

TC10   11:20–11:35
MICROWAVE SPECTRA OF Ar···AgI AND H_2O···AgI PRODUCED BY LASER ABLATION, John C Mullaney, Chris
Medcraft, Nick Walker, Anthony Legon

TD. Large amplitude motions, internal rotation
Tuesday, June 20, 2017 – 8:30 AM
Room: B102 Chemical and Life Sciences

Chair: V. Ilyushin, Institute of Radio Astronomy of NASU, Kharkiv, Ukraine

TD01   8:30 – 8:45
THE MICROWAVE SPECTROSCOPY STUDY OF 1,2-DIMETHOXYETHANE , Weixing Li, Annalisa Vigorito, Camilla Calabrese, Luca Evangelisti, Laura B. Favero, Assimo Maris, Sonia Melandri

TD02   8:47 – 9:02
AB INITIO EFFECTIVE ROVIBRATIONAL HAMILTONIANS FOR NON-RIGID MOLECULES VIA CURVILINEAR VMP2, Bryan Changala, Joshua H Baraban

TD03   9:04 – 9:19
ANOMALOUS CENTRIFUGAL DISTORTION IN NH_2, Marie-Aline Martin-Drumel, Olivier Pirali, L. H. Coudert

TD04   9:21 – 9:36
THE EFFECT OF TORSION - VIBRATION COUPLINGS ON THE ν_9 AND ν_1 BANDS IN THE $CH_3OO\cdot$ RADICAL, Meng Huang, Terry A. Miller, Anne B McCoy

TD05   9:38 – 9:53
ROVIBRATIONAL QUANTUM DYNAMICS OF THE METHANE-WATER DIMER, János Sarka, Attila Császár, Edit Mátyus

Intermission

TD06   10:12 – 10:27
A COMBINED GIGAHERTZ AND TERAHERTZ SYNCHROTRON-BASED FOURIER TRANSFORM INFRARED SPECTROSCOPIC INVESTIGATION OF ORTHO-D-PHENOL, Sieghard Albert, Ziqiu Chen, Csaba Fábri, Robert Prentner, Martin Quack, Daniel Zindel

TD07   10:29 – 10:44
ANALYSIS OF THE ν_6 ASYMMETRIC NO STRETCH BAND OF NITROMETHANE, Mahesh B. Dawadi, Lou Degliumberto, David S. Perry, Howard Mettee, Robert L. Sams

TD08   10:46 – 11:01
TORSIONAL, VIBRATIONAL AND VIBRATION-TORSIONAL LEVELS IN THE S_1 AND GROUND CATIONIC D_0^+ STATES OF PARA-FLUOROTOLUENE, Adrian M. Gardner, William Duncan Tuttle, Laura E. Whalley, Andrew Claydon, Joseph H. Carter, Timothy G. Wright

TD09   11:03 – 11:18
MOLECULAR SYMMETRY ANALYSIS OF LOW-ENERGY TORSIONAL AND VIBRATIONAL STATES IN THE S_0 AND S_1 STATES OF p-XYLENE TO INTERPRET THE REMPI SPECTRUM , Peter Groner, Adrian M. Gardner, William Duncan Tuttle, Timothy G. Wright

TD10   11:20 – 11:35
TORSIONAL, VIBRATIONAL AND VIBRATION-TORSIONAL LEVELS IN THE S_1 AND GROUND CATIONIC D_0^+ STATES OF PARA-XYLENE, Adrian M. Gardner, William Duncan Tuttle, Peter Groner, Timothy G. Wright

TE. Fundamental interest
Tuesday, June 20, 2017 – 8:30 AM
Room: 161 Noyes Laboratory

Chair: Josh Vura-Weis, University of Illinois at Urbana-Champaign, Urbana, IL, USA

TE01   8:30–8:45
COVALENT AND NONCOVALENT INTERACTIONS BETWEEN BORON AND ARGON: AN INFRARED PHOTODIS-SOCIATION SPECTROSCOPIC STUDY OF ARGON-BORON OXIDE CATION COMPLEXES, Jiaye Jin, Wei Li, Guanjun Wang, Mingfei Zhou

TE02   8:47–9:02
OBSERVATION OF QUANTUM BEATING IN Rb AT 2.1 THz AND 18.2 THz: LONG-RANGE Rb*-Rb INTERACTIONS., William Goldshlag, Brian J Ricconi, J. Gary Eden

TE03   9:04–9:19
ROTATIONAL SPECTRUM OF SACCHARINE, Elena R. Alonso, Santiago Mata, José L. Alonso

TE04   9:21–9:36
MICROWAVE OBSERVATION OF THE O_2-CONTAINING COMPLEX, O_2-HCl, Frank E Marshall, Nicole Moon, Thomas D. Persinger, Richard Dawes, G. S. Grubbs II

TE05   9:38–9:53
SUPERRADIANCE IN A STRONGLY-COUPLED MULTI-LEVEL SYSTEM, Stephen Coy, David Grimes, Timothy J Barnum, Robert W Field

TE06   9:55–10:10
ABSTRACT WITHDRAWN,

Intermission

TE07   10:29–10:44
DETECTION OF THE MW TRANSITION BETWEEN ORTHO AND PARA STATES, Hideto Kanamori, Zeinab Tafti Dehghani, Asao Mizoguchi, Yasuki Endo

TE08   10:46–11:01
NEAR-INFRARED SPECTROSCOPY OF SMALL PROTONATED WATER CLUSTERS, J. Philipp Wagner, David C McDonald II, Anne B McCoy, Michael A Duncan

TE09   11:03–11:18
NON COVALENT INTERACTIONS IN LARGE DIAMONDOID DIMERS IN THE GAS PHASE - A MICROWAVE STUDY, Cristobal Perez, Marina Sekutor, Andrey A. Fokin, Sebastian Blomeyer, Yury V. Vishnevskiy, Norbert W. Mitzel, Peter R. Schreiner, Melanie Schnell

TE10   11:20–11:35
VIBRATIONAL STARK EFFECT TO PROBE THE ELECTRIC-DOUBLE LAYER OF THE IONIC LIQUID-METAL ELECTRODES, Natalia Garcia Rey, Alexander Knight Moore, Shuichi Toyouchi, Dana Dlott

TE11   11:37–11:52
IN-SITU GENERATED GRAPHENE AS THE CATALYTIC SITE FOR VISIBLE-LIGHT MEDIATED ETHYLENE EPOXIDATION ON AG NANOCATALYSTS, Xueqiang Alex Zhang, Prashant Jain

TF. Mini-symposium: ALMA's Molecular View
Tuesday, June 20, 2017 – 1:45 PM
Room: 274 Medical Sciences Building

Chair: Susanne Aalto, Chalmers University of Technology, Onsala, Sweden

TF01   ***INVITED TALK***   **1:45 – 2:15**
PHYSICOCHEMICAL PROCESSES ON ICE DUST TOWARDS DEUTERIUM ENRICHMENT, Naoki Watanabe

TF02   **2:19 – 2:34**
CO IN PROTOSTARS (COPS): *HERSCHEL*-SPIRE SPECTROSCOPY OF EMBEDDED PROTOSTARS, Yao-Lun Yang, Joel D. Green, Neal J Evans II

TF03   **2:36 – 2:51**
THE CO TRANSITION FROM DIFFUSE MOLECULAR GAS TO DENSE CLOUDS, Johnathan S Rice, Steven Federman

TF04   **2:53 – 3:08**
HIGH RESOLUTION ROTATIONAL SPECTROSCOPY OF HCSSH: A CS_2 PROXY IN THE ISM, Domenico Prudenzano, Jacob Laas, Maria Elisabetta Palumbo, Paola Caselli

TF05   **3:10 – 3:25**
THE ASTROPHYSICAL WEEDS: ROTATIONAL TRANSITIONS IN EXCITED VIBRATIONAL STATES , José L. Alonso, Lucie Kolesniková, Elena R. Alonso, Santiago Mata

Intermission

TF06   **3:44 – 3:59**
TIME-SENSITIVE CHEMICAL TRACERS WITHIN SHOCKED ASTROPHYSICAL SOURCES, Andrew M Burkhardt, Christopher N Shingledecker, Romane Le Gal, Brett A. McGuire, Anthony Remijan, Eric Herbst

TF07   **4:01 – 4:16**
SIO OUTFLOWS AS TRACERS OF MASSIVE STAR FORMATION IN INFRARED DARK CLOUDS, Mengyao Liu

TF08   **4:18 – 4:33**
QUANTUM-CHEMICAL CALCULATIONS REVEALING THE EFFECTS OF MAGNETIC FIELDS ON METHANOL, Boy Lankhaar, Ad van der Avoird, Wouter H.T. Vlemmings, Gerrit Groenenboom, Huib Jan van Langevelde, Gabriele Surcis

TF09   **4:35 – 4:50**
ZEEMAN EFFECT IN SULFUR MONOXIDE: A PROBE TO OBSERVE MAGNETIC FIELDS IN STAR FORMING REGIONS?, Gabriele Cazzoli, Valerio Lattanzi, Sonia Coriani, Jürgen Gauss, Claudio Codella, Andrés Asensio Ramos, Jose Cernicharo, Cristina Puzzarini

TF10   **4:52 – 5:07**
MAPPING MAGNETIC FIELDS IN MOLECULAR CLOUDS WITH THE CN ZEEMAN EFFECT, Richard Crutcher

TF11   **5:09 – 5:24**
LAYING THE GROUNDWORK FOR FUTURE ALMA DIRECT MAGNETIC FIELD DETECTION IN PROTOSTELLAR ENVIRONMENTS, Erin Guilfoil Cox, Robert J Harris, Leslie Looney, Dominique M. Segura-Cox, Richard Crutcher, Zhi-Yun Li, John Tobin, Ian Stephens, Giles Novak, Manuel Fernandez-Lopez

TG. Mini-symposium: Multiple Potential Energy Surfaces
Tuesday, June 20, 2017 – 1:45 PM
Room: 116 Roger Adams Lab

Chair: John Parkhill, The University of Notre Dame, Notre Dame, IN, US

TG01 1:45 – 2:00
ROAMING ISOMERIZATION OF PHOTOEXCITED HALOGENATED ALKANES IN THE GAS AND LIQUID PHASES
, Veniamin A. Borin, Sergey M. Matveev, Darya S. Budkina, Christopher M. Hicks, Andrey S. Mereshchenko, Evgeniia V. Butaeva, Vasily V. Vorobyev, <u>Alexander N Tarnovsky</u>

TG02 2:02 – 2:17
NONRADIATIVE DECAY ROUTE OF CINNAMATE DERIVATIVES STUDIED BY FREQUENCY AND TIME DO-MAIN LASER SPECTROSCOPY IN THE GAS PHASE, MATRIX ISOLATION FTIR SPECTROSCOPY AND QUAN-TUM CHEMICAL CALCULATIONS, <u>Takayuki Ebata</u>

TG03 2:19 – 2:34
PHOTOISOMERIZATION DYNAMICS OF AZOBENZENE DERIVATIVES IN SOLUTION USING BROADBAND UL-TRAFAST SPECTROSCOPY, <u>Abdelqader Jamhawi</u>, Ishan Fursule, Qunfei Zhou, Brad J. Barron, Matthew J. Beck, Jinjun Liu

TG04 2:36 – 2:51
TIME-RESOLVED SIGNATURES ACROSS THE INTRAMOLECULAR RESPONSE IN SUBSTITUTED CYANINE DYES, <u>Muath Nairat</u>, Morgan Webb, Michael Esch, Vadim V. Lozovoy, Benjamin G Levine, Marcos Dantus

TG05 2:53 – 3:08
CAN INTERNAL CONVERSION BE CONTROLLED BY MODE-SPECIFIC VIBRATIONAL EXCITATION IN POLY-ATOMIC MOLECULES, Michael Epshtein, Alexander Portnov, <u>Ilana Bar</u>

Intermission

TG06 3:27 – 3:42
THE MOLECULAR GEOMETRIC PHASE AND LIGHT-INDUCED CONICAL INTERSECTIONS, <u>Emil J Zak</u>

TG07 3:44 – 3:59
SEMI-CLASSICAL DYNAMICS STUDIES OF THE PHOTODISSOCIATION OF ICN^- and $BrCN^-$, <u>Bernice Opoku-Agyeman</u>, Anne B McCoy

TG08 4:01 – 4:16
ULTRAFAST MOLECULAR PHOTODISSOCIATION DYNAMICS STUDIED BY FEMTOSECOND PHOTOELECTRON-PHOTOION COINCIDENCE SPECTROSCOPY, <u>Bernhard Thaler</u>, Pascal Heim, Wolfgang E. Ernst, Markus Koch

TG09 4:18 – 4:33
Cl-LOSS DYNAMICS OF VINYL CHLORIDE CATION IN B STATE: ROLE OF C STATE, <u>X Zhou</u>

TG10 4:35 – 4:50
VIBRATIONAL MODE-SPECIFIC AUTODETACHMENT AND COUPLING OF CH2CN-, <u>Justin Lyle</u>, Richard Mabbs

TG11 4:52 – 5:07
INFLUENCE OF SPIN-ORBIT QUENCHING ON THE SOLVATION OF INDIUM IN HELIUM DROPLETS, <u>Ralf Meyer</u>, Johann V. Pototschnig, Wolfgang E. Ernst, Andreas W. Hauser

TG12 5:09 – 5:24
$n \rightarrow \pi^*$ NON-COVALENT INTERACTION IS WEAK BUT STRONG IN ACTION, <u>Santosh Kumar Singh</u>, Aloke Das

TH. Instrument/Technique Demonstration
Tuesday, June 20, 2017 – 1:45 PM
Room: 1024 Chemistry Annex

Chair: Christopher F. Neese, The Ohio State University, Columbus, OH, USA

TH01 1:45 – 2:00
A NEW 2.0-6.0 GHz CHIRPED PULSE FOURIER TRANSFORM MICROWAVE SPECTROMETER: INSTRUMENTAL ANALYSIS AND INITIAL MOLECULAR RESULTS, Nathan A Seifert, Javix Thomas, Wolfgang Jäger, Yunjie Xu

TH02 2:02 – 2:17
AN 18-26 GHz SEGMENTED CHIRPED PULSE FOURIER TRANSFORM MICROWAVE SPECTROMETER FOR ASTROCHEMICAL APPLICATIONS, Amanda Steber, Mariyam Fatima, Cristobal Perez, Melanie Schnell

TH03 2:19 – 2:34
A HIGHLY-INTEGRATED SUPERSONIC-JET FOURIER TRANSFORM MICROWAVE SPECTROMETER, Qian Gou, Gang Feng, Jens-Uwe Grabow

TH04 2:36 – 2:51
LARGE OLIGOMERS STABILIZED BY WHB NETWORKS: PENTAMERS OF DIFLUOROMETHANE AND ITS WATER CLUSTERS, Emilio J. Cocinero, Iciar Uriarte, Luca Evangelisti, Camilla Calabrese, Giacomo Prampolini, Ivo Cacelli, Brooks Pate

TH05 2:53 – 3:08
MEASURING CONFORMATIONAL ENERGY DIFFERENCES USING PULSED-JET MICROWAVE SPECTROSCOPY, Cameron M Funderburk, Sydney A Gaster, Tiffany R Taylor, Gordon G Brown

TH06 3:10 – 3:25
METHOXYETHANOL, ETHOXYETHANOL, AND SPECTRAL COMPLEXITY, J. H. Westerfield, Erika Riffe, Maria Phillips, Erika Johnson, Steven Shipman

Intermission

TH07 3:44 – 3:59
LABORATORY HETERODYNE SPECTROMETERS OPERATING AT 100 AND 300 GHZ, Jakob Maßen, Nadine Wehres, Marius Hermanns, Frank Lewen, Bettina Heyne, Christian Endres, Urs Graf, Netty Honingh, Stephan Schlemmer

TH08 4:01 – 4:16
COMPLEX MOLECULES IN THE LABORATORY - A COMPARISON OF CHRIPED PULSE AND EMISSION SPECTROSCOPY, Marius Hermanns, Nadine Wehres, Jakob Maßen, Stephan Schlemmer

TH09 4:18 – 4:33
A 530-590 GHZ SCHOTTKY HETERODYNE RECEIVER FOR HIGH-RESOLUTION MOLECULAR SPECTROSCOPY WITH LILLE'S FAST-SCAN FULLY SOLID-STATE DDS SPECTROMETER, A. Pienkina, L. Margulès, R. A. Motiyenko, Martina C. Wiedner, Alain Maestrini, Fabien Defrance

TH10 4:35 – 4:50
MILLIMETER WAVE SPECTROSCOPY IN A SEMI-CONFOCAL FABRY-PEROT CAVITY, Brian Drouin, Adrian Tang, Theodore J Reck, Deacon J Nemchick, Matthew J. Cich, Timothy J. Crawford, Alexander W Raymond, M.-C. Frank Chang, Rod M. Kim

TH11 4:52 – 5:07
DETERMINING THE CONCENTRATIONS AND TEMPERATURES OF PRODUCTS IN A $CF_4/CHF_3/N_2$ PLASMA VIA SUBMILLIMETER ABSORPTION SPECTROSCOPY, Yaser H. Helal, Christopher F. Neese, Frank C. De Lucia, Paul R. Ewing, Ankur Agarwal, Barry Craver, Phillip J. Stout, Michael D. Armacost

TI. Large amplitude motions, internal rotation
Tuesday, June 20, 2017 – 1:45 PM
Room: B102 Chemical and Life Sciences

Chair: Peter Groner, University of Missouri, Kansas City, MO, USA

TI01   1:45 – 2:00
BROADBAND FTMW SPECTROSCOPY OF 2-METHYLIMIDAZOLE AND COMPLEXES WITH WATER AND AR-GON, Chris Medcraft, Juliane Heitkämper, John C Mullaney, Nick Walker

TI02   *Post-Deadline Abstract - Original Abstract Withdrawn*   2:02 – 2:17
CONNECTION BETWEEN THE SU(3) ALGEBRAIC MODEL AND CONFIGURATION SPACE FOR BENDING MODES OF LINEAR MOLECULES: APPLICATION TO ACETYLENE, Lemus Renato, Estezez-Fregozo María del Mar

TI03   2:19 – 2:34
MICROWAVE AND FIR SPECTROSCOPY OF DIMETHYLSULFIDE IN THE GROUND, FIRST AND SECOND EX-CITED TORSIONAL STATES, V. Ilyushin, Iuliia Armieieva, Olga Dorovskaya, Mykola Pogrebnyak, Igor Krapivin, E. A. Alekseev, L. Margulès, R. A. Motiyenko, F. Kwabia Tchana, Atef Jabri, Laurent Manceron, Sigurd Bauerecker, Christof Maul

TI04   2:36 – 2:51
ADVANCES IN GLOBAL MODELLING OF METHYL MERCAPTAN $CH_3{}^{32}SH$ TORSION-ROTATION SPECTRUM, V. Ilyushin, Iuliia Armieieva, Olena Zakharenko, Holger S. P. Müller, Frank Lewen, Stephan Schlemmer, Li-Hong Xu, Ronald M. Lees

Intermission

TI05   3:10 – 3:25
THE MICROWAVE SPECTROSCOPY OF CD_3SH, Kaori Kobayashi, Shozo Tsunekawa, Nobukimi Ohashi

TI06   3:27 – 3:42
QUANTUM CHEMICAL CALCULATIONS OF TORSIONALLY MEDIATED HYPERFINE SPLITTINGS IN STATES OF E SYMMETRY OF ACETALDEHYDE (CH_3CHO), Li-Hong Xu, Elias M. Reid, Bradley Guislain, Jon T. Hougen, E. A. Alekseev, Igor Krapivin

TI07   3:44 – 3:59
MICROWAVE SPECTROSCOPY OF 2-PENTANONE, Maike Andresen, Ha Vinh Lam Nguyen, Isabelle Kleiner, Wolfgang Stahl

TI08   4:01 – 4:16
COMPETITION BETWEEN TWO LARGE-AMPLITUDE MOTION MODELS: NEW HYBRID HAMILTONIAN VER-SUS OLD PURE-TUNNELING HAMILTONIAN, Isabelle Kleiner, Jon T. Hougen

TI09   4:18 – 4:33
TUNNELING EFFECTS AND CONFORMATION DETERMINATION OF THE POLAR FORMS OF 1,3,5-TRISILAPENTANE, Frank E Marshall, William Raymond Neal Tonks, David Joseph Gillcrist, Charles J. Wurrey, Gamil A Guirgis, G. S. Grubbs II

TI10   4:35 – 4:50
A COMPARISON OF THE MOLECULAR STRUCTURES OF $C_4H_9OCH_3$, $C_4H_9SCH_3$, $C_5H_{11}OCH_3$, AND $C_5H_{11}SCH_3$ USING MICROWAVE SPECTROSCOPY, Brittany E. Long, Juan Betancur, Yoon Jeong Choi, S. A. Cooke, G. S. Grubbs II, Jonathan Ogulnick, Tara Holmes

TJ. Linelists
Tuesday, June 20, 2017 – 1:45 PM
Room: 161 Noyes Laboratory

Chair: Brian Drouin, California Institute of Technology, Pasadena, CA, USA

TJ01   1:45 – 2:00
ACCURACY and COMPLETENESS of MOLECULAR LINE LISTS, Oleg Polyansky, Jonathan Tennyson

TJ02   2:02 – 2:17
A RIGOROUS COMPARISON OF THEORETICAL AND MEASURED CARBON DIOXIDE LINE INTENSITIES, Hongming Yi, Adam J. Fleisher, Lyn Gameson, Emil J Zak, Oleg Polyansky, Jonathan Tennyson, Joseph T. Hodges

TJ03   2:19 – 2:34
PRECISION CAVITY-ENHANCED DUAL-COMB SPECTROSCOPY: APPLICATION TO THE GAS METROLOGY OF CO_2, H_2O, and N_2O., Adam J. Fleisher, David A. Long, Joseph T. Hodges

TJ04   2:36 – 2:51
EXPERIMENTAL LINE LIST OF WATER VAPOR ABSORPTION LINES IN THE SPECTRAL RANGES $1850 – 2280\,CM^{-1}$ AND $2390 – 4000\,CM^{-1}$, Joep Loos, Manfred Birk, Georg Wagner

TJ05   2:53 – 3:08
PROGRESS IN THE MEASUREMENT ON TEMPERATURE-DEPENDENCE OF H_2-BROADENING OF COLD AND HOT CH_4, Keeyoon Sung, V. Malathy Devi, D. Chris Benner, Timothy J. Crawford, Arlan Mantz, Mary Ann H. Smith

TJ06   3:10 – 3:25
HIGH ACCURACY POTENTIAL ENERGY SURFACE, DIPOLE MOMENT SURFACE, ROVIBRATIONAL ENERGIES AND LINE LIST CALCULATIONS FOR $^{14}NH_3$, Phillip Coles, Sergei N. Yurchenko, Oleg Polyansky, Aleksandra Kyuberis, Roman I. Ovsyannikov, Nikolay Fedorovich Zobov, Jonathan Tennyson

TJ07   3:27 – 3:42
A NEW LINELIST FOR OH $A^2\Sigma$-$X^2\Pi$ ELECTRONIC TRANSITION, Mahdi Yousefi, Peter F. Bernath

Intermission

TJ08   4:01 – 4:16
HITRAN2016: Part I. Line lists for H_2O, CO_2, O_3, N_2O, CO, CH_4, and O_2, Iouli E Gordon, Laurence S. Rothman, Yan Tan, Roman V Kochanov, Christian Hill

TJ09   4:18 – 4:33
HITRAN2016 DATABASE PART II: OVERVIEW OF THE SPECTROSCOPIC PARAMETERS OF THE TRACE GASES, Yan Tan, Iouli E Gordon, Laurence S. Rothman, Roman V Kochanov, Christian Hill

TJ10   4:35 – 4:50
ABSORPTION CROSS-SECTIONS IN HITRAN2016: MAJOR DATABASE UPDATE FOR ATMOSPHERIC, INDUSTRIAL, AND CLIMATE APPLICATIONS, Roman V Kochanov, Iouli E Gordon, Laurence S. Rothman, Keith Shine, Steven W. Sharpe, Timothy J. Johnson, Jeremy J. Harrison, Peter F. Bernath, Timothy Wallington, Manfred Birk, Georg Wagner, Christian Hill

TJ11   4:52 – 5:07
A NEW LINE LIST FOR $A^3\Pi$ - $X^3\Sigma^-$ TRANSITION OF NH RADICAL, Anton Madushanka Fernando, Peter F. Bernath

TJ12   5:09 – 5:24
LINE LISTS FOR LiF AND LiCl IN THE $X^1\Sigma^+$ STATE, Dror M. Bittner, Peter F. Bernath

TK. Small molecules
Tuesday, June 20, 2017 – 1:45 PM
Room: 140 Burrill Hall

Chair: Vincent Boudon, CNRS / Université Bourgogne Franche-Comté, Dijon, France

TK01　　　　　　　　　　　　　　　　　　　　　　　　　　　　　　　　　　　　　1:45 – 2:00
RELATIVE INTENSITY OF A CROSS-OVER RESONANCE TO LAMB DIPS OBSERVED IN STARK SPECTROSCOPY OF METHANE, Shoko Okuda, Hiroyuki Sasada

TK02　　　　　　　　　　　　　　　　　　　　　　　　　　　　　　　　　　　　　2:02 – 2:17
CORIOLIS PERTURBATIONS TO THE $3\nu_4$ LEVEL OF THE $\tilde{A}$ STATE OF FORMALDEHYDE, Barratt Park, Bastian C. Krueger, Sven Meyer, Tim Schaefer

TK03　　　　　　　　　　　　　　　　　　　　　　　　　　　　　　　　　　　　　2:19 – 2:34
A $1 + 1'$ RESONANCE-ENHANCED MULTIPHOTON IONIZATION SCHEME FOR ROTATIONALLY STATE-SELECTIVE DETECTION OF FORMALDEHYDE VIA THE $\tilde{A}\,^1A_2 \leftarrow \tilde{X}\,^1A_1$ TRANSITION, Barratt Park, Bastian C. Krueger, Sven Meyer, Alec Wodtke, Tim Schaefer

TK04　　　　　　　　　　　　　　　　　　　　　　　　　　　　　　　　　　　　　2:36 – 2:51
AN EMPIRICAL SPECTROSCOPIC DATABASE FOR ACETYLENE IN THE REGIONS OF 5850-9415 CM^{-1} , Alain Campargue, Oleg Lyulin

TK05　　　　　　　　　　　　　　　　　　　　　　　　　　　　　　　　　　　　　2:53 – 3:08
IDENTIFICATION OF PHOTOFRAGMENTS FROM ONE-COLOR RESONANTLY-ENHANCED $(\tilde{A} - \tilde{X})$ MULTI-PHOTON PHOTODISSOCIATION OF ACETYLENE , Jun Jiang, Angelar K Muthike, Robert W Field

Intermission

TK06　　　　　　　　　　　　　　　　　　　　　　　　　　　　　　　　　　　　　3:27 – 3:42
CONFORMATIONAL ANALYSIS OF 3,3,3-TRIFLUORO-2-(TRIFLUOROMETHYL)PROPANOIC ACID , Javix Thomas, Michael J Carrillo, Agapito Serrato III, Elijah G Schnitzler, Wolfgang Jäger, Yunjie Xu, Wei Lin

TK07　　　　　　　　　　　　　　　　　　　　　　　　　　　　　　　　　　　　　3:44 – 3:59
STRUCTURE AND TUNNELING DYNAMICS OF *gauche*-1,3-BUTADIENE, Bryan Changala, Joshua H Baraban, Marie-Aline Martin-Drumel, Sandra Eibenberger, David Patterson, John F. Stanton, Barney Ellison, Michael C McCarthy

TK08　　　　　　　　　　　　　　　　　　　　　　　　　　　　　　　　　　　　　4:01 – 4:16
FIRST HIGH RESOLUTION IR SPECTRA OF 2,2-D$_2$-PROPANE. THE ν_{15} (B$_1$) A-TYPE BAND NEAR 954.709 cm^{-1}. DETERMINATION OF GROUND AND UPPER STATE CONSTANTS. , Daniel Gjuraj, S.J. Daunt, Robert Grzywacz, Walter Lafferty, Jean-Marie Flaud, Brant E. Billinghurst

TK09　　　　　　　　　　　　　　　　　　　　　　　　　　　　　　　　　　　　　4:18 – 4:33
HIGH-RESOLUTION INFRARED SPECTROSCOPY OF IMIDAZOLE CLUSTERS IN HELIUM DROPLETS USING QUANTUM CASCADE LASERS, Devendra Mani, Cihad Can, Nitish Pal, Gerhard Schwaab, Martina Havenith

TK10　　　　　　　　　　　　　　　　　　　　　　　　　　　　　　　　　　　　　4:35 – 4:50
LUMINESCENCE OF ADENINE MOLECULES IN GAS PHASE UNDER THE LOW ENERGY ELECTRON BEAM, Y.Y. Svyda, M.I. Shafranyosh, M.O. Margitych, M.I. Sukhoviya, I.I. Shafranyosh

WA. Astronomy
Wednesday, June 21, 2017 – 8:30 AM
Room: 274 Medical Sciences Building

Chair: Brett A. McGuire, National Radio Astronomy Observatory, Charlottesville, VA, USA

WA01 8:30 – 8:45
THE GIGAHERTZ AND TERAHERTZ SPECTRUM of MONO-DEUTERATED OXIRANE (c-C_2H_3DO), Sieghard Albert, Ziqiu Chen, Karen Keppler, Philippe Lerch, Martin Quack, Volker Schurig, Oliver Trapp

WA02 8:47 – 9:02
THE MICROWAVE SPECTROSCOPY OF $HCOO^{13}CH_3$ IN THE SECOND TORSIONAL EXCITED STATE, Kaori Kobayashi, Takuro Kuwahara, Yuki Urata, Nobukimi Ohashi, Masaharu Fujitake

WA03 9:04 – 9:19
ROVIBRATIONAL INTERACTIONS IN THE GROUND AND TWO LOWEST EXCITED VIBRATIONAL STATES OF METHOXY ISOCYANATE, A. Pienkina, L. Margulès, R. A. Motiyenko, J.-C. Guillemin

WA04 9:21 – 9:36
MILLIMETER AND SUBMILLIMETER WAVE SPECTROSCOPY OF HIGHER ENERGY CONFORMERS OF 1,2-PROPANEDIOL, Olena Zakharenko, Jean-Baptiste Bossa, Frank Lewen, Stephan Schlemmer, Holger S. P. Müller

WA05 9:38 – 9:53
DETERMINATION OF METHANOL PHOTOLYSIS BRANCHING RATIOS VIA ROTATIONAL SPECTROSCOPY, Carson Reed Powers, Morgan N McCabe, Susanna L. Widicus Weaver

WA06 9:55 – 10:10
PHOTOPROCESSING OF METHANOL ICE: FORMATION AND LIBERATION OF CO, Houston H Smith, AJ Mesko, Samuel Zinga, Stefanie N Milam, Susanna L. Widicus Weaver

WA07 10:12 – 10:27
INFRARED SPECTROSCOPY OF DISILICON-CARBIDE, Si_2C, Daniel Witsch, Volker Lutter, Guido W Fuchs, Jürgen Gauss, Thomas Giesen

Intermission

WA08 10:46 – 11:01
FOURIER TRANSFORM SPECTROSCOPY OF THE $A^3\Pi - X^3\Sigma^-$ TRANSITION OF OH^+, James Neil Hodges, Peter F. Bernath

WA09 11:03 – 11:18
ASTROCHEMICAL LABORATORY EXPERIMENTS AS ANALOGS TO PLUTONIAN CHEMISTRY: USING FTIR SPECTROSCOPY TO MONITOR THE SUBLIMATION OF IRRADIATED 1:1:100 $CO+H_2O+N_2$ AND 1:1:100 $CH_4+H_2O+N_2$ ICES, Kamil Bartłomiej Stelmach, Yukiko Yarnall, Paul Cooper

WA10 11:20 – 11:35
PHOTOCHEMICAL GENERATION OF H_2NCNX, H_2NNCX, $H_2NC(NX)$ (X = O, S) IN LOW-TEMPERATURE MATRICES, Tamas Voros, Gyozo Gyorgy Lajgut, Gabor Magyarfalvi, Gyorgy Tarczay

WA11 11:37 – 11:52
INFRARED SPECTRUM OF N-OXIDOHYDROXYLAMINE [•ONH(OH)] PRODUCED IN REACTION H + HONO IN SOLID *PARA*-HYDROGEN , Karolina Anna Haupa, Yuan-Pern Lee

WA12 11:54 – 12:09
INFRARED SPECTRA OF PROTONATED QUINOLINE ($1-C_9H_7NH^+$) IN SOLID *PARA*-HYDROGEN, Chih-Yu Tseng, Yuan-Pern Lee

WB. Mini-symposium: Multiple Potential Energy Surfaces
Wednesday, June 21, 2017 – 8:30 AM
Room: 116 Roger Adams Lab

Chair: Jinjun Liu, University of Louisville, Louisville, KY, USA

WB01   *INVITED TALK*   **8:30 – 9:00**
LIGHT, MOLECULES, ACTION: USING ULTRAFAST UV-VISIBLE AND X-RAY SPECTROSCOPY TO PROBE EXCITED STATE DYNAMICS IN PHOTOACTIVE MOLECULES, R.J. Sension

WB02   **9:04 – 9:19**
BLACK BOX REAL-TIME TRANSIENT ABSORPTION SPECTROSCOPY AND ELECTRON CORRELATION, John Parkhill

WB03   **9:21 – 9:36**
RESONANCE-ENHANCED EXCITED-STATE RAMAN SPECTROSCOPY OF CONJUGATED THIOPHENE DERIVATIVES: COMBINING EXPERIMENT WITH THEORY, Matthew S. Barclay, Timothy J Quincy, Marco Caricato, Christopher G. Elles

WB04   **9:38 – 9:53**
RESONANT FEMTOSECOND STIMULATED RAMAN BAND INTENSITY AND S_n STATE ELECTRONIC STRUCTURE, Timothy J Quincy, Matthew S. Barclay, Marco Caricato, Christopher G. Elles

Intermission

WB05   **10:12 – 10:27**
FEMTOSECOND ELEMENT-SPECIFIC XUV SPECTROSCOPY OF COMPLEX MOLECULES AND MATERIALS, Josh Vura-Weis

WB06   **10:29 – 10:44**
ULTRAFAST TRANSIENT ABSORPTION SPECTROSCOPY INVESTIGATION OF EXCITED-STATE DYNAMICS OF METHYL AMMONIUM LEAD BROMIDE PEROVSKITE NANOSTRUCTURES, Abdelqader Jamhawi, Hamzeh Telfah, Meghan B Teunis, Rajesh Sardar, Jinjun Liu

WB07   **10:46 – 11:01**
ULTRAFAST TRANSIENT ABSORPTION SPECTROSCOPY INVESTIGATION OF PHOTOINDUCED DYNAMICS IN POLY(3-HEXYLTHIOPHENE)-BLOCK-OLIGO(ANTHRACENE-9,10-DIYL), Jacob Strain, Hemali Rathnayake, Jinjun Liu

WB08   **11:03 – 11:18**
PHOTOCHEMICAL DYNAMICS OF INTRAMOLECULAR SINGLET FISSION, Zhou Lin, Hikari Iwasaki, Troy Van Voorhis

WB09   **11:20 – 11:35**
KEY INTERMEDIATES OF CARBON DIOXIDE REDUCTION ON SILVER FROM VIBRATIONAL NANOSPECTROSCOPY, Prashant Jain

WB10   *Post-Deadline Abstract*   **11:37 – 11:52**
TWO-PHOTON EXCITATION OF CONJUGATED MOLECULES IN SOLUTION: SPECTROSCOPY AND EXCITED-STATE DYNAMICS, Christopher G. Elles, Amanda L. Houk, Marc de Wergifosse, Anna Krylov

WC. Conformers, isomers, chirality, stereochemistry
Wednesday, June 21, 2017 – 8:30 AM
Room: 1024 Chemistry Annex

Chair: Josh Newby, Hobart and William Smith Colleges, Geneva, NY, USA

WC01 8:30 – 8:45
ROTATIONAL SPECTROSCOPY AND CONFORMATIONAL STUDIES OF 4-PENTYNENITRILE, 4-PENTENENITRILE, AND GLUTARONITRILE, Brian M Hays, Deepali Mehta-Hurt, Khadija M. Jawad, Alicia O. Hernandez-Castillo, Chamara Abeysekera, Di Zhang, Timothy S. Zwier

WC02 8:47 – 9:02
THE CONFORMER SPECIFIC ROTATIONAL SPECTRUM OF 3-PHENYLPROPIONITRILE UTILIZING STRONG FIELD COHERENCE BREAKING, Sean Fritz, Alicia O. Hernandez-Castillo, Chamara Abeysekera, Timothy S. Zwier

WC03 9:04 – 9:19
EFFECT OF INTRAMOLECULAR DISPERSION INTERACTIONS ON THE CONFORMATIONAL PREFERENCES OF MONOTERPENOIDS, Donatella Loru, Annalisa Vigorito, Andreia Santos, Jackson Tang, M. Eugenia Sanz

WC04 9:21 – 9:36
CONFORMATIONAL STUDY OF DNA SUGARS: FROM THE GAS PHASE TO SOLUTION, Iciar Uriarte, Montserrat Vallejo-López, Emilio J. Cocinero, Francisco Corzana, Benjamin G. Davis

WC05 9:38 – 9:53
FOUR STRUCTURES OF TARTARIC ACID REVEALED IN THE GAS PHASE , Vanessa Cortijo, Verónica Díez, Elena R. Alonso, Santiago Mata, José L. Alonso

Intermission

WC06 10:12 – 10:27
THE CONFORMATIONAL LANDSCAPE OF L-THREONINE: MATRIX ISOLATION INFRARED AND *AB-INITIO* STUDIES, Pankaj Dubey, Anamika Mukhopadhyay, K S Viswanathan

WC07 10:29 – 10:44
THE ROLE OF THE LOCAL CONFORMATION OF A CYCLICALLY CONSTRAINED β-AMINO ACID IN THE SECONDARY STRUCTURES OF A MIXED α/β DIASTEREOMER PAIR, Karl N. Blodgett, Timothy S. Zwier

WC08 10:46 – 11:01
CONFORMATIONAL EXPLOSION: UNDERSTANDING THE COMPLEXITY OF THE PARA-DIALKYLBENZENE POTENTIAL ENERGY SURFACES, Piyush Mishra, Daniel M. Hewett, Timothy S. Zwier

WC09 11:03 – 11:18
BEYOND THE BEND: EXPLORING THE CONFORMATIONAL LANDSCAPE OF DECYL, UNDECYL, AND DODE-CYLBENZENE, Daniel M. Hewett, Timothy S. Zwier

WC10 11:20 – 11:35
CONFORMER-SPECIFIC IR SPECTROSCOPY OF LASER-DESORBED SULFONAMIDE DRUGS: TAUTOMERIC AND CONFORMATIONAL PREFERENCES OF SULFANILAMIDE AND ITS DERIVATIVES, Thomas Uhlemann, Sebastian Seidel, Christian W. Müller

WC11 11:37 – 11:52
SODIATED SUGAR STRUCTURES: CRYOGENIC ION VIBRATIONAL SPECTROSCOPY OF Na$^+$(GLUCOSE) ADDUCTS, Jonathan Voss, Steven J. Kregel, Kaitlyn C Fischer, Etienne Garand

WD. Clusters/Complexes
Wednesday, June 21, 2017 – 8:30 AM
Room: B102 Chemical and Life Sciences

Chair: G. S. Grubbs II, Missouri University of Science and Technology, Rolla, MO, USA

WD01 8:30 – 8:45
DETECTION OF WATER BINDING TO THE OXYGEN EVOLVING COMPLEX USING LOW FREQUENCY SERS, Andrew J. Wilson, Prashant Jain

WD02 8:47 – 9:02
MICROSOLVATION AND THE EFFECTS OF NON-COVALENT INTERACTIONS ON INTRAMOLECULAR DYNAMICS, Lidor Foguel, Zachary Vealey, Patrick Vaccaro

WD03 9:04 – 9:19
THE JET-COOLED HIGH-RESOLUTION IR SPECTRUM OF FORMIC ACID CYCLIC DIMER, Manuel Goubet, Sabath Bteich, Therese R. Huet, Olivier Pirali, Pierre Asselin, Pascale Soulard, Atef Jabri, P. Roy, Robert Georges

WD04 9:21 – 9:36
ROTATIONAL SPECTRA OF 4,4,4-TRIFLUOROBUTYRIC ACID AND THE 4,4,4-TRIFLUOROBUTYRIC ACID-FORMIC ACID COMPLEX, Yoon Jeong Choi, Alex Treviño, Susanna L. Stephens, S. A. Cooke, Stewart E. Novick, Wei Lin

WD05 9:38 – 9:53
THE THz/FIR SPECTRUM OF SMALL WATER CLUSTERS IN HELIUM NANODROPLETS, Gerhard Schwaab, Raffael Schwan, Devendra Mani, Nitish Pal, Arghya Dey, Britta Redlich, Lex van der Meer, Martina Havenith

WD06 9:55 – 10:10
BROADBAND MICROWAVE SPECTROSCOPY AS A TOOL TO STUDY INTERMOLECULAR INTERACTIONS IN THE DIPHENYL ETHER - WATER SYSTEM, Mariyam Fatima, Cristobal Perez, Melanie Schnell

Intermission

WD07 10:29 – 10:44
INVESTIGATION OF THE HYDANTOIN MONOMER AND ITS INTERACTION WITH WATER MOLECULES, Sébastien Gruet, Cristobal Perez, Melanie Schnell

WD08 10:46 – 11:01
HYDRATION OF AN ACID ANHYDRIDE: THE WATER COMPLEX OF ACETIC SULFURIC ANHYDRIDE, CJ Smith, Anna Huff, Becca Mackenzie, Ken Leopold

WD09 11:03 – 11:18
VIBRATIONAL COUPLING IN SOLVATED FORM OF EIGEN PROTON, Jer-Lai Kuo

WD10 11:20 – 11:35
INFRARED PHOTODISSOCIATION CLUSTER STUDIES ON CO_2 INTERACTION WITH TITANIUM OXIDE CATALYST MODELS , Leah G Dodson, Michael C Thompson, J. Mathias Weber

WD11 11:37 – 11:52
OXALATE FORMATION IN TITANIUM–CARBON DIOXIDE ANIONIC CLUSTERS STUDIED BY INFRARED PHOTODISSOCIATION SPECTROSCOPY, Leah G Dodson, Michael C Thompson, J. Mathias Weber

WE. Spectroscopy as an analytical tool
Wednesday, June 21, 2017 – 8:30 AM
Room: 161 Noyes Laboratory

Chair: Brooks Pate, The University of Virginia, Charlottesville, VA, USA

WE01   8:30 – 8:45
PYROLYSIS AND MATRIX-ISOLATION FTIR OF ACETOIN, Sarah Cole, Martha Ellis, John Sowards, Laura R. McCunn

WE02   8:47 – 9:02
EMISSION SPECTROSCOPY OF ATMOSPHERIC-PRESSURE BALL PLASMOIDS: HIGHER ENERGY REVEALS A RICH CHEMISTRY, Scott E. Dubowsky, Amber Nicole Rose, Nick Glumac, Benjamin J. McCall

WE03   9:04 – 9:19
S-NITROSOTHIOLS OBSERVED USING CAVITY RING-DOWN SPECTROSCOPY, Mary Lynn Rad, Benjamin M Gaston, Kevin Lehmann

WE04   9:21 – 9:36
SI-TRACEABLE SCALE FOR MEASUREMENTS OF RADIOCARBON CONCENTRATION, Joseph T. Hodges, Adam J. Fleisher, Qingnan Liu, David A. Long

WE05   9:38 – 9:53
LINEAR AND NON-LINEAR THERMAL LENS SIGNAL OF THE FIFTH C-H VIBRATIONAL OVERTONE OF NAPHTHALENE IN LIQUID SOLUTIONS OF HEXANE, Carlos Manzanares, Marlon Diaz, Ann Barton, Parashu R Nyaupane

WE06   9:55 – 10:10
STUDY OF THE IMIDAZOLIUM-BASED IONIC LIQUID – Ag ELECTRIFIED INTERFACE ON THE CO_2 ELECTROREDUCTION BY SUM FREQUENCY SPECTROSCOPY., Natalia Garcia Rey, Dana Dlott

Intermission

WE07   10:29 – 10:44
SPECDATA: AUTOMATED ANALYSIS SOFTWARE FOR BROADBAND SPECTRA, Jasmine N Oliveira, Marie-Aline Martin-Drumel, Michael C McCarthy

WE08   10:46 – 11:01
IDENTIFYING BROADBAND ROTATIONAL SPECTRA WITH NEURAL NETWORKS, Daniel P. Zaleski, Kirill Prozument

WE09   11:03 – 11:18
ADVANCES IN MOLECULAR ROTATIONAL SPECTROSCOPY FOR APPLIED SCIENCE, Brent Harris, Shelby S. Fields, Robin Pulliam, Matt Muckle, Justin L. Neill

WE10   11:20 – 11:35
FOURIER TRANSFORM MICROWAVE SPECTROSCOPIC STUDIES OF DIMETHYL ETHER AND ETHYLENE FLAMES , Daniel A. Obenchain, Julia Wullenkord, Katharina Kohse-Höinghaus, Jens-Uwe Grabow, Nils Hansen

WE11   11:37 – 11:52
STRATEGIES FOR INTERPRETING TWO DIMENSIONAL MICROWAVE SPECTRA, Marie-Aline Martin-Drumel, Kyle N. Crabtree, Zachary Buchanan

WF. Mini-symposium: ALMA's Molecular View
Wednesday, June 21, 2017 – 1:45 PM
Room: 274 Medical Sciences Building

Chair: Amanda Steber, Universität Hamburg, Hamburg, Germany

WF01   *INVITED TALK*   1:45 – 2:15
PROBING CO FREEZE-OUT AND DESORPTION IN PROTOPLANETARY DISKS, Chunhua Qi

WF02   2:19 – 2:34
AN UPDATED GAS/GRAIN SULFUR NETWORK FOR ASTROCHEMICAL MODELS, Jacob Laas, Paola Caselli

WF03   2:36 – 2:51
A NEW MODEL OF THE CHEMISTRY OF IONIZING RADIATION IN SOLIDS, Christopher N Shingledecker, Eric Herbst

WF04   2:53 – 3:08
THE KEY ROLE OF NUCLEAR-SPIN ASTROCHEMISTRY, Romane Le Gal, Eric Herbst, Changjian Xie, Hua Guo, Dahbia Talbi, Sebastien Muller, Carina Persson

WF05   3:10 – 3:25
ROTATIONAL SPECTROSCOPY OF REACTIVE SPECIES AT THE CENTER FOR ASTROCHEMICAL STUDIES., Valerio Lattanzi, Silvia Spezzano, Paola Caselli

WF06   3:27 – 3:42
A PRESTELLAR CORE 3MM LINE SURVEY: MOLECULAR COMPLEXITY IN L183, Valerio Lattanzi, Luca Bizzocchi, Paola Caselli

Intermission

WF07   4:01 – 4:16
MILLIMETER WAVE SPECTRUM OF THE TWO MONOSULFUR DERIVATIVES OF METHYL FORMATE: S- AND O-METHYL THIOFORMATE, IN THE GROUND AND THE FIRST EXCITED TORSIONAL STATES, Atef Jabri, R. A. Motiyenko, L. Margulès, J.-C. Guillemin, E. A. Alekseev, Isabelle Kleiner, Belén Tercero, Jose Cernicharo

WF08   4:18 – 4:33
VIBRATIONALLY EXCITED c-C_3H_2 RE-VISITED: NEW LABORATORY MEASUREMENTS AND THEORETICAL CALCULATIONS, Harshal Gupta, J. H. Westerfield, Joshua H Baraban, Bryan Changala, Sven Thorwirth, John F. Stanton, Marie-Aline Martin-Drumel, Olivier Pirali, Carl A Gottlieb, Michael C McCarthy

WF09   4:35 – 4:50
MILLIMETER WAVE SPECTRUM OF METHYL KETENE AND ITS SEARCH IN ORION, Celina Bermúdez, L. Margulès, R. A. Motiyenko, Belén Tercero, Jose Cernicharo, J.-C. Guillemin, Y. Ellinger

WF10   4:52 – 5:07
ON THE RELATIVE STABILITY OF CUMULENONE AND ALDEHYDE ISOMERS: WHEN WE HEAT345(Q) THINGS UP, Kelvin Lee, Michael C McCarthy, John F. Stanton

WG. Mini-symposium: Chirality-Sensitive Spectroscopy
Wednesday, June 21, 2017 – 1:45 PM
Room: 116 Roger Adams Lab

Chair: Laurent Nahon, Synchrotron SOLEIL, Gif sur Yvette Cedex, France

WG01   *INVITED TALK*   1:45 – 2:15
OPTICAL ROTATORY DISPERSION: NEW TWISTS ON AN OLD TOPIC, Patrick Vaccaro

WG02   2:19 – 2:34
A CHIRAL TAG STUDY OF THE ABSOLUTE CONFIGURATION OF CAMPHOR, David Pratt, Luca Evangelisti, Taylor Smart, Martin S. Holdren, Kevin J Mayer, Channing West, Brooks Pate

WG03   2:36 – 2:51
ROTATIONAL SPECTROSCOPY OF THE METHYL GLYCIDATE-WATER COMPLEX , Jason Gall, Javix Thomas, Zhibo Wang, Wolfgang Jäger, Yunjie Xu

WG04   2:53 – 3:08
VIBRATIONAL CIRCULAR DICHROISM SPECTRA OF METHYL GLYCIDATE IN CHLOROFORM AND WATER: APPLICATION OF THE CLUSTERS-IN-A-LIQUID MODEL , Angelo Shehan Perera, Javix Thomas, Christian Merten, Yunjie Xu

WG05   3:10 – 3:25
SOLVENT, TEMPERATURE And CONCENTRATION EFFECTS On THE OPTICAL ACTIVITY Of CHIRAL FIVE-And-SIX MEMBERED RING KETONES CONFORMERS , Watheq Al-Basheer

Intermission

WG06   3:44 – 3:59
RAPID-ADIABATIC-PASSAGE CONTROL OF RO-VIBRATIONAL POPULATIONS IN POLYATOMIC MOLECULES, Emil J Zak, Andrey Yachmenev

WG07   4:01 – 4:16
CHIRAL TAGGING OF VERBENONE WITH 3-BUTYN-2-OL FOR ESTABLISHING ABSOLUTE CONFIGURATION AND DETERMINING ENANTIOMERIC EXCESS, Luca Evangelisti, Kevin J Mayer, Martin S. Holdren, Taylor Smart, Channing West, Brooks Pate, Galen Sedo, Frank E Marshall, G. S. Grubbs II

WG08   4:18 – 4:33
COHERENT POPULATION TRANSFER IN CHIRAL MOLECULES USING TAILORED MICROWAVE PULSES, Cristobal Perez, Amanda Steber, Sergio R Domingos, Anna Krin, David Schmitz, Melanie Schnell

WG09   4:35 – 4:50
COMPLEXES OF SMALL CHIRAL MOLECULES: PROPYLENE OXIDE AND 3-BUTYN-2OL, Luca Evangelisti, Channing West, Ellie Coles, Brooks Pate

WG10   4:52 – 5:07
CHIRAL PROCESS MONITORING USING FOURIER TRANSFORM MICROWAVE SPECTROSCOPY, Justin L. Neill, Matt Muckle, Brooks Pate

WG11   5:09 – 5:24
HIGH RESOLUTION FTIR SPECTROSCOPY OF TRISULFANE HSSSH: A CANDIDATE FOR DETECTING PARITY VIOLATION IN CHIRAL MOLECULES, Sieghard Albert, Irina Bolotova, Ziqiu Chen, Csaba Fábri, Martin Quack, Georg Seyfang, Daniel Zindel

WH. Dynamics and kinetics
Wednesday, June 21, 2017 – 1:45 PM
Room: 1024 Chemistry Annex

Chair: J. Gary Eden, University of Illinois, Urbana, IL, USA

WH01   1:45 – 2:00
DIRECT MEASUREMENT OF OD+CO→ *cis*-DOCO, *trans*-DOCO, AND D+CO_2 BRANCHING KINETICS USING TIME-RESOLVED FREQUENCY COMB SPECTROSCOPY, Bryce J Bjork, Thinh Quoc Bui, Bryan Changala, Ben Spaun, Kana Iwakuni, Jun Ye

WH02   2:02 – 2:17
DYNAMIC TIME-RESOLVED CHIRPED-PULSE ROTATIONAL SPECTROSCOPY OF VINYL CYANIDE PHOTO-PRODUCTS IN A ROOM TEMPERATURE FLOW REACTOR, Daniel P. Zaleski, Kirill Prozument

WH03   2:19 – 2:34
TOWARDS A QUANTUM DYNAMICAL STUDY OF THE H_2O+H_2O INELASTIC COLLISION: REPRESENTATION OF THE POTENTIAL AND PRELIMINARY RESULTS, Steve Alexandre Ndengue, Richard Dawes

WH04   2:36 – 2:51
NORMAL MODE ANALYSIS ON THE RELAXATION OF AN EXCITED NITROMETHANE MOLECULE IN ARGON BATH, Luis A. Rivera-Rivera, Albert F. Wagner

WH05   2:53 – 3:08
PROTON TRANSFER AND LOW-BARRIER HYDROGEN BONDING: A SHIFTING VIBRATIONAL LANDSCAPE DICTATED BY LARGE AMPLITUDE TUNNELING, Zachary Vealey, Lidor Foguel, Patrick Vaccaro

Intermission

WH06   3:27 – 3:42
MOLECULAR BEAM SURFACE SCATTERING OF FORMALDEHYDE FROM Au(111): CHARACTERIZATION OF THE DIRECT SCATTER AND TRAPPING-DESORPTION CHANNELS, Bastian C. Krueger, Barratt Park, Sven Meyer, Roman J. V. Wagner, Alec Wodtke, Tim Schaefer

WH07   3:44 – 3:59
ROTATIONALLY-RESOLVED SCATTERING OF FORMALDEHYDE FROM THE Au(111) SURFACE: AN AXIS SPE-CIFIC ROTATIONAL RAINBOW AND ITS ROLE IN TRAPPING PROBABILITY, Barratt Park, Bastian C. Krueger, Sven Meyer, Alexander Kandratsenka, Alec Wodtke, Tim Schaefer

WH08   *Post-Deadline Abstract - Original Abstract Withdrawn*   4:01 – 4:16
CHARACTERIZATION OF EXTENDED TIME SCALE 2D IR PROBES OF PROTEINS, Sashary Ramos, Amanda L Le Sueur, Keith J Scott, Megan Thielges

WH09   4:18 – 4:33
NONLINEAR PHOTOCHROMIC SWITCHING IN THE PLASMONIC FIELD OF A NANOPARTICLE ARRAY, Christopher J Otolski, Christos Argyropoulos, Christopher G. Elles

WH10   4:35 – 4:50
ENERGY POOLING, ION RECOMBINATION, AND REACTIONS OF RUBIDIUM AND CESIUM IN HYDROCARBON GASSES. , Sean Michael Bresler, J. Park, Michael Heaven

WH11   4:52 – 5:07
ANALYSIS OF THREE-BODY FORMATION RATES COEFFICIENTS OF Hg^*, Hg_2^*, AND Hg_3^* VIA PHOTOEXCITA-TION OF Hg VAPOR, Wenting Wendy Chen, J. Gary Eden

WI. Theory and Computation
Wednesday, June 21, 2017 – 1:45 PM
Room: B102 Chemical and Life Sciences

Chair: Tucker Carrington, Queen's University, Kingston, ON, Canada

WI01 1:45 – 2:00
A CANONICAL APPROACH TO GENERATE MULTIDIMENSIONAL POTENTIAL ENERGY SURFACES, Jay R. Walton, Luis A. Rivera-Rivera, Robert R. Lucchese

WI02 2:02 – 2:17
AB INITIO CALCULATIONS OF THE GROUND AND EXCITED STATES OF THE ZNTE MOLECULE AND ITS IONS $ZNTE^+$ AND $ZNTE^-$, Nour el Houda Bensiradj, Ourida Ouamerali, Azeddine Dekhira, Timón Vicente

WI03 2:19 – 2:34
THEORETICAL CALCULATION OF THE UV-VIS SPECTRAL BAND LOCATIONS OF PAHS WITH UNKNOWN SYNTHESES PROCEDURES AND PROSPECTIVE CARCINOGENIC ACTIVITY, Jorge Oswaldo Ona-Ruales, Yosadara Ruiz-Morales

WI04 2:36 – 2:51
THEORETICAL INVESTIGATION OF PHOTOASSOCIATIVE EXCITATION SPECTROSCOPY OF XENON MONOIODIDE, Wenting Wendy Chen, Fang Shen, J. Gary Eden

WI05 2:53 – 3:08
INSIGHT INTO THE CHARGE TRANSFER MECHANISMS OF HEAVY ATOM SUBSTITUTED MALDI MATRICES, Chelsea N Bridgmohan, Lichang Wang, Kristopher M Kirmess

WI06 3:10 – 3:25
SCALAR RELATIVISTIC EQUATION-OF-MOTION COUPLED CLUSTER CALCULATIONS OF CORE-IONIZED/EXCITED STATES, Lan Cheng

Intermission

WI07 3:44 – 3:59
MULTI-STATE EXTRAPOLATION OF UV/VIS ABSORPTION SPECTRA WITH QM/QM HYBRID METHODS, Sijin Ren, Marco Caricato

WI08 4:01 – 4:16
A PROTOCOL FOR HIGH-ACCURACY THEORETICAL THERMOCHEMISTRY, Bradley Welch, Richard Dawes

WI09 4:18 – 4:33
INVESTIGATION OF SOLVATION EFFECTS ON OPTICAL ROTATORY DISPERSION USING THE POLARIZABLE CONTINUUM MODEL, Tal Aharon, Paul M Lemler, Patrick Vaccaro, Marco Caricato

WI10 4:35 – 4:50
A CODE FOR AUTOMATED CONSTRUCTION OF POTENTIAL ENERGY SURFACES FOR VAN DER WAALS SYSTEMS, Ernesto Quintas Sánchez, Richard Dawes

WI11 4:52 – 5:07
TRIPLET TUNING – A NEW "BLACK-BOX" COMPUTATIONAL SCHEME FOR PHOTOCHEMICALLY ACTIVE MOLECULES, Zhou Lin, Troy Van Voorhis

WJ. Lineshapes, collisional effects
Wednesday, June 21, 2017 – 1:45 PM
Room: 161 Noyes Laboratory

Chair: Iouli E Gordon, Harvard-Smithsonian Center for Astrophysics, Cambridge, MA, USA

WJ01　　　　　　　　　　　　　　　　　　　　　　　　　　　**1:45–2:00**
NUMERICAL EVALUATION OF PARAMETER CORRELATION IN THE HARTMANN-TRAN LINE PROFILE, Erin M. Adkins, Zachary Reed, Joseph T. Hodges

WJ02　　　　　　　　　　　　　　　　　　　　　　　　　　　**2:02–2:17**
TEMPERATURE DEPENDENCE OF NEAR-INFRARED CO_2 LINE SHAPES MEASURED BY CAVITY RING-DOWN SPECTROSCOPY, Mélanie Ghysels, Adam J. Fleisher, Qingnan Liu, Joseph T. Hodges

WJ03　　　　　　　　　　　　　　　　　　　　　　　　　　　**2:19–2:34**
EXPERIMENTAL STUDY OF TEMPERATURE-DEPENDENCE LAWS OF NON-VOIGT ABSORPTION LINE SHAPE PARAMETERS, Jonas Wilzewski, Manfred Birk, Joep Loos, Georg Wagner

WJ04　　　　　　　　　　　　　　　　　　　　　　　　　　　**2:36–2:51**
LINE SHAPES AND INTENSITIES OF CARBON MONOXIDE TRANSITIONS IN THE (3→0) AND (4→1) BANDS, Zachary Reed, Oleg Polyansky, Joseph T. Hodges

WJ05　　　　　　　　　　　　　　　　　　　　　　　　　　　**2:53–3:08**
RELAXATION MATRICES OF THE NH_3 MOLECULE IN PARALLEL AND PERPENDICUAR BANDS, Qiancheng Ma, C. Boulet, Richard Tipping

WJ06　　　　　　　　　　　　　　　　　　　　　　　　　　　**3:10–3:25**
SATURATION DIP MEASUREMENTS OF HIGH-J TRANSITIONS IN THE $v_1 + v_3$ BAND OF C_2H_2: ABSOLUTE FREQUENCIES AND SELF-BROADENING, Trevor Sears, Sylvestre Twagirayezu, Gregory Hall

WJ07　　　　　　　　　　　　　　　　　　　　　　　　　　　**3:27–3:42**
TIME- AND FREQUENCY-DOMAIN SIGNATURES OF VELOCITY CHANGING COLLISIONS IN SUB-DOPPLER SATURATION SPECTRA AND PRESSURE BROADENING, Gregory Hall, Hong Xu, Damien Forthomme, Paul Dagdigian, Trevor Sears

Intermission

WJ08　　　　　　　　　　　　　　　　　　　　　　　　　　　**4:01–4:16**
RECENT PROGRESS ON LABFIT: A MULTISPECTRUM ANALYSIS PROGRAM FOR FITTING LINESHAPES INCLUDING THE HTP MODEL AND TEMPERATURE DEPENDENCE, Matthew J. Cich, Alexandre Guillaume, Brian Drouin, D. Chris Benner

WJ09　　　　　　　　　　　　　　　　　　　　　　　　　　　**4:18–4:33**
MULTISPECTRAL FITTING VALIDATION OF THE SPEED DEPENDENT VOIGT PROFILE AT UP TO 1300K IN WATER VAPOR WITH A DUAL FREQUENCY COMB SPECTROMETER, Paul James Schroeder, Matthew J. Cich, Jinyu Yang, Brian Drouin, Greg B Rieker

WJ10　　　　　　　　　　　　　　　　　　　　　　　　　　　**4:35–4:50**
HIGH PRECISION MEASUREMENTS OF LINE MIXING AND COLLISIONAL INDUCED ABSORPTION IN THE O_2 A-BAND , Erin M. Adkins, Mélanie Ghysels, David A. Long, Joseph T. Hodges

WJ11　　　　　　　　　　　　　　　　　　　　　　　　　　　**4:52–5:07**
COLLISON-INDUCED ABSORPTION OF OXYGEN MOLECULE AS STUDIED BY HIGH SENSITIVITY SPECTROSCOPY, Wataru Kashihara, Atsushi Shoji, Akio Kawai

WJ12　　　　　　　　　　　　　　　　　　　　　　　　　　　**5:09–5:24**
THEORY OF COLLISION-INDUCED ABSORPTION FOR ELECTRONIC TRANSITIONS IN THE ATMOSPHERICALLY RELEVANT O_2-O_2 AND O_2-N_2 PAIRS., Tijs Karman, Ad van der Avoird, Gerrit Groenenboom

WK. Metal containing
Wednesday, June 21, 2017 – 1:45 PM
Room: 140 Burrill Hall

Chair: Leah C O'Brien, Southern Illinois University, Edwardsville, IL, USA

WK01   1:45 – 2:00
LASER INDUCED FLUORESCENCE SPECTROSCOPY OF JET-COOLED MgOMg, <u>Michael N. Sullivan</u>, Daniel J. Frohman, Michael Heaven, Wafaa M Fawzy

WK02   2:02 – 2:17
HIGH RESOLUTION LASER SPECTROSCOPY OF THE $[15.45]0 - a^3\Delta_1$ TRANSITION OF TANTALUM MONONITRIDE, TaN, <u>Colan Linton</u>, Timothy Steimle, Damian L Kokkin

WK03   2:19 – 2:34
ELECTRONIC TRANSITIONS OF TUNGSTEN MONOSULFIDE , L. Г. Tsang, Man-Chor Chan, Wenli Zou, <u>Allan S.C. Cheung</u>

WK04   2:36 – 2:51
RE-VISITING THE ELECTRONIC ENERGY MAP OF THE COPPER DIMER BY DOUBLE-RESONANT FOUR-WAVE MIXING, Bradley Visser, Peter Bornhauser, Martin Beck, Gregor Knopp, Roberto Marquardt, Christophe Gourlaouen, Jeroen A. van Bokhoven, <u>Peter Radi</u>

WK05   2:53 – 3:08
THE LOW-LYING ELECTRONIC STATES OF SCANDIUM MONOCARBIDE, ScC, Chiao-Wei Chen, <u>Anthony Merer</u>, Yen-Chu Hsu

Intermission

WK06   3:27 – 3:42
THRESHOLD IONIZATION AND SPIN-ORBIT COUPLING OF CERIUM MONOXIDE, <u>Wenjin Cao</u>, Yuchen Zhang, Lu Wu, Dong-Sheng Yang

WK07   3:44 – 3:59
A DATABASE FOR TRANSITION METAL DIATOMICS, <u>Corinne Duperrouzel</u>, Nikesh S. Dattani, Paul W. Ayers

WK08   4:01 – 4:16
REANALYSIS OF THE $a\ ^4\Sigma^- - X\ ^2\Pi_r$ TRANSITION OF GeH USING INTRACAVITY LASER SPECTROSCOPY, <u>Jack C Harms</u>, Leah C O'Brien, James J O'Brien

WK09   4:18 – 4:33
A REEXAMINATION OF THE RED BAND OF CuO: ANALYSIS OF THE $[16.5]\ ^2\Sigma^- - X\ ^2\Pi_i$ TRANSITION OF ^{63}CuO and ^{65}CuO, <u>Jack C Harms</u>, Ethan M Grames, Sirkhoo Yun, Bushra Ahmed, Leah C O'Brien, James J O'Brien

WK10   4:35 – 4:50
ANALYSIS OF SOME NEW ELECTRONIC TRANSITIONS OBSERVED USING INTRACAVITY LASER SPECTROSCOPY (ILS): POSSIBLE IDENTIFICATION OF HCuN, <u>Jack C Harms</u>, Ethan M Grames, Leah C O'Brien, James J O'Brien

WK11   4:52 – 5:07
THE PURE ROTATIONAL SPECTRUM OF KO, <u>Mark Burton</u>, Benjamin Russ, Phillip M. Sheridan, Matthew Bucchino, Lucy M. Ziurys

RA. Plenary
Thursday, June 22, 2017 – 8:30 AM
Room: Foellinger Auditorium

Chair: Anthony Remijan, NRAO, Charlottesville, VA, USA

RA01   8:30 – 9:10
PRECISION SPECTROSCOPY OF MOLECULAR HYDROGEN AND THE SEARCH FOR NEW PHYSICS,
Wim Ubachs

RA02   9:15 – 9:55
EXPLORING THE DETAILS OF INTERMOLECULAR INTERACTIONS VIA A SYSTEMATIC CHARACTERIZATION
OF THE STRUCTURES OF THE BIMOLECULAR HETERODIMERS FORMED BETWEEN PROTIC ACIDS AND
HALOETHYLENES, Helen O. Leung

Intermission

RAO AWARDS   **10:35**
Presentation of Awards by Gary Douberly, University of Georgia

2016 Rao Award Winners
David Grimes, Massachusetts Institute of Technology
Kasper Mackeprang, University of Copenhagen
Josey E. Topolski, Indiana University Bloomington

MILLER PRIZE   **10:45**
Introduction by Susanna Widicus Weaver, Emory University

RA03   *Miller Prize Lecture*   10:50 – 11:05
AUTOMATED MICROWAVE DOUBLE RESONANCE SPECTROSCOPY: A TOOL TO IDENTIFY AND CHARACTER-
IZE CHEMICAL COMPOUNDS, Marie-Aline Martin-Drumel, Michael C McCarthy, David Patterson, Brett A. McGuire,
Kyle N. Crabtree

COBLENTZ AWARD   **11:10**
Presentation of Award by Linda Kidder, Coblentz Society

RA04   *Coblentz Society Award Lecture*   11:15 – 11:55
BIOORTHOGONAL CHEMICAL IMAGING FOR BIOMEDICINE , Wei Min

RF. Mini-symposium: ALMA's Molecular View
Thursday, June 22, 2017 – 1:45 PM
Room: 274 Medical Sciences Building

Chair: Marie-Aline Martin-Drumel, Institut des Sciences Moléculaires d'Orsay, Orsay, France

RF01   *INVITED TALK*   1:45 – 2:15
FEEDING, FEEDBACK AND THE GROWTH OF GALAXIES – MOLECULES AS TOOLS FOR PROBING GALAXY EVOLUTION, Susanne Aalto

RF02   2:19 – 2:34
WATER EMISSION FROM EARLY UNIVERSE, Sreevani Jarugula, Joaquin Vieira

RF03   2:36 – 2:51
THE SPT+ALMA CO REDSHIFT SURVEY OF DUSTY GALAXIES, Joaquin Vieira

RF04   2:53 – 3:08
PHOSPHORUS CHEMISTRY IN OXYGEN RICH STARS, Jacob Bernal, Deborah Schmidt, Julie Anderson, Lucy M. Ziurys

RF05   3:10 – 3:25
THE HIGH RESOLUTION VIBRATION-ROTATION SPECTRUM OF SiH^+, Jose Luis Domenech

Intermission

RF06   *Journal of Molecular Spectroscopy Review Lecture*   3:44 – 4:14
BUILDING BLOCKS OF DUST AND LARGE ORGANIC MOLECULES: A COORDINATED LABORATORY AND ASTRONOMICAL STUDY OF AGB STARS, Michael C McCarthy, Carl A Gottlieb, Jose Cernicharo

RF07   4:18 – 4:33
THE (SUB-)MILLIMETER-WAVE SPECTRUM OF PROPANAL, Oliver Zingsheim, Holger S. P. Müller, Frank Lewen, Stephan Schlemmer

RF08   4:35 – 4:50
ROTATIONAL SPECTRA AND STRUCTURAL DETERMINATION OF HCCNCS, Wenhao Sun, Rebecca Davis, Jennifer van Wijngaarden

RF09   4:52 – 5:07
THE JET-COOLED ROTATIONAL SPECTRUM OF GLYCINAMIDE, AN AMINOACID PRECURSOR, Elena R. Alonso, Lucie Kolesniková, Zbigniew Kisiel, J.-C. Guillemin, José L. Alonso

RF10   5:09 – 5:24
THE MICROWAVE SPECTRUM OF LACTALDEHYDE, THE SIMPLEST CHIRAL SUGAR. , Elena R. Alonso, Lucie Kolesniková, Carlos Cabezas, Santiago Mata, J.-C. Guillemin, José L. Alonso

RF11   5:26 – 5:41
EXTENDING THE MILLIMETER-SUBMILLIMETER SPECTRUM OF PROTONATED FORMALDEHYDE, Kevin Roenitz, Luyao Zou, Susanna L. Widicus Weaver

RG. Mini-symposium: Chirality-Sensitive Spectroscopy
Thursday, June 22, 2017 – 1:45 PM
Room: 116 Roger Adams Lab

Chair: Yunjie Xu, University of Alberta, Edmonton, AB, Canada

RG01   *INVITED TALK*   1:45–2:15
WHAT CAN WE LEARN ON GAS PHASE CHIRAL COMPOUNDS BY PHOTOELECTRON CIRCULAR DICHROISM ?, Laurent Nahon

RG02   2:19–2:34
INTERNAL DYNAMICS AND CHIRAL ANALYSIS OF PULEGONE, USING MICROWAVE BROADBAND SPECTROSCOPY, Anna Krin, Cristobal Perez, Melanie Schnell, María del Mar Quesada-Moreno, Juan Jesús López-González, Juan Ramón Avilés-Moreno, Pablo Pinacho, Susana Blanco, Juan Carlos Lopez

RG03   2:36–2:51
A CHIRAL TAGGING STRATEGY FOR DETERMINING ABSOLUTE CONFIGURATION AND ENANTIOMERIC EXCESS BY MOLECULAR ROTATIONAL SPECTROSCOPY, Luca Evangelisti, Walther Caminati, David Patterson, Javix Thomas, Yunjie Xu, Channing West, Brooks Pate

RG04   2:53–3:08
HIGH SENSITIVITY 1-D AND 2-D MICROWAVE SPECTROSCOPY VIA CRYOGENIC BUFFER GAS COOLING, David Patterson, Sandra Eibenberger

RG05   3:10–3:25
NATURAL OPTICAL ACTIVITY OF CHIRAL EPOXIDES: THE INFLUENCE OF STRUCTURE AND ENVIRONMENT ON THE INTRINSIC CHIROPTICAL RESPONSE, Paul M Lemler, Clayton L. Craft, Patrick Vaccaro

RG06   3:27–3:42
CHARACTERIZATION OF INTERMOLECULAR INTERACTIONS AT PLAY IN THE 2,2,2-TRIFLUOROETHANOL TRIMERS USING CAVITY AND CHIRPED-PULSE MICROWAVE SPECTROSCOPY, Nathan A Seifert, Javix Thomas, Wolfgang Jäger, Yunjie Xu

Intermission

RG07   *INVITED TALK*   4:01–4:31
ADVANCED APPLICATIONS OF VIBRATIONAL CIRCULAR DICHROISM: FROM SMALL CHIRAL MOLECULES TO FIBRILS, Rina K. Dukor

RG08   4:35–4:50
THE MICROWAVE SPECTRA AND MOLECULAR STRUCTURES OF 2-(TRIFLUOROMETHYL)-OXIRANE AND 2-VINYLOXIRANE, TWO CANDIDATES FOR CHIRAL ANALYSIS VIA NONCOVALENT CHIRAL TAGGING, Mark D. Marshall, Helen O. Leung, Desmond Acha, Kevin Wang

RG09   4:52–5:07
INTRINSIC OPTICAL ACTIVITY AND CONFORMATIONAL FLEXIBILITY: NEW INSIGHTS ON THE ROLE OF RING MORPHOLOGY FROM CYCLIC AMINES, Clayton L. Craft, Paul M Lemler, Patrick Vaccaro

RG10   5:09–5:24
ABSOLUTE CONFIGURATION OF 3-METHYLCYCLOHEXANONE BY CHIRAL TAG ROTATIONAL SPECTROSCOPY AND VIBRATIONAL CIRCULAR DICHROISM, Luca Evangelisti, Martin S. Holdren, Kevin J Mayer, Taylor Smart, Channing West, Brooks Pate

RG11   *Post-Deadline Abstract*   5:26–5:41
CHIRALITY RECOGNITION IN CAMPHOR - 1,2-PROPANEDIOL COMPLEXES, Cristobal Perez, Mariyam Fatima, Anna Krin, Melanie Schnell

RG12   5:43–5:58
THE COMPLETE HEAVY-ATOM STRUCTURE OF A CP-FTMW CHIRAL TAG PRECURSOR, VERBENONE, Frank E Marshall, Channing West, Galen Sedo, Brooks Pate, G. S. Grubbs II

RH. Clusters/Complexes
Thursday, June 22, 2017 – 1:45 PM
Room: 1024 Chemistry Annex

Chair: Jer-Lai Kuo, Academia Sinica, Taipei, Taiwan

RH01   1:45 – 2:00
MILLIMETER-WAVE SPECTROSCOPY OF He-HCN AND He-DCN: ENERGY LEVELS NEAR THE DISSOCIATION LIMIT., Kensuke Harada, Keiichi Tanaka

RH02   2:02 – 2:17
THEORETICAL STUDY OF GROUP 14 $M^+(^2P_J)$-RG COMPLEXES (M^+ = C^+, Si^+; RG = He - Ar), William Duncan Tuttle, Rebecca L. Thorington, Timothy G. Wright, Larry A. Viehland

RH03   2:19 – 2:34
ROVIBRATIONAL SPECTRUM OF THE Ar-NO COMPLEX IN 5.3 μm REGION , Chuanxi Duan

RH04   2:36 – 2:51
WEAK INTERACTIONS AND CO_2 MICROSOLVATION IN THE CIS-1,2-DIFLUOROETHYLENE...CO_2 COMPLEX, William Trendell, Rebecca A. Peebles, Sean A. Peebles

RH05   2:53 – 3:08
MICROWAVE SPECTROSCOPIC STUDY OF THE ATMOSPHERIC OXIDATION PRODUCT *m*-TOLUIC ACID AND ITS MONOHYDRATE, Mohamad Al-Jabiri, Elijah G Schnitzler, Nathan A Seifert, Wolfgang Jäger

RH06   3:10 – 3:25
THE ETHANOL-CO_2 DIMER IS AN ELECTRON DONOR-ACCEPTOR COMPLEX, Brett A. McGuire, Marie-Aline Martin-Drumel, Michael C McCarthy

Intermission

RH07   3:44 – 3:59
MICROWAVE SPECTRUM AND STRUCTURE OF THE METHANE-PROPANE COMPLEX, Karen I. Peterson, Wei Lin, Eric A. Arsenault, Yoon Jeong Choi, Stewart E. Novick

RH08   4:01 – 4:16
NON-COVALENT INTERACTIONS AND INTERNAL DYNAMICS IN PYRIDINE-AMMONIA: A COMBINED QUANTUM-CHEMICAL AND MICROWAVE SPECTROSCOPY STUDY, Lorenzo Spada, Nicola Tasinato, Fanny Vazart, Vincenzo Barone, Walther Caminati, Cristina Puzzarini

RH09   4:18 – 4:33
SPECTROSCOPIC CHARACTERIZATION OF N_2O_5 HALIDE CLUSTERS AND THE FORMATION OF HNO_3 , Joanna K. Denton, Patrick J Kelleher, Fabian Menges, Mark Johnson

RH10   4:35 – 4:50
ETHANOL DIMER: OBSERVATION OF THREE NEW CONFORMERS BY BROADBAND ROTATIONAL SPECTROSCOPY, Donatella Loru, Isabel Peña, M. Eugenia Sanz

RH11   *Post-Deadline Abstract*   4:52 – 5:07
BROADBAND FTMW SPECTROSCOPY OF THE UREA-ARGON AND THIOUREA-ARGON COMPLEXES, Chris Medcraft, Dror M. Bittner, Graham A. Cooper, John C Mullaney, Nick Walker

RI. Instrument/Technique Demonstration
Thursday, June 22, 2017 – 1:45 PM
Room: B102 Chemical and Life Sciences

Chair: Kyle N. Crabtree, University of California, Davis, CA, USA

RI01 1:45–2:00
DOPPLER-FREE TWO-PHOTON ABSORPTION SPECTROSCOPY OF VIBRONIC EXCITED STATES OF NAPHTHA-LENE ASSISTED BY AN OPTICAL FREQUENCY COMB, Akiko Nishiyama, Kazuki Nakashima, Masatoshi Misono, Masaaki Baba

RI02 2:02–2:17
TWO-PHOTON ABSORPTION SPECTROSCOPY OF RUBIDIUM WITH A DUAL-COMB TEQUNIQUE, Akiko Nishiyama, Satoru Yoshida, Takuya Hariki, Yoshiaki Nakajima, Kaoru Minoshima

RI03 2:19–2:34
SPIN POLARIZATION SPECTROSCOPY OF ALKALI-NOBLE GAS INTERATOMIC POTENTIALS, Andrey E. Mironov, William Goldshlag, J. Gary Eden

RI04 2:36–2:51
MOLECULAR STRUCTURE AND DYNAMICS PROBED BY PHOTOIONIZATION OUT OF RYDBERG STATES, Fedor Rudakov

RI05 2:53–3:08
LASER-MILLIMETER-WAVE TWO-PHOTON RABI OSCILLATIONS EN ROUTE TO COHERENT POPULATION TRANSFER, David Grimes, Timothy J Barnum, Yan Zhou, Tony Colombo, Robert W Field

Intermission

RI06 3:27–3:42
HIGH HARMONIC GENERATION XUV SPECTROSCOPY FOR STUDYING ULTRAFAST PHOTOPHYSICS OF CO-ORDINATION COMPLEXES, Elizabeth S Ryland, Ming-Fu Lin, Kristin Benke, Max A Verkamp, Kaili Zhang, Josh Vura-Weis

RI07 3:44–3:59
EXTENDING TABLETOP XUV SPECTROSCOPY TO THE LIQUID PHASE TO EXAMINE TRANSITION METAL CATALYSTS, Kristin Benke, Elizabeth S Ryland, Josh Vura-Weis

RI08 4:01–4:16
ULTRAFAST EXTREME ULTRAVIOLET SPECTROSCOPY OF METHYLAMMONIUM LEAD IODIDE PEROVSKITE FOR CARRIER SPECIFIC PHOTOPHYSICS, Max A Verkamp, Ming-Fu Lin, Elizabeth S Ryland, Kristin Benke, Josh Vura-Weis

RI09 4:18–4:33
LIQUID PHASE SUPERCONTINUUM FIBER-LOOP CAVITY ENHANCED ABSORPTION SPECTROSCOPY FOR H_2O IN ORGANICS, Mingyun Li, Kevin Lehmann

RI10 4:35–4:50
MULTIPLEXED SATURATION SPECTROSCOPY WITH ELECTRO-OPTIC FREQUENCY COMBS, David A. Long, Adam J. Fleisher, David F. Plusquellic, Joseph T. Hodges

RI11 4:52–5:07
DIRECT ABSORPTION SPECTROSCOPY WITH ELECTRO-OPTIC FREQUENCY COMBS, Adam J. Fleisher, David A. Long, David F. Plusquellic, Joseph T. Hodges

RJ. Radicals
Thursday, June 22, 2017 – 1:45 PM
Room: 161 Noyes Laboratory

Chair: Neil J Reilly, University of Massachusetts Boston, Boston, MA, USA

RJ01   1:45 – 2:00
INFRARED SPECTRUM OF THE CYCLOBUTYL RADICAL IN He DROPLETS, <u>Alaina R. Brown</u>, Peter R. Franke, Gary E. Douberly

RJ02   2:02 – 2:17
$O(^3P)$ DOPED HELIUM DROPLETS, <u>Joseph T. Brice</u>, Gary E. Douberly

RJ03   2:19 – 2:34
THE O_2 + ETHYL REACTION IN HELIUM NANODROPLETS: INFRARED SPECTROSCOPY OF THE ETHYLPEROXY RADICAL, <u>Peter R. Franke</u>, Gary E. Douberly

RJ04   2:36 – 2:51
INFRARED SPECTRA OF THE 1-CHLOROMETHYL-1-METHYLALLYL AND 1-CHLOROMETHYL-2-METHYLALLYL RADICALS ISOLATED IN SOLID $PARA$-HYDROGEN, <u>Jay C. Amicangelo</u>, Yuan-Pern Lee

RJ05   2:53 – 3:08
THERMAL DECOMPOSITION OF METHYL ACETATE (CH_3COOCH_3) IN A FLASH-PYROLYSIS MICRO-REACTOR, <u>Jessica P Porterfield</u>, David H. Bross, Branko Ruscic, James H. Thorpe, Thanh Lam Nguyen, Joshua H Baraban, John F. Stanton, John W Daily, Barney Ellison

RJ06   3:10 – 3:25
BROADBAND MICROWAVE STUDY OF REACTION INTERMEDIATES AND PRODUCTS THROUGH THE PYROLYSIS OF OXYGENATED BIOFUELS , <u>Chamara Abeysekera</u>, Alicia O. Hernandez-Castillo, Sean Fritz, Timothy S. Zwier

Intermission

RJ07   3:44 – 3:59
HIGH-RESOLUTION THz MEASUREMENTS OF BrO GENERATED IN AN INDUCTIVELY COUPLED PLASMA, <u>Deacon J Nemchick</u>, Brian Drouin

RJ08   4:01 – 4:16
DETECTION AND CHARACTERIZATION OF THE STANNYLENE (SnH_2) RADICAL IN THE GAS PHASE, <u>Tony Smith</u>, Dennis Clouthier

RJ09   4:18 – 4:33
FOURIER TRANSFORM ABSORPTION SPECTROSCOPY OF C_3 IN THE ν_3 ANTISYMMETRIC STRETCH MODE REGION, Michel Vervloet, Marie-Aline Martin-Drumel, Dennis W. Tokaryk, <u>Olivier Pirali</u>

RJ10   4:35 – 4:50
IDENTIFICATION OF A JAHN-TELLER ACTIVE GAS PHASE SILOXY FREE RADICAL (Cl_3SiO) BY LIF SPECTROSCOPY, <u>Tony Smith</u>, Dennis Clouthier

RJ11   4:52 – 5:07
PHOTOELECTRON IMAGING SPECTROSCOPY AS A WINDOW TO UNEXPECTED MOLECULES, <u>Christopher C Blackstone</u>

RJ12   5:09 – 5:24
LASER SPECTROSCOPY OF THE JET-COOLED SiCF FREE RADICAL, <u>Tony Smith</u>, Dennis Clouthier

FA. Planetary atmospheres
Friday, June 23, 2017 – 8:30 AM
Room: 274 Medical Sciences Building

Chair: James Neil Hodges, Old Dominion University, Norfolk, Virginia, United States

FA01 8:30–8:45
SPECTRAL LINE SHAPES IN THE ν_3 Q BRANCH OF $^{12}CH_4$ NEAR 3.3 μm, V. Malathy Devi, D. Chris Benner, Robert R. Gamache, Mary Ann H. Smith, Robert L. Sams

FA02 8:47–9:02
LINE POSITIONS OF CENTRIFUGAL DISTORSION INDUCED ROTATIONAL TRANSITIONS OF METHANE MEASURED UP TO 2.6 THZ AT SUB-MHZ ACCURACY WITH A CW-THZ PHOTOMIXING SPECTROMETER, Cédric Bray, Arnaud Cuisset, Francis Hindle, Gaël Mouret, Robin Bocquet, Vincent Boudon

FA03 9:04–9:19
INFRARED ABSORPTION CROSS SECTIONS OF COLD PROPANE IN THE LOW FREQUENCY REGION BETWEEN 600 - 1300 cm^{-1}, Andy Wong, Robert J. Hargreaves, Brant E. Billinghurst, Peter F. Bernath

FA04 9:21–9:36
FIRST HIGH RESOLUTION IR SPECTRA OF 2-^{13}C-PROPANE. THE ν_9 B-TYPE BAND NEAR 366.767 cm^{-1} AND THE ν_{26} C-TYPE BAND NEAR 746.615 cm^{-1}. DETERMINATION OF GROUND AND UPPER STATE CONSTANTS., S.J. Daunt, Robert Grzywacz, Walter Lafferty, Jean-Marie Flaud, Brant E. Billinghurst

FA05 9:38–9:53
FIRST HIGH RESOLUTION IR SPECTRA OF 1-^{13}C-PROPANE. THE ν_9 B-TYPE BAND NEAR 366.404 cm^{-1} AND THE ν_{26} C-TYPE BAND NEAR 748.470 cm^{-1}. DETERMINATION OF GROUND AND UPPER STATE CONSTANTS., S.J. Daunt, Robert Grzywacz, Walter Lafferty, Jean-Marie Flaud, Brant E. Billinghurst

FA06 9:55–10:10
FIRST HIGH RESOLUTION IR STUDY OF THE ν_{14} (A') A-TYPE BAND NEAR 421.847 cm^{-1} OF 2-^{13}C-PROPENE, S.J. Daunt, Robert Grzywacz, Brant E. Billinghurst

Intermission

FA07 10:29–10:44
HIGH RESOLUTION INFRARED CAVITY ENHANCED ABSORPTION OF PROPYNE, Parashu R Nyaupane, Marlon Diaz, Ann Barton, Carlos Manzanares

FA08 10:46–11:01
AB INITIO CHARACTERIZATION OF SULFUR COMPOUNDS AND THEIR CHEMISTRY FOR VENUS AND THE INTERSTELLAR MEDIUM, David E. Woon

FA09 11:03–11:18
INFRARED SPECTROSCOPIC AND THEORETICAL STUDY OF THE HC_nO^+(N=5-12) CATIONS, Wei Li, Jiaye Jin, Guanjun Wang, Mingfei Zhou

FA10 11:20–11:35
IMPACT OF INSERTION REACTION OF $O(^1D)$ INTO THE CARBONIC ACID MOLECULE IN THE ATMOSPHERE OF EARTH AND MARS, Sourav Ghoshal, Montu K. Hazra

FA11 11:37–11:52
PHOTOCHEMICAL FORMATION OF SULFUR-CONTAINING AEROSOLS, Jay A Kroll, Veronica Vaida

FB. (Hyper)fine structure, tunneling
Friday, June 23, 2017 – 8:30 AM
Room: 116 Roger Adams Lab

Chair: Isabelle Kleiner, CNRS et Universités Paris-Est et Paris Diderot, Créteil, France

FB01　　　　　　　　　　　　　　　　　　　　　　　　　　　　　　　　　　　　　8:30 – 8:45
MICROWAVE SPECTRUM OF 1-SILA-1-ISOCYANOCYCLOPENT-3-ENE, Frank E Marshall, Daniel V. Hickman, Gamil A Guirgis, Michael H. Palmer, Charles J. Wurrey, Nicole Moon, Thomas D. Persinger, G. S. Grubbs II

FB02　　　　　　　　　　　　　　　　　　　　　　　　　　　　　　　　　　　　　8:47 – 9:02
NUCLEAR QUADRUPOLE COUPLING IN SiH_2I_2 DUE TO THE PRESENCE OF TWO IODINE NUCLEI, Eric A. Arsenault, Daniel A. Obenchain, W. Orellana, Stewart E. Novick

FB03　　　　　　　　　　　　　　　　　　　　　　　　　　　　　　　　　　　　　9:04 – 9:19
MICROWAVE SPECTRUM OF THE H_2S DIMER: OBSERVATION OF $K_a=1$ LINES, Arijit Das, Pankaj Mandal, Frank J Lovas, Chris Medcraft, Elangannan Arunan

FB04　　　　　　　　　　　　　　　　　　　　　　　　　　　　　　　　　　　　　9:21 – 9:36
ROTATIONAL SPECTRA AND NUCLEAR QUADRUPOLE COUPLING CONSTANTS OF 4-HALOPYRAZOLES $C_3N_2H_3X$ (X = Br, I), Graham A. Cooper, Chris Medcraft, Anthony Legon, Nick Walker

FB05　　　　　　　　　　　　　　　　　　　　　　　　　　　　　　　　　　　　　9:38 – 9:53
AN INVESTIGATION OF THE DIPOLE FORBIDDEN TRANSITION EFFECTS IN BROMOFLUOROCARBONS AS IT PERTAINS TO 3 BROMO 1,1,1,2,2-PENTAFLUOROPROPANE USING CP-FTMW SPECTROSCOPY, Frank E Marshall, Nicole Moon, Thomas D. Persinger, David Joseph Gillcrist, N. E. Shreve, William C. Bailey, G. S. Grubbs II

Intermission

FB06　　　　　　　　　　　　　　　　　　　　　　　　　　　　　　　　　　　　10:12 – 10:27
A REINVESTIGATION OF THE ELECTRONIC PROPERTIES OF 2-BROMOPYRIDINE WITH HIGH-RESOLUTION MICROWAVE SPECTROSCOPY, Angela Y. Chung, Eric A. Arsenault, Stewart E. Novick

FB07　　　　　　　　　　　　　　　　　　　　　　　　　　　　　　　　　　　　10:29 – 10:44
USING HYPERFINE STRUCTURE TO QUANTIFY THE EFFECTS OF SUBSTITUTION ON THE ELECTRON DISTRIBUTION WITHIN A PYRIDINE RING: A STUDY OF 2-, 3-, AND 4-PICOLYLAMINE, Lindsey M McDivitt, Korrina M Himes, Josiah R Bailey, Timothy J McMahon, Ryan G Bird

FB08　　　　　　　　　　　　　　　　　　　　　　　　　　　　　　　　　　　　10:46 – 11:01
PURE ROTATIONAL SPECTRUM OF THE "NON-POLAR" DIMER OF FORMIC ACID, Luca Evangelisti, Weixing Li, Qian Gou, Rolf Meyer, Walther Caminati

FB09　　　　　　　　　　　　　　　　　　　　　　　　　　　　　　　　　　　　11:03 – 11:18
CROSS-CONTAMINATION OF FITTING PARAMETERS IN MULTIDIMENSIONAL TUNNELING TREATMENTS, Nobukimi Ohashi, Jon T. Hougen

FB10　　　　　　　　　　　　　　　　　　　　　　　　　　　　　　　　　　　　11:20 – 11:35
SPIN-SPIN AND SPIN-ROTATION FINE STRUCTURE OF THE METASTABLE $a\,^3\Sigma_u^+$ STATES OF MOLECULAR HELIUM, Paul Jansen, Luca Semeria, Frederic Merkt

FB11　　　　　　　　　　　　　*Post-Deadline Abstract*　　　　　　　　　11:37 – 11:52
ROTATIONAL SPECTRA AND NUCLEAR QUADRUPOLE COUPLING CONSTANTS OF IODOIMIDAZOLES, Graham A. Cooper, Cara J Anderson, Chris Medcraft, Anthony Legon, Nick Walker

FC. Vibrational structure/frequencies
Friday, June 23, 2017 – 8:30 AM
Room: 1024 Chemistry Annex

Chair: Melanie A.R. Reber, University of Georgia, Athens, GA, USA

FC01   **8:30 – 8:45**
LASER SPECTROSCOPY OF VINYL ALCOHOL EMBEDDED IN HELIUM DROPLETS , Hayley Bunn, Paul Raston, Gary E. Douberly

FC02   **8:47 – 9:02**
INFRARED SPECTRA OF THE n-PROPYL AND i-PROPYL RADICALS IN SOLID PARA-HYDROGEN, Gregory T. Pullen, Peter R. Franke, Gary E. Douberly, Yuan-Pern Lee

FC03   **9:04 – 9:19**
THE INFLUENCE OF ANHARMONIC, DISPERSION AND SOLVATION EFFECTS IN THE IR SPECTRA OF PROTO-NATED NEUROTRANSMITTERS SEROTONIN AND DOPAMINE, Vipin Bahadur Singh

FC04   **9:21 – 9:36**
TRANSIENT RAMAN SPECTRA, STRUCTURE AND THERMOCHEMISTRY OF THE THIOCYANATE DIMER RADICAL ANION IN WATER, Ireneusz Janik, G. N. R. Tripathi, Ian Carmichael

FC05   **9:38 – 9:53**
INTRAMOLECULAR VIBRATIONAL ENERGY REDISTRIBUTION (IVR) IN SELECTED S_1 LEVELS ABOVE 1000 cm^{-1} IN PARA-FLUOROTOLUENE., Laura E. Whalley, Adrian M. Gardner, William Duncan Tuttle, Julia A Davies, Katharine L Reid, Timothy G. Wright

Intermission

FC06   **10:12 – 10:27**
VIBRATION AND VIBRATION-TORSION LEVELS OF THE S_1 AND GROUND CATIONIC $D_0{}^+$ STATES OF PARA-FLUOROTOLUENE AND PARA-XYLENE BELOW 1000 cm^{-1}, William Duncan Tuttle, Adrian M. Gardner, Laura E. Whalley, Timothy G. Wright

FC07   **10:29 – 10:44**
CO-ASSIGNMENTS OF THE CALCULATED QUANTUM-MECHANICAL MOLECULAR VIBRATIONAL FREQUENCIES OF CIS-ACROLEIN IN THE GROUND S_0 AND LOWEST EXCITED T_1 AND S_1 ELECTRONIC STATES, V.A. Bataev, Yurii Panchenko, Alexander Abramenkov

FC08   **10:46 – 11:01**
AB INITIO CALCULATION OF THE INFRARED SPECTRUM FOR XeF6 MOLECULE, Lan Cheng

FC09   **11:03 – 11:18**
MOLECULAR AND ELECTRONIC STRUCTURES OF CERIUM AND CERIUM SUBOXIDE CLUSTERS, Jared O. Kafader, Josey E Topolski, Caroline Chick Jarrold

FD. Clusters/Complexes
Friday, June 23, 2017 – 8:30 AM
Room: B102 Chemical and Life Sciences

Chair: Gerhard Schwaab, Ruhr University Bochum, Bochum, Germany

FD01   8:30 – 8:45
CHARACTERIZATION OF A CARBON DIOXIDE-HEXAFLOUROBENZENE COMPLEX USING MATRIX ISOLATION INFRARED SPECTROSCOPY, Jay C. Amicangelo, Bradley K. Gall, Maryn N. Horn

FD02   8:47 – 9:02
VIBRATIONAL PREDISSOCIATION OF THE Ã STATE OF THE C_3Ar COMPLEX IN THE EXCITATION ENERGY REGION OF 25410-25535 CM^{-1}, Yi-Jen Wang, Yen-Chu Hsu

FD03   9:04 – 9:19
THE ν_3 FUNDAMENTAL VIBRATIONAL BAND OF SCCCS REVISITED, Thomas Salomon, John B Dudek, Sven Thorwirth

FD04   9:21 – 9:36
CORE ION STRUCTURES AND SOLVATION EFFECTS IN GAS PHASE $[Sn(CO_2)_n]^-$ CLUSTERS, Michael C Thompson, J. Mathias Weber

Intermission

FD05   9:55 – 10:10
EXPERIMENTAL INSIGHT ON THE CONFORMATIONAL LANDSCAPE OF THE SF_6 DIMER: EVIDENCE FOR THREE CONFORMERS, Pierre Asselin, Alexey Potapov, Vincent Boudon, Laurent Bruel, Marc-André Gaveau, Michel Mons

FD06   10:12 – 10:27
SAMARIUM DOPED CERIUM OXIDE CLUSTERS: A STUDY ON THE MODULATION OF ELECTRONIC STRUCTURE , Josey E Topolski, Jared O. Kafader, Vicmarie Marrero-Colon, Caroline Chick Jarrold

FD07   10:29 – 10:44
JET-COOLED INFRARED LASER SPECTROSCOPY IN THE UMBRELLA ν_2 VIBRATION REGION OF NH_3: IMPROVING THE POTENTIAL ENERGY SURFACE MODEL OF THE $NH_3 - Ar$ VAN DER WAALS COMPLEX, Pierre Asselin, Atef Jabri, Alexey Potapov, Jérome Loreau, Ad van der Avoird

FD08   *Post-Deadline Abstract*   10:46 – 11:01
VIBRATIONALLY EXCITED CARBON MONOXIDE PRODUCED VIA A CHEMICAL REACTION BETWEEN CARBON VAPOR AND OXYGEN, Elijah R Jans, Zakari Eckert, Kraig Frederickson, Bill Rich, Igor V. Adamovich

FE. Ions
Friday, June 23, 2017 – 8:30 AM
Room: 161 Noyes Laboratory

Chair: Timothy S. Zwier, Purdue University, West Lafayette, IN, USA

FE01 8:30 – 8:45
CHARGE OSCILLATION IN C−O STRETCHING VIBRATIONS: A COMPARISON OF CO_2^- ANION AND CARBOXYLATE FUNCTIONAL GROUPS, Michael C Thompson, J. Mathias Weber

FE02 8:47 – 9:02
THRESHOLD IONIZATION SPECTROSCOPY OF $La(CH_3CN)$ AND $La(C_4H_9CN)$ RADICALS FORMED BY La REACTIONS WITH ALKANE NITRILES, Ahamed Ullah, Jong Hyun Kim, Wenjin Cao, Dong-Sheng Yang

FE03 9:04 – 9:19
MASS-ANALYZED THRESHOLD IONIZATION SPECTROSCOPY AND SPIN-ORBIT COUPLING OF CERIUM-HYDROCARBON COMPLEXES , Yuchen Zhang, Sudesh Kumari, Michael W Schmidt, Mark S Gordon, Dong-Sheng Yang

FE04 9:21 – 9:36
CONFORMATION-SPECIFIC INFRARED AND ULTRAVIOLET SPECTROSCOPY OF COLD $[YAPAA+H]^+$ AND $[YGPAA+H]^+$ IONS: A STEREOCHEMICAL "TWIST" ON THE β-HAIRPIN TURN, Andrew F DeBlase, Christopher P Harrilal, John T Lawler, Nicole L Burke, Scott A McLuckey, Timothy S. Zwier

FE05 9:38 – 9:53
PROBING IR-INDUCED ISOMERIZATION OF A MODEL PENTAPEPTIDE IN A CRYO-COOLED ION TRAP USING IR-UV DOUBLE RESONANCE , Christopher P Harrilal, Andrew F DeBlase, Joshua L Fischer, John T Lawler, Scott A McLuckey, Timothy S. Zwier

FE06 9:55 – 10:10
CHARACTERIZING PEPTIDE β-HAIRPIN LOOPS VIA COLD ION SPECTROSCOPY OF MODEL COMPOUNDS, John T Lawler, Andrew F DeBlase, Christopher P Harrilal, Joshua L Fischer, Scott A McLuckey, Timothy S. Zwier

Intermission

FE07 10:29 – 10:44
FELIX SPECTROSCOPY OF LIKELY ASTRONOMICAL MOLECULAR IONS: HC_3O^+, $C_2H_3CNH^+$, and $C_2H_5CNH^+$, Sven Thorwirth, Oskar Asvany, Sandra Brünken, Pavol Jusko, Stephan Schlemmer, Marie-Aline Martin-Drumel, Michael C McCarthy

FE08 10:46 – 11:01
SUB-DOPPLER ROVIBRATIONAL SPECTROSCOPY OF THE H_3^+ CATION AND ISOTOPOLOGUES, Charles R. Markus, Jefferson E. McCollum, Thomas S Dieter, Philip A Kocheril, Benjamin J. McCall

FE09 11:03 – 11:18
SPECTROSCOPY OF THE LOW LYING STATES OF CaO^+, Robert A. VanGundy, Michael Heaven

FE10 11:20 – 11:35
SPECTROSCOPY OF HIGHLY CHARGED TIN IONS FOR AN EXTREME ULTRAVIOLET LIGHT SOURCE FOR LITHOGRAPHY, Francesco Torretti, Alexander Windberger, Wim Ubachs, Ronnie Hoekstra, Oscar Versolato, Alexander Ryabtsev, Anastasia Borschevsky, Julian Berengut, Jose Crespo Lopez-Urrutia

MA. Plenary
Monday, June 19, 2017 – 8:30 AM
Room: Foellinger Auditorium

Chair: Martin Gruebele, University of Illinois at Urbana-Champaign, Urbana, IL, USA

Welcome   8:30
Peter Schiffer, Vice Chancellor for Research
University of Illinois at Urbana-Champaign

MA01   8:40 – 9:20

MOLECULES FROM CLOUDS TO PLANETS: SWEET RESULTS FROM ALMA

EWINE VAN DISHOECK, *Leiden Observatory, University of Leiden, Leiden, Netherlands.*

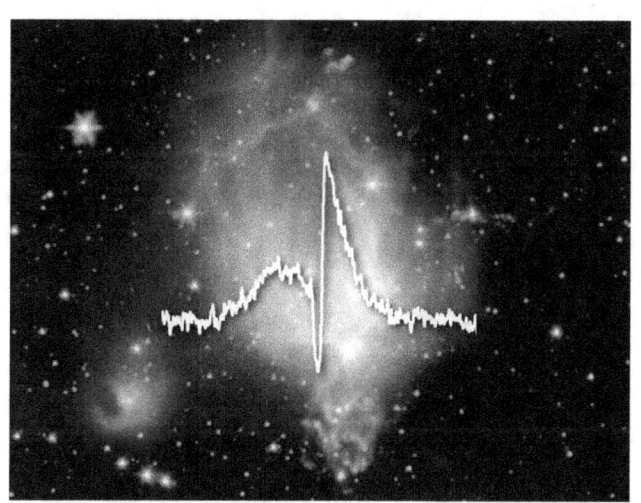

One of the most exciting developments in astronomy is the discovery of thousands of planets around stars other than our Sun. But how do these exo-planets form, and which chemical ingredients are available to build them? Thanks to powerful new telescopes, especially the Atacama Large Millimeter/submillimeter Array (ALMA), astronomers are starting to address these age-old questions scientifically. Stars and planets are born in the cold and tenuous clouds between the stars in the Milky Way. In spite of the extremely low temperatures and densities, a surprisingly rich and interesting chemistry occurs in these interstellar clouds, as evidenced by the detection of more than 180 different molecules. Highly accurate spectroscopic data are key to their identification, and examples of the continued need and close interaction between laboratory work and astronomical observations will be given.

ALMA now allows us to zoom in on solar system construction for the first time. Spectral scans of the birth sites of young stars contain tens of thousands of rotational lines. Water and a surprisingly rich variety of organic materials are found, including simple sugars and high abundances of deuterated species. How are these molecules formed? Can these pre-biotic molecules end up on new planets and form the basis for life elsewhere in the universe? Stay tuned for the latest analyses and also a comparison with recent results from the Rosetta mission to comet 67 P/C-G in our own Solar System.

MA02   9:25 – 10:05

INFRARED SPECTROSCOPY OF NEW MOLECULES AND CLUSTERS

MINGFEI ZHOU, *Fudan University, Department of Chemistry, Shanghai, China.*

Gas phase infrared photodissociation spectroscopy and matrix isolation infrared absorption spectroscopy have proven to be effective spectroscopic methods to investigate novel molecular and cluster species. Vibrational spectroscopy combined with state-of-the-art quantum chemical calculations provides detailed information on geometric and electronic structures as well as chemical bonding of the observed species. In this presentation, I will highlight our recent studies on the formation and infrared spectroscopic characterization of a number of neutral and charged metal-containing compounds including high oxidation state transition metal and lanthanide oxide species and metal carbonyl clusters featuring unprecedented metal-metal multiple bonds. These findings help to expand chemical understanding of the behavior of elements and their compounds.

Intermission

MA03 10:40 – 11:20

ELECTRONIC WAVE PACKET INTERFEROMETRY OF GAS PHASE SAMPLES: HIGH RESOLUTION SPECTRA AND COLLECTIVE EFFECTS

FRANK STIENKEMEIER, *Institute of Physics, University of Freiburg, Freiburg, Germany.*

Time-resolved coherent spectroscopy has opened many new directions to study ultrafast dynamics in complex quantum systems. While most applications have been achieved in the condensed phase, we are focusing on dilute gas phase samples, in particular, on doped helium droplet beams. Isolation in such droplets at millikelvin temperatures provides unique opportunities to synthesize well-defined complexes, to prepare specific ro-vibronic states, and study their dynamics. To account for the small densities in our samples, we apply a phase modulation technique in order to reach enough sensitivity and a high spectral resolution in electronic wave packet interferometry experiments. The combination with mass-resolved ion detection enabled us e.g. to characterize vibrational structures of excimer molecules. By extending this technique we have observed collective resonances in samples of very low density ($10^8\,\mathrm{cm}^{-3}$). With a variant of this method, we are currently elaborating the implementation of nonlinear all-XUV spectroscopy.

FLYGARE AWARDS **11:25**
Introduction by Trevor Sears, Brookhaven National Laboratory

MA04 11:30 – 11:45

DUAL CRYOGENIC ION TRAP SPECTROMETER FOR SPECTROSCOPY OF COLD ION-MOLECULES COMPLEXES

ETIENNE GARAND, *Department of Chemistry, University of Wisconsin–Madison, Madison, WI, USA.*

Ion traps provide a great environment for carrying out controlled ion-neutral molecular reactions. They not only allow for low-temperature chemistry but also for the formation of weakly-bound clusters suitable for vibrational predissociation spectroscopy. Here we present a novel dual cryogenic ion trap spectrometer which combines both capabilities. The first ion trap allows for temperature controlled (77-300K) ion-neutral reaction and clustering, while the second ion trap further thermalize (10K) the reacted complexes and prepare them for subsequent infrared vibrational predissociation characterization.

Our studies show that at 80K, large solvated clusters with more than 50 water molecules can be formed around almost any ions inside the first ion trap. This opens the door for studies of peptide structures as a function of solvation. Preliminary data on the microsolvation of model protonated $(Gly)_n$ peptides will be presented. One complication in these studies is the presence of multiple conformations and resulting spectral congestion which hinders the spectral analysis. We approached this issue in two different ways. First, taking advantage of temperature dependent H-D exchange, we formed D_2O solvated all H peptides and thus separated the spectral signatures of the solvent and solute into two different regions. Second, implementation of a simplified IR-IR double resonance scheme allowed us to efficiently extract conformation-specific spectrum from complex mixture of isobaric molecules. The combination of these two approaches opens the possibility of studying very complex clusters with high structural specificity.

MA05   **11:50 – 12:05**

HIGH-RESOLUTION LASER SPECTROSCOPY OF FREE RADICALS IN NEARLY DEGENERATE ELECTRONIC STATES

<u>JINJUN LIU</u>, *Department of Chemistry, University of Louisville, Louisville, KY, USA.*

Rovibronic structure of molecules in orbitally degenerate electronic states including Renner-Teller (RT) and Jahn-Teller (JT) active molecules has been extensively studied. Less is known about rotational structure of polyatomic molecules in *nearly* degenerate states, especially those with low (e.g., C_s) symmetry that are subject to the pseudo-Jahn-Teller (pJT) effect. In the case of free radicals, the unpaired electron further complicates energy levels by inducing spin-orbit (SO) and spin-rotation (SR) splittings. Asymmetric deuteration or methyl substitution of C_{3v} free radicals such as CH_3O, $CaCH_3$, and $CaOCH_3$ lowers the molecular symmetry, lifts the vibronic degeneracy, and reduces the JT effect to the pJT effect. New spectroscopic models are required to reproduce the rovibronic structure and simulate the experimentally obtained spectra of pJT-active free radicals. It has been found that rotational and fine-structure analysis of spectra involving nearly degenerate states may aid in vibronic analysis and interpretation of *effective* molecular constants. Especially, SO and Coriolis interactions that couple the two states can be determined accurately from fitting the experimental spectra. Coupling between the two electronic states also affects the intensities of rotational and vibronic transitions. The study on free radicals in nearly degenerate states provides a promising avenue of research which may bridge the gap between symmetry-induced degenerate states and the Born-Oppenheimer (BO) limit of unperturbed electronic states.

MF. Mini-symposium: ALMA's Molecular View

Monday, June 19, 2017 – 1:45 PM

Room: 274 Medical Sciences Building

Chair: Susanna L. Widicus Weaver, Emory University, Atlanta, GA, USA

MF01   1:45–2:00

THE ATACAMA LARGE MILLIMETER/SUBMILLIMETER ARRAY - FROM EARLY SCIENCE TO FULL OPERATIONS.

ANTHONY REMIJAN, *ALMA, National Radio Astronomy Observatory, Charlottesville, VA, USA.*

The Atacama Large Millimeter/Submillimeter Array (ALMA) is now entering its 6th cycle of scientific observations. Starting with Cycle 3, science observations were no longer considered "Early Science" or "best efforts". Cycle 5 is now the third cycle of "steady state" observations and Cycle 7 is advertised to begin ALMA "full science" operations. ALMA Full Science Operations will include all the capabilities that were agreed upon by the international consortium after the ALMA re-baselining effort. In this talk, I will detail the upcoming ALMA Cycle 5 observing capabilities, describe the process of selecting new observing modes for upcoming cycles and provide an update on the status of the ALMA Full Science capabilities.

MF02   ***INVITED TALK***   2:02–2:32

EXPLORING MOLECULAR COMPLEXITY IN THE INTERSTELLAR MEDIUM WITH ALMA

ARNAUD BELLOCHE, *Millimeter- und Submillimeter-Astronomie, Max-Planck-Institut für Radioastronomie, Bonn, NRW, Germany.*

The search for complex organic molecules (COMs) in the interstellar medium (ISM) relies heavily on the progress made in the laboratory to record and characterize the rotational spectra of these molecules. Almost 200 different molecules have been identified in the ISM so far, in particular thanks to millimeter-wavelength observations of the star-forming molecular cloud core Sgr B2(N) in the Galactic Center region. The advent of the Atacama Large Millimeter/submillimeter Array (ALMA) has recently opened a new door to explore the molecular complexity of the ISM. Thanks to its high angular resolution, the spectral confusion of star-forming cores can be reduced, and its tremendous sensitivity allows astronomers to detect molecules of low abundance that could not be probed by previous generations of telescopes.

I will present results of the EMoCA survey conducted toward Sgr B2(N) with ALMA. The main goal of this spectral line survey is to decipher the molecular content of Sgr B2(N) in order to test the predictions of astrochemical numerical simulations and gain insight into the chemical processes at work in the ISM. I will in particular report on the tentative detection of N-methylformamide[a], on the deuterium fractionation of COMs[b], and on the detection of a branched alkyl molecule in the ISM[cd]. The latter detection has unveiled a new domain in the structures available to the chemistry of star-forming regions and established a further connection to the COMs found in meteorites.

[a]A. Belloche, A. A. Meshcheryakov, R. T. Garrod et al. 2017, A&A, in press, DOI: 10.1051/0004-6361/201629724
[b]A. Belloche, H. S. P. Müller, R. T. Garrod, and K. M. Menten 2016, A&A, 587, A91
[c]A. Belloche, R. T. Garrod, H. S. P. Müller, and K. M. Menten 2014, Science, 345, 1584
[d]R. T. Garrod, A. Belloche, H. S. P. Müller, and K. M. Menten 2017, A&A, in press, DOI: 10.1051/0004-6361/201630254

MF03

ROTATIONAL SPECTROSCOPY OF THE LOW ENERGY CONFORMER OF 2-METHYLBUTYRONITRILE AND SEARCH FOR IT TOWARD SAGITTARIUS B2(N2)

HOLGER S. P. MÜLLER, NADINE WEHRES, OLIVER ZINGSHEIM, FRANK LEWEN, STEPHAN SCHLEMMER, *I. Physikalisches Institut, Universität zu Köln, Köln, Germany*; JENS-UWE GRABOW, *Institut für Physikalische Chemie und Elektrochemie, Gottfried-Wilhelm-Leibniz-Universität, Hannover, Germany*; ROBIN T. GARROD, *Departments of Chemistry and Astronomy, The University of Virginia, Charlottesville, VA, USA*; ARNAUD BELLOCHE, KARL M. MENTEN, *Millimeter- und Submillimeter-Astronomie, Max-Planck-Institut für Radioastronomie, Bonn, NRW, Germany.*

Quite recently, some of us detected *iso*-propyl cyanide as the first branched alkyl molecule in space.[a] The identification was made in an ALMA Cycle 0 and 1 molecular line survey of Sagittarius B2(N) at 3 mm. The branched isomer was only slightly less abundant than its straight-chain isomer with a ratio of about $2 : 5$. While initial chemical models favored the branched isomer somewhat, more recent models[b] are able to reproduce the observed ratio. Moreover, the models predicted that among the next longer butyl cyanides (BuCNs) 2-methylbutyronitrile (2-MBN) should be more abundant than both n-BuCN and 3-MBN by factors of around 2, with t-BuCN being almost negligible.

With the rotational spectra of t- and n-BuCN studied,[c] we investigated those of 2-MBN and 3-MBN betwen ~40 and ~400 GHz by conventional absorption spectroscopy and by chirped-pulse and resonator Fourier transform microwave (FTMW) spectroscopy. The analyses were guided by quantum-chemical calculations.

Here we report the analysis of the low-energy conformer of 2-MBN and a search for it in our current ALMA data. Two additional conformers are higher by ~250 and ~280 cm^{-1}. The low-energy conformer displays a very rich rotational spectrum because of its great asymmetry ($\kappa \approx 0.14$) and large a- and b-dipole moment components. Accurate ^{14}N quadrupole coupling parameters were obtained from the FTMW spectral recordings.

[a] A. Belloche, R. T. Garrod, H. S. P. Müller, and K. M. Menten, *Science* **345** 1584.

[b] R. T. Garrod, A. Belloche, H. S. P. Müller, and K. M. Menten, *Astron. Astrophys.*, in press; doi: 10.1051/0004-6361/201630254.

[c] Z. Kisiel, *Chem. Phys. Lett.* **118** 134; M. H. Ordu et al., *Astron. Astrophys.* **541** A121.

MF04

SUBMILLIMETER WAVE SPECTRUM OF THIOACETALDEHYDE AND ITS SEARCH IN SgrB2

L. MARGULÈS, R. A. MOTIYENKO, *Laboratoire PhLAM, UMR 8523 CNRS - Université Lille 1, Villeneuve d'Ascq, France*; J.-C. GUILLEMIN, *Institut des Sciences Chimiques de Rennes, UMR 6226 CNRS - Université de Rennes 1, Rennes, France*; Y. ELLINGER, *Laboratoire de Chimie Théorique (UMR 7616), Université Paris 6, Paris, FRANCE*; BRETT A. McGUIRE, *NAASC, National Radio Astronomy Observatory, Charlottesville, VA, USA*; ANTHONY REMIJAN, *ALMA, National Radio Astronomy Observatory, Charlottesville, VA, USA.*

Sulfur chemistry in the interstellar medium is clearly misunderstood, the tenth most abundant element in the Galaxy, it is depleted in molecular clouds by a factor of 1000 [a]. This suggests that sulfur chemistry is important on icy grain mantles, and that sulfur-bearing molecules may be not detected yet due to the lack of laboratory data. The present study is about thioacetaldehyde (CH_3CHS), the analog of acetaldehyde. Previously, the rotational spectra were recorded up to 40 GHz [b],[c]. New spectra were recorded from 150 to 660 GHz using the Lille solid-state based spectrometer. The new fast version of the spectrometer using DDS component is particulary suitable for reactive species like thioacetaldehyde. Thioacetaldehyde was produced in-situ by pyrolisis at $750°C$ and introduced in a 1m long pyrex cell in a flow mode. Analysis of the spectra is not obvious, like in acetaldehyde, as this molecule exhibits internal rotation of the methyl group. The internal rotation barrier is higher in thioacetaldehyde, 542 cm-1, than in acetaldehyde, 408 cm-1. However, the coupling between the internal rotation and the overall rotation in thioacetaldehyde is strong, the coupling parameter ρ value is 0.261 just slightly smaller than the acetaldehyde value of 0.329. The spectroscopic results and its searches in SgrB2 will be presented.

This work was supported by the CNES and the Action sur Projets de l'INSU, PCMI. This work was also done under ANR-13-BS05-0008-02 IMOLABS

[a] A. Tieftrunk; *et al.*, 1994, *A&A* **289**, 579

[b] H. Kroto; *et al.*, 1974, *Chem. Phys. Lett.* **29**, 265

[c] H. Kroto; *et al.*, 1976, *J. Mol. Spectrosc.* **62**, 346

MF05   3:10 – 3:25

THE SURPRISING COMPLEXITY OF DIFFUSE AND TRANSLUCENT CLOUDS TOWARD SGR B2: DIATOMICS AND COMs FROM 4 GHz TO 1.2 THz

BRETT A. McGUIRE, *NAASC, National Radio Astronomy Observatory, Charlottesville, VA, USA*; JOANNA F. CORBY, *Physics, University of South Florida, Tampa, FL, USA*; MARIE-ALINE MARTIN-DRUMEL, *CNRS, Institut des Sciences Moléculaires d'Orsay, Orsay, France*; P. SCHILKE, *I. Physikalisches Institut, Universität zu Köln, Köln, Germany*; MICHAEL C McCARTHY, *Atomic and Molecular Physics, Harvard-Smithsonian Center for Astrophysics, Cambridge, MA, USA*; ANTHONY REMIJAN, *ALMA, National Radio Astronomy Observatory, Charlottesville, VA, USA.*

Many diffuse and translucent clouds lie along the line of sight between Earth and the Galactic Center that can be probed through molecular absorption at characteristic velocities. We highlight results of a study of diffuse and translucent clouds along the line of sight to Sgr B2, including SOFIA observations of SH near 1.4 THz and GBT PRIMOS observations from 4 to 50 GHz. We find significant variation in the chemical conditions within these clouds, and the abundances do not appear to correlate with the total optical depth. Additionally, from the GBT observations, we report the first detections of multiple complex organic molecules (COMs) in diffuse and translucent clouds, including CH_3CN, HC_3N, CH_3CHO, and NH_2CHO. We compare the GBT results to complementary observations of SH, H_2S, and others at mm, sub-mm, and THz frequencies from the NRAO 12m, Herschel HIFI, and SOFIA facilities, and comment on the insights into interstellar sulfur chemistry which is currently not well constrained.

MF06   3:27 – 3:42

SIO EMISSION FROM THE INNER PC OF SGR A*

FARHAD YUSEF-ZADEH, *Physics and Astronomy, Northwestern University, Evanston, IL, USA.*

A critical question regarding star formation near supermassive black holes (SMBHs) is whether tidal shear completely suppresses star formation or whether it induces star formation. The circumnuclear molecular ring orbiting the 4 million solar mass black hole Sgr A* in the inner few parsecs of the Galactic center is an excellent testing ground to study star formation in extreme tidal environments. We have carried out ALMA observations of SiO (5-4) line emission to resolve protostellar outflow candidates in the molecular ring and its interior. We will describe preliminary results of these observations. In addition, we will present ALMA and VLA observations of continuum sources that show bow-shock structures. The characteristics of these mm sources suggest the presence of protoplanetary disks. These continuum measurements suggest on-going low-mass star formation with the implication that gas clouds can survive near the strong tidal and radiation fields of the Galactic center.

Intermission

MF07

ASTROCHEMICALLY RELEVANT MOLECULES IN THE W-BAND REGION

BENJAMIN E ARENAS, *CoCoMol, Max-Planck-Institut für Struktur und Dynamik der Materie, Hamburg, Germany*; AMANDA STEBER, SÉBASTIEN GRUET, *CUI, The Hamburg Centre for Ultrafast Imaging, Hamburg, Germany*; MELANIE SCHNELL, *CoCoMol, Max-Planck-Institut für Struktur und Dynamik der Materie, Hamburg, Germany*.

The interplay between laboratory spectroscopy and observational astronomy has allowed for the chemical complexity of the interstellar medium (ISM) to be explored. Our laboratory studies involve the measurement of the rotational spectra of commercially available samples in the region 75-110 GHz, thus covering a portion of Band 3 of the Atacama Large Millimeter/submillimeter Array (ALMA). Up until recently, we have concentrated on medium-sized (5 to 9 heavy atoms) nitrogen- and oxygen-containing molecules and their vibrationally excited states. Examples include amino alcohols, such as alaninol (2-amino-1-propanol), and cyanides. Further, we have extended the capabilities of our segmented chirped-pulse spectrometer [1] with electrical discharge apparatus. We present here the recent results from our set-up, including the typical rotational spectra of astrochemically relevant samples and the discharge-enabled rotational spectroscopy of mixtures of simple organic molecules. These experimental results have yielded transitions that will facilitate the detection of these molecules in the ISM with ALMA, and the discharge experiments should allow us to consider formation pathways of organic molecules from smaller building blocks.

[1] B.E. Arenas, S. Gruet, A.L. Steber, B.M. Giuliano, M. Schnell, Phys. Chem. Chem. Phys. 19 (2017) 1751-1756.

MF08

ASTRONOMICAL TRIPLETS: ALMA OBSERVATIONS OF C2H4O2 ISOMERS IN SGR B2 (N)

CI XUE, *Department of Chemistry, The University of Virginia, Charlottesville, VA, USA*; ANTHONY REMIJAN, *ALMA, National Radio Astronomy Observatory, Charlottesville, VA, USA*; ANDREW M BURKHARDT, *Department of Astronomy, The University of Virginia, Charlottesville, VA, USA*; ERIC HERBST, *Department of Chemistry, The University of Virginia, Charlottesville, VA, USA*.

The $C_2H_4O_2$ triplet found in the interstellar medium (ISM) consists of glycolaldehyde (CH_2OHCHO), acetic acid (CH_3COOH) and methyl formate ($HCOOCH_3$). The forming mechanism of their HCO-bearing component involves both gas-phase and grain-surface processes whose relative roles plays into fundamental questions within the fields of astrochemistry and astrobiology. Glycolaldehyde is closely related to ribose and deoxyribose, the primary components of genetic materials. The first detection of Glycolaldehyde was toward Sgr B2 with using NRAO 12 m telescope in 2000 (J. M. Hollis et al). A new careful search for glycolaldehyde toward the hot dense core Sgr B2 (N) is needed. While methyl formate has a large number of detected transitions throughout the ISM, the detection of acetic acid, the least abundant of these isomers, is more tentative. Mehringer et al. (1997) reported only 4 lines of acetic acid toward Sgr B2 Large Molecule Heimat source. Here, we confirm these detections of each species toward Sgr B2 (N) with the more sensitive and larger bandwidth from ALMA Band 3 observations (A. Belloche, 2012), providing us more transitions and more accurate continuum subtraction. Based on these results, the abundances and spatial distributions of the $C_2H_4O_2$ triplet species would be obtained and compared.

MF09

DIMETHYL ETHER BETWEEN 214.6 AND 265.3 GHZ: THE COMPLETE, TEMPERATURE RESOLVED SPECTRUM

JAMES P. McMILLAN, CHRISTOPHER F. NEESE, FRANK C. DE LUCIA, *Department of Physics, The Ohio State University, Columbus, OH, USA*.

We have studied dimethyl ether, one of the so-called 'astronomical weeds', in the 214.6–265.3 GHz band. We have experimentally gathered a set of intensity calibrated, complete, and temperature resolved spectra from across the temperature range of 238–391 K. Using our previously reported method of analysis[a], the point by point method, we are capable of generating the complete spectrum at astronomically significant temperatures. Many lines, of nontrivial intensity, which were previously not included in the available astrophysical catalogs have been found. Lower state energies and line strengths have been found for a number of lines which are not currently present in the catalogs. The extent to which this may be useful in making assignments will be discussed.

[a] J. McMillan, S. Fortman, C. Neese, F. DeLucia, ApJ. 795, 56 (2014)

MF10 4:52 – 5:07

H$_2$S: A NEW PROBE OF HIDDEN LUMINOSITY IN ORION KL

<u>MARIA KLESHCHEVA</u>, *Division of Geological and Planetary Sciences, California Institute of Technology, Pasadena, CA, USA*; NATHAN CROCKETT, *Geological and Planetary Sciences , California Institute of Techonolgy, Pasadena, CA, USA*; GEOFFREY BLAKE, *Division of Chemistry and Chemical Engineering, California Institute of Technology, Pasadena, CA, USA.*

We present 20" X 20" interferometric maps toward the Orion Kleinmann-Low nebula (Orion KL) for the 2(0,2) - 1(1,1), and 9(7,2) - 9(6,3) transition of H$_2$S. These lines were among more than 90 H$_2$S transitions detected toward Orion KL using Herschel/HIFI. Modeling the line excitation reveals that the J=9 transition is tracing extremely dense gas that is likely irradiated by a mid-IR continuum which is stronger than observed. The source of this intense radiation field is presumably hot dust heated by massive embedded protostars. ALMA maps of the J=9 transition therefore trace those regions exposed to the strongest sources of hidden mid-IR radiation, while maps of the J=2 transition trace the global distribution of H$_2$S. The sources identified by the J=9 transition may trace embedded protostars, representing the origin of Orion KL's tremendous luminosity.

MF11 5:09 – 5:24

TRACING THE ORIGINS OF NITROGEN BEARING ORGANICS TOWARD ORION KL WITH ALMA

<u>BRANDON CARROLL</u>, *Division of Chemistry and Chemical Engineering, California Institute of Technology, Pasadena, CA, USA*; NATHAN CROCKETT, *Geological and Planetary Sciences , California Institute of Techonolgy, Pasadena, CA, USA*; OLIVIA H. WILKINS, *Chemistry, California Institute of Technology, Pasadena, CA, USA*; EDWIN BERGIN, *Department of Astronomy, University of Michigan, Ann Arbor, MI, USA*; GEOFFREY BLAKE, *Division of Chemistry and Chemical Engineering, California Institute of Technology, Pasadena, CA, USA.*

A comprehensive analysis of a broadband 1.2 THz wide spectral survey of the Orion Kleinmann-Low nebula (Orion KL) has shown that nitrogen bearing complex organics trace systematically hotter gas than O-bearing organics toward this source. The origin of this O/N dichotomy remains a mystery. If complex molecules originate from grain surfaces, N-bearing species may be more difficult to remove from grain surfaces than O-bearing organics. Theoretical studies, however, have shown that hot (T=300 K) gas phase chemistry can produce high abundances of N-bearing organics while suppressing the formation of O-bearing complex molecules. In order to distinguish these distinct formation pathways we have obtained extremely high angular resolution observations of methyl cyanide (CH$_3$CN) using the Atacama Large Millimeter/Submillimeter Array (ALMA) toward Orion KL. By simultaneously imaging ^{13}CH$_3$CN and CH$_2$DCN we map the temperature structure and D/H ratio of CH$_3$CN. We will present updated results of these observations and discuss their implications for the formation of N-bearing organics in the interstellar medium.

MG. Mini-symposium: Multiple Potential Energy Surfaces
Monday, June 19, 2017 – 1:45 PM
Room: 116 Roger Adams Lab

Chair: Christopher G. Elles, University of Kansas, Lawrence, KS, USA

MG01   *INVITED TALK*   1:45 – 2:15

THE ELECTRONIC GROUND STATE OF THE NITRATE RADICAL: A DECADE OF CONTROVERSY

JOHN F. STANTON, *Physical Chemistry, University of Florida, Gainesville, FL, USA.*

In the ten years since the traditional assignment of its degenerate stretching fundamental became controversial, a great deal of work - both theoretical and experimental - has been done on the NO_3 molecule. A brief review of these developments will be given, and results of very high-level calculations of the dispersed fluorescence and negative ion photoelectron spectra of this molecule will be presented together with the corresponding experimental results. In addition, the question of "what is next to do" on the ground state - from a theoretical point of view - will be addressed. Time permitting, some discussion will also be devoted to the strongly Jahn-Teller active $^2E''$ (first excited) electronic state, where the level of understanding and agreement thus far obtained from experiment and theory is still at a rather primitive stage.

MG02   2:19 – 2:34

A ZERO-ORDER PICTURE OF THE INFRARED SPECTRA OF CH_3O AND CD_3O: ASSIGNMENT OF STATES

BRITTA JOHNSON, EDWIN SIBERT, *Department of Chemistry, University of Wisconsin–Madison, Madison, WI, USA.*

Experimentalists and theorists alike have been intrigued by the infrared spectra of the methoxy radical; due to the presence of a conical intersection at the C_{3v} molecular geometry, methoxy's IR spectrum is strongly influenced by Jahn-Teller vibronic coupling which leads to large amplitude vibrations and extensive mixing of the two lowest electronic states. This radical's complex IR spectra, which also contains moderate mixing from spin-orbit and Fermi couplings, serves as an important test for models which seek to understand complex molecular vibrations.

The assignment of the IR spectra in methoxy, and its partially and fully deuterated analogues, is considered. All vibronic states below 2575 cm^{-1} in CH_3O and 2035 cm^{-1} in CD_3O are assigned. The mixing between the zero-order normal modes complicates the assignment using this representation. Alternative zero-order representations, that include specific Jahn-Teller couplings, are explored and used to create definitive assignments for the low lying vibronic states. In many instances it is possible to plot the wavefunctions on which the assignments are based. The plots also enable one to visualize the conical seam and its effect on the wavefunctions. The first and second order Jahn-Teller coupling in the rocking motion dominates the spectral features in CH_3O, while first order and modulated first order couplings dominate the spectral features in CD_3O. The methods described here are general and can be applied to other Jahn-Teller systems.

MG03   2:36 – 2:51

DVR3DUV: A SUITE FOR HIGH ACCURACY CALCULATIONS OF RO-VIBRONIC SPECTRA OF TRIATOMIC MOLECUlES

EMIL J ZAK, JONATHAN TENNYSON, *Department of Physics and Astronomy, University College London, Gower Street, London WC1E 6BT, United Kingdom.*

We present a computer code (DVR3DUV) for calculations of high-accuracy ro-vibronic spectra of triatomic molecules. The current implementation is an extension to the DVR3D suite [1], which operates with the exact kinetic energy operator, a single potential energy surface and a single dipole moment surface (ro-vibrational transitions only). The main function of the new code is calculation of transition intensities between different electronic states in the rotational-vibrational resolution. As a case study, two electronic states of SO_2 molecule are considered. Ro-vibrational wavefunctions and energy levels for the ground $\tilde{X}^1A_1$ state of SO_2 are calculated using Ames PES [2], while energy levels and wavefunctions of the $\tilde{C}^1B_2$ state are calculated using *ab initio* PES (MRCI-F12-AVTZ). Transition intensities are computed using a) Franck-Condon approximation; b) *ab initio* dipole moment surface between the two electronic states. Results are compared to the latest theoretical and experimental works. Future applications of the DVR3DUV code will focus on highly accurate electronic spectra for atmospherically important species, such as ozone molecule.

[1] J. Tennyson, M A. Kostin, P. Barletta, G. J. Harris, O L. Polyansky, J. Ramanlal, N. F. Zobov, Computer Physics Communications 163 (2004) 85–116.

[2] X. Huang, D. W. Schwenke, T. J. Lee, J Chem Phys. 2014 ;140(11):11431

MG04

THE ROVIBRONIC SPECTRA OF THE CYCLOPENTADIENYL RADICAL

KETAN SHARMA, TERRY A. MILLER, *Department of Chemistry and Biochemistry, The Ohio State University, Columbus, OH, USA*; JOHN F. STANTON, *Physical Chemistry, University of Florida, Gainesville, FL, USA*; DAVID NESBITT, *Department of Chemistry, JILA CU-NIST, Boulder, CO, USA*.

Cyclopentadienyl (Cp) radical has been subject to numerous studies for the greater part of half a century. Experimental work has involved photo-electron spectroscopy,[a] laser induced fluorescence excitation[b] and emission,[c] infrared absorption spectroscopy,[d] and recently rotationally resolved spectra in the CH stretch region taken at JILA. Even more theoretical works appear in the literature, but substantial advances in computation have occurred since their completion. Cp's highly symmetric (D_{5h}) structure and doubly degenerate electronic ground ($\tilde{X}^2 E_1''$), which is subject to linear Jahn-Teller distortion, have been a great motivation for work on it. We have commenced new computational work to obtain a broad understanding of the electronic, vibrational, and rotational, i.e. rovibronic, structure of the Cp radical as revealed by its spectra, with particular emphasis on the new infrared spectra. The goal is to guide experiments and their analyses and reconcile results from spectroscopy and quantum chemistry calculations.

[a] T. ICHINO, *et al. J. Chem. Phys.* 129, 084310 (2008)

[b] L. YU, S. C. FOSTER, J. M. WILLIAMSON, M. C. HEAVEN AND T. A. MILLER *J. Phys. Chem.* 92, 4263 (1988)

[c] B. E. APPLEGATE, A. J. BEZANT AND T. A. MILLER *J. Chem. Phys* 114, 4869 (2001)

[d] D. LEICHT, M. KAUFMANN, G. SCHWAAB, AND M. HAVENITH *J. Chem. Phys. 145*, 7 (2016), 074304.

MG05

ROTATIONAL PARAMETERS FROM VIBRONIC EIGENFUNCTIONS OF JAHN-TELLER ACTIVE MOLECULES

SCOTT M. GARNER, TERRY A. MILLER, *Department of Chemistry and Biochemistry, The Ohio State University, Columbus, OH, USA*.

The structure in rotational spectra of many free radical molecules is complicated by Jahn-Teller distortions. Understanding the magnitudes of these distortions is vital to determining the equilibrium geometric structure and details of potential energy surfaces predicted from electronic structure calculations. For example, in the recently studied $\tilde{A}^2 E''$ state of the NO_3 radical, the magnitudes of distortions are yet to be well understood as results from experimental spectroscopic studies of its vibrational and rotational structure disagree with results from electronic structure calculations of the potential energy surface. By fitting either vibrationally resolved spectra or vibronic levels determined by a calculated potential energy surface, we obtain vibronic eigenfunctions for the system as linear combinations of basis functions from products of harmonic oscillators and the degenerate components of the electronic state. Using these vibronic eigenfunctions we are able to predict parameters in the rotational Hamiltonian such as the Watson Jahn-Teller distortion term, h_1, and compare with the results from the analysis of rotational experiments.

Intermission

MG06

RYDBERG STATES OF ALKALI METAL ATOMS ON SUPERFLUID HELIUM DROPLETS - THEORETICAL CONSIDERATIONS

JOHANN V. POTOTSCHNIG, FLORIAN LACKNER, ANDREAS W. HAUSER, WOLFGANG E. ERNST, *Institute of Experimental Physics, Graz University of Technology, Graz, Austria.*

The bound states of electrons on the surface of superfluid helium have been a research topic for several decades. One of the first systems treated was an electron bound to an ionized helium cluster.[a] Here, a similar system is considered, which consists of a helium droplet with an ionized dopant inside and an orbiting electron on the outside. In our theoretical investigation we select alkali metal atoms (AK) as central ions, stimulated by recent experimental studies of Rydberg states for Na,[b] Rb,[c] and Cs[d] attached to superfluid helium nanodroplets. Experimental spectra , obtained by electronic excitation and subsequent ionization, showed blueshifts for low lying electronic states and redshifts for Rydberg states.

In our theoretical treatment the diatomic AK^+-He potential energy curves are first computed with *ab initio* methods. These potentials are then used to calculate the solvation energy of the ion in a helium droplet as a function of the number of atoms. Additional potential terms, derived from the obtained helium density distribution, are added to the undisturbed atomic pseudopotential in order to simulate a 'modified' potential felt by the outermost electron. This allows us to compute a new set of eigenstates and eigenenergies, which we compare to the experimentally observed energy shifts for highly excited alkali metal atoms on helium nanodroplets.

[a] A. Golov and S. Sekatskii, Physica B, 1994, 194, 555-556

[b] E. Loginov, C. Callegari, F. Ancilotto, and M. Drabbels, J. Phys. Chem. A, 2011, 115, 6779-6788

[c] F. Lackner, G. Krois, M. Koch, and W. E. Ernst, J. Phys. Chem. Lett., 2012, 3, 1404-1408

[d] F. Lackner, G. Krois, M. Theisen, M. Koch, and W. E. Ernst, Phys. Chem. Chem. Phys., 2011, 13, 18781-18788

MG07

OBSERVATION OF HEAVY RYDBERG STATES IN H_2 AND HD

MAXIMILIAN BEYER, FREDERIC MERKT, *Laboratorium für Physikalische Chemie, ETH Zurich, Zurich, Switzerland.*

The binding energies of the hydrogen atom are given by the Rydberg formula

$$E_n = -\frac{\mathcal{R}_\infty \mu/m_e}{(n-\delta)^2},$$

where the quantum defect δ vanishes in the case of a pure Coulomb potential.

Heavy Rydberg systems can be realized when the electron is replaced by an anion, which leads in the case of H^+H^- to an almost 1000 times larger Rydberg constant and to an infinite number of vibrational states. In the diabatic molecular basis, these ion-pair states are described by long-range Coulomb potentials with $^1\Sigma_g^+$ and $^1\Sigma_u^+$ symmetry. In this basis, the level energies are described by an almost energy-independent, nonzero quantum defect, reflecting the finite size of H^-. Strong interactions at small internuclear distances lead to strong variation of δ with n.

Gerade [2] and ungerade [3] ion-pair states have been observed in H_2 with principal quantum numbers up to $n = 240$. The quantum defects in this range were found to vary with energy, indicating the inadequacy of a pure diabatic picture.

Spectra of ungerade heavy Rydberg states of H_2 with $n = 160 - 520$ showing that the quantum defect only becomes energy independent for $n > 350$ will be presented, supporting the description using a diabatic basis.

I will also present first observations of ion-pair states in HD, showing two series of heavy Rydberg states, H^+D^- and H^-D^+, which have different series limits. The experimental results will be discussed and compared with calculations using both an adiabatic and a diabatic basis.

[1] S. Pan, and F. H. Mies, *J. Chem. Phys.* **89**, 3096 (1988).

[2] M. O. Vieitez, T. I. Ivanov, E. Reinhold, C. A. de Lange, and W. Ubachs, *Phys. Rev. Lett.* **101**, 163001 (2008).

[3] R. C. Ekey, and E. F. McCormack, *Phys. Rev. A* **84**, 020501(R) (2011).

MG08 4:18 – 4:33

THE ROLE OF PERTURBATIONS IN THE B-X UV SPECTRUM OF S_2 IN A TEMPERATURE-DEPENDENT MECHANISM FOR SULFUR MASS INDEPENDENT FRACTIONATION

ALEXANDER W. HULL, ROBERT W FIELD, *Department of Chemistry, MIT, Cambridge, MA, USA*; SHUHEI ONO, *Earth, Atmospheric, and Planetary Sciences, MIT, Cambridge, MA, USA.*

Sulfur mass independent fractionation (S-MIF) describes anomalous sulfur isotope ratios commonly found in sedimentary rocks older than 2.45 billion years. These anomalies likely originate from photochemistry of small, sulfur-containing molecules in the atmosphere, and their sudden disappearance from rock samples younger than 2.45 years is thought to be correlated with a sharp rise in atmospheric oxygen levels. The emergence of atmospheric oxygen is an important milestone in the development of life on Earth, but the mechanism for sulfur MIF in an anoxic atmosphere is not well understood. In this context, we present an analysis of the B-X UV spectrum of S_2, an extension of work presented last year. The B state of S_2 is strongly perturbed by the nearby B" state, as originally described by Green and Western (1996). Our analysis suggests that a doorway-mediated transfer mechanism shifts excited state population from the short-lifetime B state to the longer-lifetime B" state. Furthermore, access to the perturbed doorway states is strongly dependent on the population distribution in the ground state. This suggests that the temperature of the Achaean atmosphere may have played a significant role in determining the extent of S-MIF.

MG09 4:35 – 4:50

THE ORGIN OF UNEQUAL BOND LENGTHS IN THE $\tilde{C}^1B_2$ STATE OF SO_2: SIGNATURES OF HIGH-LYING POTENTIAL ENERGY SURFACE CROSSINGS IN THE LOW-LYING VIBRATIONAL STRUCTURE

BARRATT PARK, *Dynamics at Surfaces, Max Planck Institute for Biophysical Chemistry, Göttingen, Germany*; JUN JIANG, ROBERT W FIELD, *Department of Chemistry, MIT, Cambridge, MA, USA.*

The $\tilde{C}^1B_2$ state of SO_2 has a double-minimum potential in the antisymmetric stretch coordinate, such that the minimum energy geometry has nonequivalent SO bond lengths. The asymmetry in the potential energy surface is expressed as a staggering in the energy levels of the ν'_3 progression. We have recently made the first observation of low-lying levels with odd quanta of ν'_3, which allows us to characterize the origins of the level staggering. Our work demonstrates the usefulness of low-lying vibrational level structure, where the character of the wavefunctions can be relatively easily understood, to extract information about dynamically important potential energy surface crossings that occur at much higher energy. The measured staggering pattern is consistent with a vibronic coupling model for the double-minimum, which involves direct coupling to the bound 2^1A_1 state and indirect coupling with the repulsive 3^1A_1 state. The degree of staggering in the ν'_3 levels increases with quanta of bending excitation, which is consistent with the approach along the $\tilde{C}$-state potential energy surface to a conical intersection with the 2^1A_1 surface at a bond angle of $\sim 145°$.

MG10 4:52 – 5:07

COMPUTATIONAL MODELING OF ELECTRONIC SPECTROSCOPY OF 3-PHENYL-2-PROPYNENITRILE

CLAUDIA I VIQUEZ ROJAS, KHADIJA M. JAWAD, TIMOTHY S. ZWIER, LYUDMILA V SLIPCHENKO, *Department of Chemistry, Purdue University, West Lafayette, IN, USA.*

3-phenyl-2-propynenitrile (PPN) is a potentially important component of Titan's atmosphere. This molecule exhibits intriguing patterns in high-resolved absorption and fluorescence spectra. To better understand PPN's photochemistry, we employ computational tools to examine its electronic structure and excited states. The presence of vibronic coupling is evaluated by mapping potential energy surfaces of the first four electronic excitations along different vibrational modes. The parameters that describe the interactions between vibrational and electronic states are used to build the vibronic Hamiltonian and predict the absorption and emission spectra of PPN with the multi configuration time dependent Hartree (MCTDH) algorithm.

MG11

INFLUENCE OF THE RENNER-TELLER COUPLING IN CO+H COLLISION DYNAMICS

STEVE ALEXANDRE NDENGUE, RICHARD DAWES, *Department of Chemistry, Missouri University of Science and Technology, Rolla, MO, USA.*

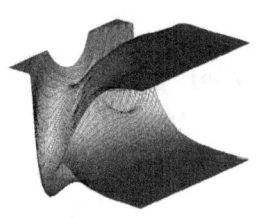

Carbon monoxide is after molecular hydrogen the second most abundant molecule in the universe and an important molecule for processes occurring in the atmosphere, hydrocarbon combustion and the interstellar medium. The rate coefficients of CO in collision with dominant species like H, H_2, He, etc are necessary to understand the CO emission spectrum or to model combustion chemistry processes. The inelastic scattering of CO with H has been intensively studied theoretically in the past decades,[1] mostly using the so-called WKS PES[6] developed by Werner et al. or recently a modified version by Song et al.[2] Though the spectroscopic agreement of the WKS surface with experiment is quite good, so far the studies of scattering dynamics have neglected coupling to an electronic excited state. We present new results on a set of HCO surfaces of the ground and the excited Renner-Teller coupled electronic states[3] with the principal objective of studying the influence of the Renner-Teller coupling on the inelastic scattering of CO+H. Our calculations done using the MCTDH[4] algorithm in the 0-2 eV energy range allow evaluation of the contribution of the Renner-Teller coupling on the rovibrationally inelastic scattering and discuss the relevance and reliability of the calculations.

References:

1. N. Balakrishnan, M. Yan and A. Dalgarno, Astrophys. J. 568, 443 (2002); B.C. Shepler et al, Astron. & Astroph. 475, L15 (2007); L. Song et al, J. Chem. Phys. 142, 204303 (2015); K.M. Walker et al, Astroph. J. 811, 27 (2015).
2. L. Song et al, Astrophys. J. 813, 96 (2015).
3. H.-M. Keller et al, J. Chem. Phys. 105, 4983 (1996).
4. S. Ndengue, R. Dawes and H. Guo, J. Chem. Phys. 144, 244301 (2016).
5. M.H. Beck et al., Phys. Rep. 324, 1 (2000).

MH. Small molecules

Monday, June 19, 2017 – 1:45 PM

Room: 1024 Chemistry Annex

Chair: Wei Lin, University of Texas Rio Grande Valley, Brownsville, TX, USA

MH01   1:45 – 2:00

A SEMI-CLASSICAL APPROACH TO THE CALCULATION OF HIGHLY EXCITED ROTATIONAL ENERGIES FOR ASYMMETRIC-TOP MOLECULES

HANNO SCHMIEDT, STEPHAN SCHLEMMER, *I. Physikalisches Institut, University of Cologne, Cologne, Germany*; SERGEI N. YURCHENKO, *Department of Physics and Astronomy, University College London, Gower Street, London WC1E 6BT, United Kingdom*; ANDREY YACHMENEV, *Center for Free-Electron Laser Science (CFEL), Deutsches Elektronen-Synchrotron (DESY), Hamburg, Germany*; PER JENSEN, *Faculty of Mathematics and Natural Sciences, University of Wuppertal, Wuppertal, Germany*.

We report a new semi-classical method to compute highly excited rotational energy levels of an asymmetric-top molecule. The method forgoes the idea of a full quantum mechanical treatment of the ro-vibrational motion of the molecule. Instead, it employs a semi-classical Green's function approach to describe the rotational motion, while retaining a quantum mechanical description of the vibrations. Similar approaches have existed for some time, but the method proposed here has two novel features. First, inspired by the path integral method, periodic orbits in the phase space and tunneling paths are naturally obtained by means of molecular symmetry analysis. Second, the rigorous variational method is employed for the first time to describe the molecular vibrations. In addition, we present a new robust approach to generating rotational energy surfaces for vibrationally excited states; this is done in a fully quantum-mechanical, variational manner. The semi-classical approach of the present work is applied to calculating the energies of very highly excited rotational states and it reduces dramatically the computing time as well as the storage and memory requirements when compared to the fully quantum-mechanical variational approach. Test calculations for excited states of SO_2 yield semi-classical energies in very good agreement with the available experimental data and the results of fully quantum-mechanical calculations. We hope to be able to present at the meeting also semi-classical calculations of transition intensities.

See also the open-access paper *Phys. Chem. Chem. Phys.* **19**, 1847–1856 (2017). DOI: 10.1039/C6CP05589C

MH02   2:02 – 2:17

NEW WAYS OF TREATING DATA FOR DIATOMIC MOLECULE 'SHELF' AND DOUBLE-MINIMUM STATES

ROBERT J. LE ROY, JASON TAO, SHIRIN KHANNA, *Department of Chemistry, University of Waterloo, Waterloo, ON, Canada*; ASEN PASHOV, *Department of Physics, Sofia University, Sofia, Bulgaria*; JOEL TELLINGHUISEN, *Department of Chemistry, Vanderbilt University, Nashville, TN, USA*.

Electronic states whose potential energy functions have 'shelf' or double-minimum shapes have always presented special challenges because, as functions of vibrational quantum number, the vibrational energies/spacings and inertial rotational constants either have an abrupt change of character with discontinuous slope, or past a given point, become completely chaotic. The present work shows that a 'traditional' methodology developed for deep 'regular' single-well potentials can also provide accurate 'parameter-fit' descriptions of the v-dependence of the vibrational energies and rotational constants of shelf-state potentials that allow a conventional RKR calculation of their Potential energy functions. It is also shown that a merging of Pashov's uniquely flexible 'spline point-wise' potential function representation with Le Roy's 'Morse/Long-Range' (MLR) analytic functional form which automatically incorporates the correct theoretically known long-range form, yields an analytic function that incorporates most of the advantages of both approaches. An illustrative application of this method to data to a double-minimum state of Na_2 will be described.

MH03 $\quad$ 2:19 – 2:34

HIGH-RESOLUTION INFRARED SPECTROSCOPY AND ANALYSIS OF THE ν_2/ν_4 BENDING DYAD AND ν_3 STRETCHING FUNDAMENTAL OF RUTHENIUM TETROXIDE

MBAYE FAYE, *AILES beamline, Synchrotron SOLEIL, Saint Aubin, France*; SÉBASTIEN REYMOND-LARUINAZ, *Département de Physico-chimie, CEA/Saclay, CEA, DEN, Gif-sur-Yvette, France*; JEAN VANDER AUWERA, *Laboratoire de Chimie Quantique et Photophysique, Universite Libre de Bruxelles, Brussels, Belgium*; VINCENT BOUDON, *Laboratoire ICB, CNRS/Université de Bourgogne, DIJON, France*; DENIS DOIZI, *Département de Physico-chimie, CEA/Saclay, CEA, DEN, Gif-sur-Yvette, France*; LAURENT MANCERON, *AILES Beamline, Synchrotron SOLEIL, Saint-Aubin, France.*

RuO_4 is a heavy tetrahedral molecule which has practical uses for several industrial fields. Due to its chemical toxicity and the radiological impact of its 103 and 106 isotopologues, the possible remote sensing of this compound in the atmosphere has renewed interest in its spectroscopic properties. We investigate here for the first time at high resolution the bending dyad region in the far IR and the line intensities in the ν_3 stretching region. Firstly, new high resolution FTIR spectra of the bending modes region in the far infrared have been recorded at room temperature, using a specially constructed cell and an isotopically pure sample of $^{102}RuO_4$. New assignments and effective Hamiltonian parameter fits for this main isotopologue have been performed, treating the whole ν_2/ν_4 bending mode dyad. We provide precise effective Hamiltonian parameters, including band centers and Coriolis interaction parameters. Secondly, we investigate the line intensities for the strongly infrared active stretching mode ν_3, in the mid infrared window near 10 μm. New high resolution FTIR spectra have also been recorded at room temperature, using the same cell and sample. Using assignments and effective Hamiltonian parameter for $^{102}RuO_4$, line intensities have been retrieved and the dipole moment parameters fitted for the ν_3 fundamental. A frequency and intensity line list is proposed.

MH04 $\quad$ 2:36 – 2:51

A MOLECULAR FOUNTAIN

CUNFENG CHENG, AERNOUT P.P. VAN DER POEL, WIM UBACHS, HENDRICK BETHLEM, *Department of Physics and Astronomy, VU University , Amsterdam, Netherlands.*

The resolution of any spectroscopic experiment is limited by the coherent interaction time between the probe radiation and the particle that is being studied. The introduction of cooling techniques for atoms and ions has resulted in a dramatic increase of interaction times and accuracy, it is hoped that molecular cooling techniques will lead to a similar increase. Here we demonstrate the first molecular fountain, a development which permits hitherto unattainably long interrogation times with molecules. In our experiment, beams of ammonia molecules are decelerated, trapped and cooled using inhomogeneous electric fields and subsequently launched. Using a combination of quadrupole lenses and buncher elements, the beam is shaped such that it has a large position spread and a small velocity spread (corresponding to a transverse temperature of less than 10μK and a longitudinal temperature of less than 1μK) while the molecules are in free fall, but strongly focused at the detection region. The molecules are in free fall for up to 266 milliseconds, making it possible, in principle, to perform sub-Hz measurements in molecular systems and paving the way for stringent tests of fundamental physics theories.

MH05 2:53 – 3:08

COMB-ASSISTED CAVITY RING DOWN SPECTROSCOPY OF ^{17}O ENRICHED WATER BETWEEN 7443 AND 7921 CM^{-1}

DIDIER MONDELAIN, *UMR5588 LIPhy, Université Grenoble Alpes/CNRS, Saint Martin d'Hères, France*; SEMEN MIKHAILENKO, *Atmospheric Spectroscopy Div., Institute of Atmospheric Optics, Tomsk, Russia*; EKATERINA KARLOVETS, *Laboratory of Quantum Molecular Mechanics and Radiation Processes, Tomsk State University, Tomsk, Russia*; MAGDALENA KONEFAL, SERGE BÉGUIER, SAMIR KASSI, ALAIN CAMPARGUE, *UMR5588 LIPhy, Université Grenoble Alpes/CNRS, Saint Martin d'Hères, France*.

The room temperature absorption spectrum of water vapour highly enriched in ^{17}O has been recorded by Cavity Ring Down Spectroscopy (CRDS) between 7443 and 7921 cm^{-1}. Three series of recordings were performed with pressure values around 0.1, 1 and 10 Torr. The frequency calibration of the present spectra benefited of the combination of the CRDS spectrometer to a self-referenced frequency comb. The resulting CRD spectrometer combines excellent frequency accuracy over a broad spectral region with a high sensitivity (Noise Equivalent Absorption, $\alpha_{min} \sim 10^{-11} - 10^{-10}$ cm^{-1}). The investigated spectral region corresponds to the high energy range of the first hexade. The assignments were performed using known experimental energy levels as well as calculated line lists based on the results of Partridge and Schwenke. Overall about 4150 lines were measured and assigned to 4670 transitions of six water isotopologues ($H_2^{16}O$, $H_2^{17}O$, $H_2^{18}O$, $HD^{16}O$, $HD^{17}O$ and $HD^{18}O$). Their intensities span six orders of magnitude from 10^{-28} to 10^{-22} cm/molecule. Most of the new results concern the $H_2^{17}O$ and $HD^{17}O$ isotopologues for which about 1600 and 400 transitions were assigned leading to the determination of 329 and 207 new energy levels, respectively. For comparison only about 300 and four transitions of $H_2^{17}O$ and $HD^{17}O$ were previously known in the region, respectively. By comparison to highly accurate $H_2^{16}O$ line positions available in the literature, the average accuracy on our line centers is checked to be on the order of 3 MHz (10^{-4} cm^{-1}) or better for unblended lines. This small uncertainty represents a significant improvement of the line center determination of many $H_2^{16}O$ lines in the considered region.

MH06 3:10 – 3:25

MILLIMETER-WAVE SPECTROSCOPY OF MgI ($\tilde{X}^2\Sigma^+$)

MARK BURTON, K. M. KILCHENSTEIN, *Department of Chemistry and Biochemistry, University of Arizona, Tucson, AZ, USA*; LUCY M. ZIURYS, *Department of Chemistry and Biochemistry; Department of Astronomy, Arizona Radio Observatory, University of Arizona, Tuscon, AZ, USA*.

The pure rotational spectrum of MgI in its ground electronic state ($\tilde{X}^2\Sigma^+$) has been measured using millimeter/submillimeter wave direct-absorption techniques. Rotational transitions arising from the v = 0, 1, and 2 vibrational states of ^{24}MgI, as well as the v = 0 state for the isotopologues (^{25}MgI and ^{26}MgI), have been measured in their natural abundance in the region of 200 – 300 GHz. Rotational, centrifugal distortion, and spin-rotation constants were determined for each isotopologue and the excited vibrational states of ^{24}MgI. Equilibrium parameters B_e, α_e, D_e, and γ_e were ascertained and used to calculate an equilibrium bond length (r_e) for MgI. The spin-rotation coupling constants of several magnesium monohalides were examined and the contribution to the spin-rotation parameter appears to be dominated by second order spin-orbit coupling of the nearby excited $A^2\Pi$ electronic state.

Intermission

MH07

QUANTIFICATION OF FLUORESCENCE FROM THE LYMAN-ALPHA PHOTOLYSIS OF WATER FOR SPACE-CRAFT PLUME CHARACTERIZATION.

JUSTIN W. YOUNG, CHRISTOPHER ANNESLEY, JAIME A. STEARNS, *Space Vehicles Directorate, Air Force Research Lab, Kirtland AFB, NM, USA.*

A quantified characterization of a spacecraft's thruster plume is achievable through measurements of fluorescence from the plume. Fluorescence is present in a spacecraft's plume due to electronic excitation from solar photons, primarily Lyman-alpha (121.6 nm). Excitation of water with Lyman-alpha leads to photodissociation through four possible channels, one of which produces fluorescent hydroxyl radicals (OH(A)). Dependent on the rovbirational state, this species is either predissociative or fluorescent. Here, dispersed fluorescence from water photolysis at Lyman-alpha has been recorded. Comparing our current florescent data with previous H-atom Rydberg tagging results, the ratio of predissociation to fluorescence of OH(A) is quantified.

MH08

LIF SPECTROSCOPY OF ThF AND THE PREPARATION OF ThF$^+$ FOR THE JILA eEDM EXPERIMENT

KIA BOON NG, YAN ZHOU, DAN GRESH, WILLIAM CAIRNCROSS, *JILA, National Institute of Standards and Technology and Univ. of Colorado Department of Physics, University of Colorado, Boulder, Boulder, CO, USA;* TANYA ROUSSY, *JILA, National Institute of Standards and Technology and Univ. of Colorado Department of Physics, University of Colorado, Boulder, Colorado, Boulder, CO, USA;* YUVAL SHAGAM, *JILA, National Institute of Standards and Technology and Univ. of Colorado Department of Physics, University of Colorado, Boulder, Boulder, CO, USA;* LAN CHENG, *Department of Chemistry, Johns Hopkins University, Baltimore, MD, USA;* JUN YE, ERIC CORNELL, *JILA, National Institute of Standards and Technology and Univ. of Colorado Department of Physics, University of Colorado, Boulder, Boulder, CO, USA.*

ThF$^+$ is a promising candidate for a second-generation molecular ion-based measurement of the permanent electric dipole moment of the electron (eEDM). Compared to the current HfF$^+$ eEDM experiment, ThF$^+$ has several advantages: (i) the eEDM-sensitive $^3\Delta_1$ electronic state is the ground state, which facilitates a long measurement coherence time; (ii) its effective electric field (38 GV/cm) is 50% larger than that of HfF+, which promises a direct increase of the eEDM sensitivity; and (iii) the ionization energy of neutral ThF is lower than its dissociation energy, which introduces a greater flexibility for rotational state-selective photoionization via core-nonpenetrating Rydberg states. We use laser-induced fluorescence (LIF) spectroscopy to find suitable intermediate states required for the state selective ionization process. We present the results of our LIF spectroscopy of ThF, and our current progress on efficient ThF ionization and on ThF$^+$ dissociation.

MH09

LASER-INDUCED FLUORESCENCE SPECTROSCOPY OF TWO RUTHENIUM-BEARING MOLECULES: RuF AND RuCl

HANIF ZARRINGHALAM, *Department of Physics, University of New Brunswick, Fredericton, NB, Canada;* ALLAN G. ADAM, *Department of Chemistry, University of New Brunswick, Fredericton, NB, Canada;* COLAN LINTON, DENNIS W. TOKARYK, *Department of Physics, University of New Brunswick, Fredericton, NB, Canada.*

This work extends the electronic spectroscopy of RuF, and reports on what we believe is the first observation of RuCl. Both molecules have been created in a laser-ablation molecular beam apparatus at UNB, and their spectra have been detected by laser-induced fluorescence. In the low-resolution survey of RuF from 400 to 770 nm, five bands were detected in the blue, green and infrared regions of the electromagnetic spectrum. Four of them were rotationally analyzed from high-resolution data. The three bands in the green region are associated with the $^4\Gamma_{11/2} - X^4\Phi_{9/2}$ system first observed by Steimle et al.[a] A new $^4\Delta_{7/2} - X^4\Phi_{9/2}$ transition in the blue region was also detected. Two high-resolution bands of RuCl were rotationally analyzed, and the ground state was also found to be $X^4\Phi_{9/2}$. The data provide detailed structural information about the molecules, such as bond lengths, vibrational frequencies, isotopic structure, spin-orbit interactions and hyperfine interactions.

[a]T. C. Steimle, W. Virgo and T. Ma, J. Chem. Phys. **124** 024309 (2006).

MH10 4:35 – 4:50

OBSERVATIONS OF LOW-LYING ELECTRONIC STATES OF NiD, AND MULTI-ISOTOPE ANALYSIS

MAHDI ABBASI, ALIREZA SHAYESTEH, *School of Chemistry, University of Tehran, Tehran, Iran*; PATRICK CROZET, <u>AMANDA J. ROSS</u>, *UMR 5306, ILM University Lyon 1 and CNRS, Villeurbanne, France.*

Resolved laser induced fluorescence spectra of NiD, recorded at Doppler resolution between 11500 and 18000 cm^{-1}, have defined some 200 term energies in two of the three strongly-interacting, low-lying ($X\,^2\Delta$, $W\,^2\Pi$ and $V\,^2\Sigma^+$) states of NiD associated with an Ni$^+(3d^9)$–D$^-$ configuration. Our observations span $v = 0$ - 5 in the lowest spin-orbit component of the ground state, $X_1\,^2\Delta_{5/2}$, $v = 0$ - 3 in $X_2\,^2\Delta_{3/2}$ and $v = 0$ - 1 in $W_1\,^2\Pi_{3/2}$, the lower component of the $W\,^2\Pi$ state. Spin-orbit and rotation-electronic interactions are strong in NiD. Large parity splittings are seen, due to interactions with the unobserved $^2\Sigma^+$ state. We have attempted a global, multi-isotope fit to reproduce observed term energies up to 6000 cm^{-1} in NiD and ^{58,60,62}NiH, in an extension of the 'Supermultiplet' model proposed by Gray and co-workers ^{a}, because fits with NiD term energies alone failed to converge to sensible solutions. Dunham-type parameters have been used to represent the unperturbed $X\,^2\Delta$, $W\,^2\Pi$ and $V\,^2\Sigma^+$ states, with off-diagonal matrix elements (treating spin-orbit, L- and S-uncoupling effects) based on Ni$^+$ atomic properties. Some electronic Born-Oppenheimer breakdown terms were included in the model.

The spectra show emission from several excited states close to the unique level populated by the single-mode laser. Bands of collisionally-induced fluorescence identify three levels (A ($\Omega = 5/2$) $v = 1$, E ($\Omega = 3/2$) $v = 1$ and I ($\Omega = 3/2$) $v = 0$) that have not been reported before.

---

^{a}Gray, Li, Nelis, and Field, *J. Chem. Phys.* <u>95</u>, 7164 (1991)

MH11 4:52 – 5:07

$B\,^1\Pi \rightarrow A\,^1\Sigma^+$ ELECTRONIC TRANSFER IN NaK

<u>AMANDA J. ROSS</u>, HEATHER HARKER, MAXIME GIRAUD, ELLA WYLLIE^a, *UMR 5306, ILM University Lyon 1 and CNRS, Villeurbanne, France.*

We investigate collisionally-induced $A\,^1\Sigma^+ \rightarrow X\,^1\Sigma^+$ fluorescence in NaK, observed following ro-vibrationally selective excitation of the $B\,^1\Pi$ state. NaK molecules are formed in a heatpipe oven, and excited with a single-mode dye laser operating around 17000 cm^{-1}. Direct $B\,^1\Pi \rightarrow X\,^1\Sigma^+$ fluorescence is dominated by $\Delta J = 0$ or ± 1 transitions, with rotational satellites whose intensities vary according to collisional partner (rare gas or K atoms). The $B \rightarrow X$ emission also shows even weaker $\Delta v \pm 1$ vibrational relaxation bands, with a more even spread of rotational population. Some 5000 cm^{-1} to lower wavenumber, we observe apparently non-selective $A\,^1\Sigma^+ \rightarrow X\,^1\Sigma^+$ emission, with an oscillating bound-free $c\,^3\Sigma^+ \rightarrow a\,^3\Sigma^+$ contribution to the baseline that becomes increasingly important as higher vibrational levels of $B\,^1\Pi$ are populated by the laser. Earlier work defining analytical functions describing the $A\,^1\Sigma^+$ and $b\,^3\Pi$ states of NaK, and the spin-orbit functions coupling them^b, allowed us to assign this dense and irregular $A \rightarrow X$ spectrum, and to see how upper state level populations change as a function of B state excitation and heatpipe conditions (He, Ar or N$_2$ as buffer gas). Our spectra reveal a propensity to conserve vibrational quantum number, at least from low v in the B state. This is in contradiction with A-X emission in Na$_2$ observed^c following laser excitation of the (corresponding) $B\,^1\Pi_u$ state. The non-Boltzmann populations in the $A\,^1\Sigma_u^+$ state of Na$_2$ were explained^d by near-resonant transfer from $B\,^1\Pi_u$ to $2\,^1\Sigma_g^+$ (with no counterpart in heteronuclear NaK), followed by spontaneous emission to the $A\,^1\Sigma_u^+$ state.

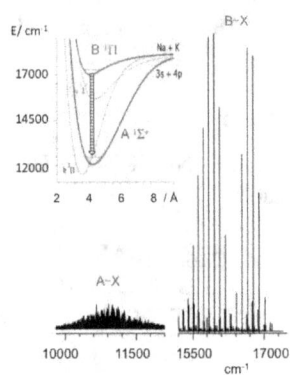

---

^{a}visiting Erasmus student, from Physics Department, University of Strathclyde, Glasgow.
^{b}Harker *et al.*, Phys. Rev. A 92 (1), 012506 (2015)
^{c}Astill *et al.*, Chem. Phys. Lett. <u>125</u> 33 (1986); Camacho *et al.*, J. Phys B. <u>39</u> 2665 (2006)
^{d}Hussein *et al.*, J. Mol Spectrosc. <u>114</u> 105 (1985)

QUANTUM CONTROLLED NUCLEAR FUSION

MARTIN GRUEBELE, *Department of Chemistry, University of Illinois at Urbana-Champaign, Urbana, IL, USA.*

Laser-assisted nuclear fusion is a potential means for providing short, well-controlled particle bursts in the lab, such as neutron or alpha particle pulses. I will discuss computational results of how coherent control by shaped, amplified 800 nm laser pulses can be used to enhance the nuclear fusion cross section of diatomic molecules such as BH or DT. Quantum dynamics simulations show that a strong laser pulse can simultaneously field-bind the diatomic molecule after electron ejection, and increase the amplitude of the vibrational wave function at small internuclear distances. When VUV shaped laser pulses become available, coherent laser control may also be extended to muonic molecules such as D-mu-T, held together by muons instead of electrons. Muonic fusion has been extensively investigated for many decades, but without coherent laser control it falls slightly short of the break-evne point.

MI. Comparing theory and experiment

Monday, June 19, 2017 – 1:45 PM

Room: B102 Chemical and Life Sciences

Chair: Luis A. Rivera-Rivera, Ferris State University , Big Rapids , MI, USA

EXPERIMENTAL AND COMPUTATIONAL INVESTIGATIONS OF THE THRESHOLD PHOTOELECTRON SPECTRUM OF THE HCCN RADICAL

B. GANS, CYRIL FALVO, <u>L. H. COUDERT</u>, *Institut des Sciences Moléculaires d'Orsay, Université Paris-Sud, Orsay, France*; GUSTAVO A. GARCIA, J. KÜGER, *L'Orme des Merisiers; Saint Aubin BP 48, Synchrotron SOLEIL, Gif sur Yvette, France*; J.-C. LOISON, *Institut des Sciences Moléculaires, Université de Bordeaux, Talence, France.*

The HCCN radical, already detected in the interstellar medium, is also important for nitrile chemistry in Titan's atmosphere.[a] Quite recently the photoionization spectrum of the radical has been recorded[b] using mass selected threshold photoelectron (TPE) spectroscopy and this provided us with the first spectroscopic information about the HCCN$^+$ cation. Modeling such a spectrum requires accounting for the non-rigidity of HCCN and for the Renner-Teller effect in HCCN$^+$.

In its $^3A''$ electronic ground state, HCCN is a non-rigid molecule as the potential for the $\angle$HCC bending angle is very shallow.[c] Vibronic couplings with the same bending angle leads, in the $^2\Pi$ electronic ground state of HCCN$^+$, to a strong Renner-Teller effect giving rise to a bent $^2A'$ and a quasi-linear $^2A''$ state.[d]

In this paper the photoionization spectrum of the HCCN radical is simulated. The model developed treats the $\angle$HCC bending angle as a large amplitude coordinate in both the radical and the cation and accounts for the overall rotation and the Renner-Teller couplings. Gaussian quadrature are used to calculate matrix elements of the three potential energy functions retrieved through *ab initio* calculations and rovibrational operators going to infinity for the linear configuration are treated rigorously.

The HCCN TPE spectrum is computed with the above model calculating all rotational components and choosing the appropriate lineshape. This synthetic spectrum will be shown in the paper and compared with the experimental one.[b]

[a]Guélin and Cernicharo, *A&A* **244** (1991) L21; Loison *et al.*, *Icarus* **247** (2015) 218

[b]Garcia, Krüger, Gans, Falvo, Coudert, and Loison, *J. Chem. Phys.* (2017) submitted

[c]Koput, *J. Phys. Chem. A* **106** (2002) 6183

[d]Zhao, Zhang, and Sun, *J. Phys. Chem. A* **112** (2008) 12125

MI02

EXPERIMENTAL AND NUMERICAL CHARACTERIZATION OF A PULSED SUPERSONIC UNIFORM FLOW FOR KINETICS AND SPECTROSCOPY

NICOLAS SUAS-DAVID, SHAMEEMAH THAWOOS, BERNADETTE M. BRODERICK, ARTHUR SUITS, *Department of Chemistry, University of Missouri, Columbia, MO, USA.*

The current CPUF[a] (Chirped Pulse Uniform Flow) and the new UF-CRDS (Uniform Flow Cavity Ring-Down Spectroscopy) setups relie mostly on the production of a good quality supersonic uniform flow.

A supersonic uniform flow is produced by expanding a gas through a Laval nozzle - similar to the nozzles used in aeronautics - linked to a vacuum chamber. The expansion is characterized by an isentropic core where constant very low kinetic temperature (down to 20K) and constant density are observed. The relatively large diameter of the isentropic core associated with homogeneous thermodynamic conditions makes it a relevant tool for low temperature spectroscopy. On the other hand, the length along the axis of the flow of this core (could be longer than 50cm) allows kinetic studies which is one of the main interest of this setup (CRESU technique[b]).

The formation of a uniform flow requires an extreme accuracy in the design of the shape of the nozzle for a set of defined temperature/density. The design is based on a Matlab program which retrieves the shape of the isentropic core according to the method of characteristics prior to calculate the thickness of the boundary layer[c].

Two different approaches are used to test the viability of a new nozzle derived from the program. First, a computational fluid dynamic software (OpenFOAM) models the distribution of the thermodynamic properties of the expansion. Then, fabricated nozzles using 3-D printing are tested based on Pitot measurements and spectroscopic analyses[d]. I will present comparisons of simulation and measured performance for a range of nozzles. We will see how the high level of accuracy of numerical simulations provides a deeper knowledge of the experimental conditions.

[a]J. M. Oldham, C. Abeysekera, J. Joalland, L. N. Zack, K. Prozument, I. R. Sims, G. Barrat Park, R. W. Filed and A. G. Suits, J. Chem. Phys. 141, 154202, (2014).

[b]I. Sims, J. L. Queffelec, A. Defrance, C. Rebrion-Rowe, D. Travers, P. Bocherel, B. Rowe, I. W. Smith, J. Chem. Phys. 100, 4229-4241, (1994).

[c]D. B. Atkinson and M. A. Smith, Rev. Sci. Instrum. 66, 4434, (1995).

[d]N. Suas-David, V. Kulkarni, A. Benidar, S. Kassi and R. Georges, Chem. Phys. Lett. 659, 209-215, (2016)

MI03

DETERMINATION OF THE OSCILLATOR STRENGTHS FOR THE THIRD AND FOURTH VIBRATIONAL OVERTONE TRANSITIONS IN SIMPLE ALCOHOLS

JENS WALLBERG, HENRIK G. KJAERGAARD, *Department of Chemistry, University of Copenhagen, Copenhagen, Denmark.*

Absolute measurements of the weak transitions require sensitive spectroscopic techniques.

With our recently constructed pulsed cavity ring down (CRD) spectrometer, we have recorded the third and fourth vibrational overtone of the OH stretching vibration in a series of simple alcohols: methanol (MeOH), ethanol (EtOH), 1-propanol (1-PrOH), 2-propanol (2-PrOH) and tert-butanol (tBuOH). The CRD setup (in a flow cell configuration) is combined with a conventional FTIR spectrometer to determine the partial pressure of the alcohols from the fundamental transitions of the OH-stretching vibration. The oscillator strengths of the overtone transitions are determined from the integrated absorbances of the overtone spectra and the partial pressures.

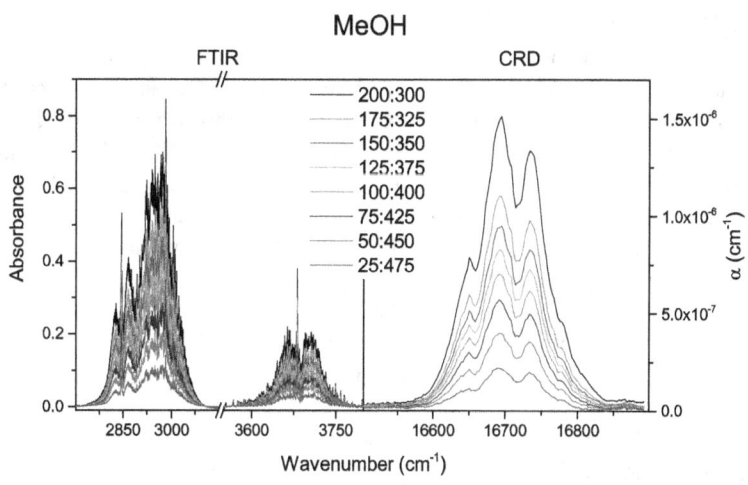

Furthermore, the oscillator strengths were calculated using vibrational local mode theory with energies and dipole moments calculated at CCSD(T)/aug-cc-pVTZ level of theory. We find a good agreement between the observed and calculated oscillator strengths across the series of alcohols.

MI04

AUTOMATED SPECTROSCOPIC ANALYSIS USING THE PARTICLE SWARM OPTIMIZATION ALGORITHM: IMPLEMENTING A GUIDED SEARCH ALGORITHM TO AUTOFIT

KATHERINE ERVIN, STEVEN SHIPMAN, *Department of Chemistry, New College of Florida, Sarasota, FL, USA.*

While rotational spectra can be rapidly collected, their analysis (especially for complex systems) is seldom straightforward, leading to a bottleneck. The AUTOFIT program[a] was designed to serve that need by quickly matching rotational constants to spectra with little user input and supervision. This program can potentially be improved by incorporating an optimization algorithm in the search for a solution. The Particle Swarm Optimization Algorithm (PSO) was chosen for implementation. PSO is part of a family of optimization algorithms called heuristic algorithms, which seek approximate best answers. This is ideal for rotational spectra, where an exact match will not be found without incorporating distortion constants, etc., which would otherwise greatly increase the size of the search space. PSO was tested for robustness against five standard fitness functions and then applied to a custom fitness function created for rotational spectra. This talk will explain the Particle Swarm Optimization algorithm and how it works, describe how Autofit was modified to use PSO, discuss the fitness function developed to work with spectroscopic data, and show our current results.

---

[a]Seifert, N.A., Finneran, I.A., Perez, C., Zaleski, D.P., Neill, J.L., Steber, A.L., Suenram, R.D., Lesarri, A., Shipman, S.T., Pate, B.H., J. Mol. Spec. 312, 13-21 (2015)

MI05

SUBSTITUTION STRUCTURES OF LARGE MOLECULES AND MEDIUM RANGE CORRELATIONS IN QUANTUM CHEMISTRY CALCULATIONS

LUCA EVANGELISTI, *Dipartimento di Chimica G. Ciamician, Università di Bologna, Bologna, Italy*; BROOKS PATE, *Department of Chemistry, The University of Virginia, Charlottesville, VA, USA.*

A study of the minimally exciting topic of agreement between experimental and measured rotational constants of molecules was performed on a set of large molecules with 16-18 heavy atoms (carbon and oxygen). The molecules are: nootkatone ($C_{15}H_{22}O$), cedrol ($C_{15}H_{26}O$), ambroxide ($C_{16}H_{28}O$), sclareolide ($C_{16}H_{22}O_2$), and dihydroartemisinic acid ($C_{15}H_{24}O_2$). For this set of molecules we obtained 13C-subsitution structures for six molecules (this includes two conformers of nootkatone). A comparison of theoretical structures and experimental substitution structures was performed in the spirit of the recent work of Grimme and Steinmetz.[1] Our analysis focused the center-of-mass distance of the carbon atoms in the molecules. Four different computational methods were studied: standard DFT (B3LYP), dispersion corrected DFT (B3LYP-D3BJ), hybrid DFT with dispersion correction (B2PLYP-D3), and MP2. A significant difference in these theories is how they handle medium range correlation of electrons that produce dispersion forces. For larger molecules, these dispersion forces produce an overall contraction of the molecule around the center-of-mass. DFT poorly treats this effect and produces structures that are too expanded. MP2 calculations overestimate the correction and produce structures that are too compact. Both dispersion corrected DFT methods produce structures in excellent agreement with experiment. The analysis shows that the difference in computational methods can be described by a linear error in the center-of-mass distance. This makes it possible to correct poorer performing calculations with a single scale factor. We also reexamine the issue of the "Costain error" in substitution structures and show that it is significantly larger in these systems than in the smaller molecules used by Costain to establish the error limits.

[1] Stefan Grimme and Marc Steinmetz, "Effects of London dispersion correction in density functional theory on structures of organic molecules in the gas phase", Phys. Chem. Chem. Phys. 15, 16031-16042 (2013).

Intermission

NEW VARIATIONAL METHODS FOR COMPUTING VIBRATIONAL SPECTRA OF MOLECULES WITH UP TO 11 ATOMS

JAMES BROWN, PHILLIP THOMAS, <u>TUCKER CARRINGTON</u>, *Department of Chemistry, Queen's University, Kingston, ON, Canada.*

I shall present two new variational methods for computing vibrational spectra. Both rely on the Hamiltonian being a sum of products (SOP). To use a variational method one represents wavefunctions in a basis and uses methods of numerical linear algebra to determine the basis function coefficients. A direct product basis has the advantage that it enables one to efficiently calculate the eigenvalues and eigenvectors of the Hamiltonian matrix using an iterative eigensolver. A direct product basis has the crucial disadvantage that the memory cost of a calculation scales exponentially with the number of atoms in the molecule. One of the new methods uses an expanding basis of products of 1D functions and an iterative eigensolver. For ethylene oxide (7 atoms), converged results are obtained with a basis that is many orders of magnitude smaller than the direct product basis with which similar results would be obtained. The second new method uses sum-of-product basis functions stored in canonical polyadic (CP) tensor format and generated by evaluating matrix-vector products. The memory cost scales linearly with the number of atoms in the molecule. Recent improvements make it possible to compute the spectrum of cyclopentadiene (11 atoms).

A NUMERICALLY EXACT FULL-DIMENSIONAL CALCULATION OF RO-VIBRATIONAL LEVELS OF WATER DIMER

XIAO-GANG WANG, <u>TUCKER CARRINGTON</u>, *Department of Chemistry, Queen's University, Kingston, ON, Canada.*

We have developed a new method for computing numerically exact rovibrational levels of a Van der Waals dimer with flexible monomers and applied it to water dimer, a 12-dimensional cluster. The method uses basis functions that are products of an inter-monomer function and an intra-monomer function. The inter-monomer function is a product of Wigner functions, used to study dimers within the rigid monomer approximation. The intra-monomer functions are monomer vibrational wavefunctions. When the coupling between inter- and intra-monomer coordinates is weak, this new basis is very efficient and only a few monomer vibrational wavefunctions are necessary. The product structure of the basis makes it efficient to use the Lanczos algorithm to calculate eigenvalues and eigenfunctions of the Hamiltonian matrix. In particular, potential matrix-vector products are evaluated, without storing the potential on a full-dimensional grid, by adapting the F-matrix idea previously used to compute rovibrational levels of 5-atom and 6-atom molecules with a contracted basis and an iterative eigensolver.[a] We have obtained numerically exact and converged inter-monomer energy levels and compare these with results obtained using the 6D + 6D adiabatic approach on the CCpol-8sf *ab initio* potential energy surface.[b] We have also obtained the water bend levels and their shifts. We compare with results of the previous adiabatic calculation and experiment.

[a] X.-G. Wang and T. Carrington Jr. J. Chem. Phys. **119**, 101 (2003) and **129**, 234102 (2008).
[b] C. Leforestier, K. Szalewicz, and A. van der Avoid, J. Chem. Phys. **137**, 014305 (2012).

MI08

COMPUTER SPECTROMETERS

NIKESH S. DATTANI, *Department of Chemistry, Kyoto University, Kyoto, Japan.*

Ideally, the cataloguing of spectroscopic linelists would not demand laborious and expensive experiments. Whatever an experiment might achieve, the same information would be attainable by running a calculation on a computer. Kolos and Wolniewicz were the first to demonstrate that calculations on a computer can outperform even the most sophisticated molecular spectroscopic experiments of the time, when their 1964 calculations of the dissociation energies of H_2 and D_2 were found to be more than 1 cm^{-1} larger than the best experiments by Gerhard Herzberg, suggesting the experiment violated a strict variational principle. As explained in his Nobel Lecture, it took 5 more years for Herzberg to perform an experiment which caught up to the accuracy of the 1964 calculations.

Today, numerical solutions to the Schrödinger equation, supplemented with relativistic and higher-order quantum electrodynamics (QED) corrections can provide ro-vibrational spectra for molecules that we strongly believe to be correct, even in the absence of experimental data. Why do we believe these calculated spectra are correct if we do not have experiments against which to test them? All evidence seen so far suggests that corrections due to gravity or other forces are not needed for a computer simulated QED spectrum of ro-vibrational energy transitions to be correct at the precision of typical spectrometers. Therefore a computer-generated spectrum can be considered to be as good as one coming from a more conventional spectrometer, and this has been shown to be true not just for the H_2 energies back in 1964, but now also for several other molecules.

So are we at the stage where we can launch an array of calculations, each with just the atomic number changed in the input file, to reproduce the NIST energy level databases? Not quite. But I will show that for the 6e$^-$ molecule Li_2, we have reproduced the vibrational spacings to within 0.001 cm^{-1} of the experimental spectrum, and I will discuss present-day prospects for replacing laborious experiments for spectra of certain systems within the reach of today's "computer spectrometers".

MI09

JET-COOLED INFRARED LASER SPECTROSCOPY OF DIMETHYL SULFIDE: HIGH RESOLUTION ANALYSIS OF THE ν_{14} CH$_3$-BENDING MODE

ATEF JABRI, *Department of Chemistry, MONARIS, CNRS, UMR 8233, Sorbonne Universités, UPMC Univ Paris 06, Paris, France*; ISABELLE KLEINER, *Laboratoire Interuniversitaire des Systèmes Atmosphériques (LISA), CNRS et Universités Paris Est et Paris Diderot, Créteil, France*; PIERRE ASSELIN, *Department of Chemistry, MONARIS, CNRS, UMR 8233, Sorbonne Universités, UPMC Univ Paris 06, Paris, France.*

The rovibrational spectrum of the ν_{14} CH$_3$-bending mode of dimethyl sulfide (CH$_3$)$_2$S was recorded in the 963-987 cm^{-1} spectral region using our sensitive tunable quantum cascade laser spectrometer coupled to a pulsed slit jet[a][b]. The combined use of a high dilution (CH$_3$)$_2$S/Ar gas mixture expanded at high backing pressure through a slit nozzle enabled to obtain an efficient rovibrational cooling which narrows the rotational distribution and eliminates hot bands arising from three low frequency modes below 300 cm^{-1}[c]. The characteristic PQR band contour of a b$_1$ symmetry mode centered at 975.29 cm^{-1} was observed and will be compared with theoretical calculations at the CCSD(T)/VTZ level[c] (ν_{14} mode at 986 cm^{-1}) and room temperature experiments at low resolution (974 cm^{-1})[d]. Starting from the accurate set of ground state parameters derived from microwave, millimeter and far-infrared measurements, the rovibrational analysis will be presented and discussed.

[a] P. Asselin, Y. Berger, T. R. Huet, R. Motiyenko, L. Margulès, R. J. Hendricks, M. R. Tarbutt, S. Tokunaga, B. Darquié, Phys. Chem. Chem. Phys. 19, 4576 (2017)

[b] P. Asselin, A. Potapov, A. Turner, V. Boudon, L. Bruel, M. A. Gaveau and M. Mons, submitted to J. Phys. Chem. Lett. (2017)

[c] M. L. Senent, C. Puzzarini, R. Domínguez-Gómez, M. Carvajal, and M. Hochlaf, J. Chem. Phys., 140, 124302 (2014)

[d] J. W. Ypenburg & H. Gerding, Recueil des Travaux Chimiques des Pays-Bas, 90, 885 (1971)

MI10

WEAK HYDROGEN BONDS FROM ALIPHATIC AND FLUORINATED ALOCOHOLS TO MOLECULAR NITROGEN DETECTED BY SUPERSONIC JET FTIR SPECTROSCOPY

SOENKE OSWALD, MARTIN A. SUHM, *Institute of Physical Chemistry, Georg-August-Universität Göttingen, Göttingen, Germany.*

Complexes of organic molecules with the main component of earth's atmosphere are of interest,[a] also for a stepwise understanding of the phenomenon of matrix isolation.[b] Via its large quadrupole moment, nitrogen binds strongly to polarized OH groups in hydrogen-bonded dimers. Further complexation leads to a smooth spectral transition from free to embedded molecules which we probe in supersonic jets. Results for 1,1,1,3,3,3-hexafluoro-2-propanol,[c] methanol,[d] t-butyl alcohol,[e] and the conformationally more complex ethanol[f] are presented and assigned with the help of quantum chemical calculations.

[a]Kuma, S., Slipchenko, M. N., Kuyanov, K. E., Momose, T., Vilesov, A. F., Infrared Spectra and Intensities of the H_2O and N_2 Complexes in the Range of the ν_1- and ν_3-Bands of Water, *J. Phys. Chem. A*, **2006**, *110*, 10046–10052.

[b]Coussan, S., Bouteiller, Y., Perchard, J. P., Zheng, W. Q., Rotational Isomerism of Ethanol and Matrix Isolation Infrared Spectroscopy, *J. Phys. Chem. A*, **1998**, *102*, 5789–5793.

[c]Suhm, M. A., Kollipost, F., Femtisecond single-mole infrared spectroscopy of molecular clusters, *Phys. Chem. Chem. Phys.*, **2013**, *15*, 10702–10721.

[d]Larsen, R. W., Zielke, P., Suhm, M. A., Hydrogen bonded OH stretching modes of methanol clusters: a combined IR and Raman isotopomer study, *J. Chem. Phys.*, **2007**, *126*, 194307.

[e]Zimmermann, D., Häber, T., Schaal, H., Suhm, M. A., Hydrogen bonded rings, chains and lassos: The case of t-butyl alcohol clusters, *Mol. Phys.*, **2001**, *99*, 413–425.

[f]Wassermann, T. N., Suhm, M. A., Ethanol Monomers and Dimers Revisited: A Raman Study of Conformational Preferences and Argon Nanocoating Effects, *J. Phys. Chem. A*, **2010**, *114*, 8223–8233.

MI11

PHOTOELECTRON VELOCITY MAP IMAGING SPECTROSCOPY OF BeS^-

AMANDA REED DERMER, MALLORY THEIS, KYLE MASCARITOLO, MICHAEL HEAVEN, *Department of Chemistry, Emory University, Atlanta, GA, USA.*

The BeS^- anion to neutral ground state transition, $X^2\Sigma^+ \rightarrow X^1\Sigma^+$ has been studied using photoelectron velocity map imaging spectroscopy. Rotational constants, vibrational intervals and the electron binding energy of BeS^- have been determined for the first time. Rotational constants were derived from band contour analyses, as the contours exhibited band head features associated with changes in the rotational angular momenta of $\Delta N=0$, ± 1, and ± 2. A dipole bound state (DBS) of BeS^- was observed 130 cm^{-1} below the detachment threshold. Autodetachment spectra for the transition to the DBS were rotationally resolved, providing an accurate rotational constant for BeS^-, v=0. The experimental results were found to be in reasonable agreement with the predictions of high level ab initio calculations.

5:09 – 5:24

EXTENDING THE LOCAL MODE HAMILTONIAN INTO THE CONDENSED PHASE: USING VIBRATIONAL SUM
FREQUENCY GENERATION TO STUDY THE BENZENE-AIR INTERFACE

BRITTA JOHNSON, EDWIN SIBERT, *Department of Chemistry, University of Wisconsin–Madison, Madison,
WI, USA.*

Surfaces and interfaces play an important role in understanding many chemical process; they also contain molecular
configurations and vibrations that are unique compared to those seen in the bulk and gas phases. Sum frequency generated
(SFG) vibrational spectroscopy provides an incredibly detailed picture of these interfaces. In particular, the CH stretch region
of the spectrum contains an extensive degree of information about the molecular vibrations and arrangements at the surface
or interface. The presence of a strong bandwidth SFG signal for the benzene/air interface has generated controversy since
it was discovered; since benzene is centrosymmetric, no SFG signal is expected. It has been hypothesized that this signal
is primarily a result of bulk contributions that results from electric quadrupole transitions. Our work focuses on testing this
conclusion by calculating a theoretical VSF spectrum from pure surface contributions using a mixed quantum/classical local
mode Hamiltonian.

We take as a starting point our local mode CH/OH stretch Hamiltonian, that was previously used to study alkylbenzenes,
benzene-$(H_2O)_n$, and DPOE-water clusters, and extend it to the condensed phase by including shifts in the intensities and
frequencies as a function of the environment. This environment is modeled using a SAPT-based force-field that accurately
reproduces the quadrupole for the benzene dimer. A series of independent time-dependent trajectories are used to obtain an
ensemble of surface configurations and calculate the appropriate correlation functions. These correlations functions allow us
to determine the origins of the VSF signal. Our talk will focus on the challenges of extending our local mode Hamiltonian
into the condensed phase.

MJ. Atmospheric science
Monday, June 19, 2017 – 1:45 PM
Room: 161 Noyes Laboratory

Chair: Jacob Stewart, Connecticut College, New London, CT, USA

MJ01   1:45 – 2:00

SPECTROSCOPIC CHARACTERIZATION OF THE REACTION PRODUCTS BETWEEN HCl AND THE SIMPLEST CRIEGEE INTERMEDIATE CH_2OO

<u>CARLOS CABEZAS</u>, YASUKI ENDO, *Department of Applied Chemistry, National Chiao Tung University, Hsinchu, Taiwan.*

Carbonyl oxides (R_1R_2COO), also known as Criegee intermediates (CIs), react quickly with many trace atmospheric gases, including inorganic gases such as HCl, which are present in polluted urban atmospheres. A theoretical investigation of the reaction between the simplest CI, CH_2OO, with HCl suggests the formation of chloromethyl hydroperoxide (CMHP) through an insertion mechanism. To gain some insight, we have interrogated the reaction system containing CH_2OO and HCl through pure rotational spectroscopy. In our experiment, CH_2OO molecules have been generated in the discharged plasma of a CH_2I_2/O_2 mixture, which containg a small amount of HCl enough to react with CH_2OO. The resulting products (including CH_2OO) were characterized by Fourier-transform microwave (FTMW) spectroscopy. Rotational transitions in the 6-40 GHz frequency range were observed by FTMW spectroscopy together with FTMW-mmW and MW-MW double-resonance techniques. The observed species was identified with the help of quantum chemical calculations as the most stable conformer of CMHP. The non observation of other different reaction products together with the absence of spectral features of the complex between HCl and CH_2OO enable us to understand the pathway of the $HCl+CH_2OO$ reaction.

MJ02   2:02 – 2:17

PROBING THE CONFORMATIONAL BEHAVIOR OF THE C_3 ALKYL-SUBSTITUTED CRIEGEE INTERMEDIATES BY FTMW SPECTROSCOPY

<u>CARLOS CABEZAS</u>, *Department of Applied Chemistry, National Chiao Tung University, Hsinchu, Taiwan;* J.-C. GUILLEMIN, *Institut des Sciences Chimiques de Rennes, UMR 6226 CNRS - ENSCR, Rennes, France;* YASUKI ENDO, *Department of Applied Chemistry, National Chiao Tung University, Hsinchu, Taiwan.*

Carbonyl oxides (R_1R_2COO), often called Criegee intermediates (CIs), have been assumed as intermediates generated by the ozonolysis reaction of alkenes, and are thought to play important roles in atmospheric chemistry. After the first laboratory observation of the simplest CI, CH_2OO, their experimental characterization has been drastically progressing. Especially alkyl-substituted CIs have attracted much attention. Here we report rotational spectra of alkyl-substituted CIs with three carbon atoms in the substituent groups, named C_3 alkyl-substituted CIs. This group includes methyl-ethyl-ketone oxide or 2-butanone oxide ($C_2H_5CCH_3OO$) and its structural isomers n-butyraldehyde oxide (C_3H_7CHOO) and isobutyraldehyde oxide (($CH_3)_2CHCHOO$). These molecules have been produced in the discharge plasma of diiodo-alkyl-derivative/O_2 gas mixtures, and characterized by Fourier-transform microwave spectroscopy. For the first of them, $C_2H_5CCH_3OO$, four different conformers were observed coexisting in the supersonic expansion. Spectra of the four species show small splittings due to the threefold methyl internal rotation which made possible to determine their respective barrier heights of the hindered methyl rotation. Preliminary results of the ongoing investigation of C_3H_7CHOO and ($CH_3)_2CHCHOO$ molecules are also presented.

MJ03   2:19 – 2:34

MICROWAVE CHARACTERIZATION OF PROPIOLIC SULFURIC ANHYDRIDE AND TWO CONFORMERS OF ACRYLIC SULFURIC ANHYDRIDE

<u>CJ SMITH</u>, ANNA HUFF, BECCA MACKENZIE, KEN LEOPOLD, *Chemistry Department, University of Minnesota, Minneapolis, MN, USA.*

Sulfur trioxide reacts with propiolic acid and acrylic acid to form propiolic sulfuric anhydride ($HC{\equiv}C\text{-}COOSO_2OH$) and acrylic sulfuric anhydride ($H_2C{=}CH\text{-}COOSO_2OH$), respectively. Both species have been observed by chirped-pulse and conventional cavity microwave spectroscopy. In the case of acrylic acid, two conformers derived from the cis and trans form of the acid have been observed. The reaction mechanism and energetics are investigated by density functional theory and CCSD calculations. These results add to a growing body of evidence that establishes carboxylic sulfuric anhydrides, $RCOOSO_2OH$, as a class of molecules formed readily from $SO_3 + RCOOH$ in the gas phase and which, therefore, may be of significance in the nucleation and growth of atmospheric aerosol particles.

MJ04

FACILE FORMATION OF ACETIC SULFURIC ANHYDRIDE IN A SUPERSONIC JET: CHARACTERIZATION BY MICROWAVE SPECTROSCOPY AND COMPUTATIONAL CHEMISTRY

ANNA HUFF, CJ SMITH, BECCA MACKENZIE, KEN LEOPOLD, *Chemistry Department, University of Minnesota, Minneapolis, MN, USA.*

Sulfur trioxide and acetic acid are shown to react under supersonic jet conditions to form acetic sulfuric anhydride, CH_3COOSO_2OH. Rotational spectra of the parent, ^{34}S, methyl ^{13}C, and fully deuterated isotopologues have been observed by chirped-pulse and conventional cavity microwave spectroscopy. A and E internal rotation states have been observed for each isotopologue studied and the methyl group internal rotation barriers have been determined (241.043(65) cm^{-1} for the parent species). The reaction is analogous to that of our previous report on the reaction of sulfur trioxide and formic acid. DFT and CCSD calculations are also presented which indicate that the reaction proceeds via a $\pi_2 + \pi_2 + \sigma_2$ cycloaddition reaction. These results support our previous conjecture that the reaction of SO_3 with carboxylic acids is both facile and general. Possible implications for atmospheric aerosol formation are discussed.

MJ05

LINE SHAPE PARAMETERS OF WATER VAPOR TRANSITIONS IN THE 3645-3975 cm^{-1} REGION

V. MALATHY DEVI, D. CHRIS BENNER, *Department of Physics, College of William and Mary, Williamsburg, VA, USA*; ROBERT R. GAMACHE, BASTIEN VISPOEL, CANDICE L. RENAUD, *Department of Environmental, Earth, and Atmospheric Sciences, University of Massachusetts Lowell, Lowell, MA, USA*; MARY ANN H. SMITH, *Science Directorate, NASA Langley Research Center, Hampton, VA, USA*; ROBERT L. SAMS, THOMAS A. BLAKE, *Chemical Physics, Pacific Northwest National Laboratory, Richland, WA, USA.*

A Bruker IFS 120HR Fourier transform spectrometer (FTS) at the Pacific Northwest National Laboratory (PNNL) in Richland, Washington was used to record a series of spectra in the regions of the ν_1 and ν_3 bands of H_2O. The samples included low pressures of pure H_2O as well as H_2O broadened by air at different pressures, temperatures and volume mixing ratios. We fit simultaneously 16 high-resolution (0.008 cm^{-1}), high S/N ratio absorption spectra recorded at 268, 296 and 353 K (L=19.95 cm), employing a multispectrum fitting technique[a] to retrieve accurate line positions, relative intensities, Lorentz air-broadened half-width and pressure-shift coefficients and their temperature dependences for more than 220 H_2O transitions. Self-broadened half-width and self-shift coefficients were measured for over 100 transitions. For select sets of transition pairs for the H_2O-air system we determined collisional line mixing coefficients via the off-diagonal relaxation matrix element formalism[b], and we also measured speed dependence parameters for 85 transitions. Modified Complex Robert Bonamy (MCRB) calculations of the half-widths, line shifts, and temperature dependences were made for self-, N_2-, O_2-, and air-broadening. The measurements and calculations are compared with each other and with similar parameters reported in the literature.

[a] D. C. Benner, C. P. Rinsland, V. Malathy Devi, M. A. H. Smith, D. Atkins, *JQSRT* **53** (1995) 705-721.
[b] A. Levy, N. Lacome, C. Chackerian, Collisional line mixing, in *Spectroscopy of the Earth's Atmosphere and Interstellar Medium*, Academic Press, Inc., Boston (1992) 261-337.

Intermission

MJ06 3:27 – 3:42

SPECTROSCOPIC STUDY OF AIR-BROADENED NITROUS OXIDE IN THE ν_3 BAND

ROBAB HASHEMI, HOSSEIN NASERI, ADRIANA PREDOI-CROSS, *Department of Physics and Astronomy, University of Lethbridge, Lethbridge, Canada*; <u>MARY ANN H. SMITH</u>, *Science Directorate, NASA Langley Research Center, Hampton, VA, USA*; V. MALATHY DEVI, *Department of Physics, College of William and Mary, Williamsburg, VA, USA*.

We present results of a recent analysis of laboratory spectra to determine line positions, intensities, air-broadened half-widths and pressure-induced shifts and their temperature dependences in the ν_3 fundamental band of N_2O. The spectra used in this study were recorded using the 1-m McMath-Pierce Fourier transform spectrometer while it was located at the National Solar Observatory on Kitt Peak, AZ. Multispectrum analysis software[a] was used to retrieve the line parameters using the Voigt and speed-dependent Voigt line profiles. The line mixing coefficients were calculated using the Exponential Power Gap scaling law. Comparisons with similar published results will be presented.

[a]D. C. Benner, C. P. Rinsland, V. Malathy Devi, M. A. H. Smith, D. Atkins, *JQSRT* **53** (1995) 705-721.

MJ07 3:44 – 3:59

ATMOSPHERIC ISOTOPOLOGUES OBSERVED WITH ACE-FTS AND MODELED WITH WACCM

<u>ERIC M. BUZAN</u>, *Department of Chemistry and Biochemistry, Old Dominion University, Norfolk, VA, USA*; CHRISTOPHER A. BEALE, *Department of Ocean, Earth and Atmospheric Sciences, Old Dominion University, Norfolk, VA, USA*; MAHDI YOUSEFI, *Department of Physics, Old Dominion University, Norfolk, VA, USA*; CHRIS BOONE, *Department of Chemistry, University of Waterloo, Waterloo, ON, Canada*; PETER F. BERNATH, *Department of Chemistry and Biochemistry, Old Dominion University, Norfolk, VA, USA*.

Atmospheric isotopologues are useful tracers of dynamics and chemistry and can be used to constrain budgets of gases in the atmosphere. The Atmospheric Chemistry Experiment (ACE) routinely measures vertical profiles of over 35 molecules and 20 isotopologues via solar occultation from a satellite in low Earth orbit. The primary instrument is an infrared Fourier transform spectrometer with a spectral range of $750 - 4400$ cm^{-1} and a resolution of 0.02 cm^{-1}. ACE began taking measurements in 2004 and is still active today. This talk focuses on isotopic measurements of CH_4, CO, CO_2, and N_2O from ACE-FTS. To complement ACE-FTS data, modeling using the Whole Atmosphere Community Climate Model (WACCM) was performed for each molecule.

MJ08 4:01 – 4:16

THE ATMOSPHERIC CHEMISTRY EXPERIMENT (ACE): LATEST RESULTS

<u>PETER F. BERNATH</u>, *Department of Chemistry and Biochemistry, Old Dominion University, Norfolk, VA, USA*.

ACE (also known as SCISAT) is making a comprehensive set of simultaneous measurements of numerous trace gases, thin clouds, aerosols and temperature by solar occultation from a satellite in low earth orbit. A high inclination orbit gives ACE coverage of tropical, mid-latitudes and polar regions. The primary instrument is a high-resolution (0.02 cm^{-1}) infrared Fourier Transform Spectrometer (FTS) operating in the $750-4400$ cm^{-1} region, which provides the vertical distribution of trace gases, and the meteorological variables of temperature and pressure. Aerosols and clouds are being monitored through the extinction of solar radiation using two filtered imagers as well as by their infrared spectra. After 14 years in orbit, the ACE-FTS is still operating well. A short introduction and overview of the ACE mission will be presented (see http://www.ace.uwaterloo.ca for more information). This talk will focus on recent ACE results and comparisons with chemical transport models.

MJ09

ACCURATE LASER MEASUREMENTS OF THE WATER VAPOR SELF-CONTINUUM ABSORPTION IN FOUR NEAR INFRARED ATMOSPHERIC WINDOWS. A TEST OF THE MT_CKD MODEL.

ALAIN CAMPARGUE, SAMIR KASSI, DIDIER MONDELAIN, DANIELE ROMANINI, *UMR5588 LIPhy, Université Grenoble Alpes/CNRS, Saint Martin d'Hères, France*; LOÏC LECHEVALLIER, *Institut des Géosciences de l'Environnement, Université Grenoble Alpes, Saint Martin d'Hères, France*; SEMYON VASILCHENKO, *UMR5588 LIPhy, Université Grenoble Alpes/CNRS, Saint Martin d'Hères, France*.

The semi empirical MT_CKD model of the absorption continuum of water vapor is widely used in atmospheric radiative transfer codes of the atmosphere of Earth and exoplanets but lacks of experimental validation in the atmospheric windows. Recent laboratory measurements by Fourier transform Spectroscopy have led to self-continuum cross-sections much larger than the MT_CKD values in the near infrared transparency windows. In the present work, we report on accurate water vapor absorption continuum measurements by Cavity Ring Down Spectroscopy (CRDS) and Optical-Feedback-Cavity Enhanced Laser Spectroscopy (OF-CEAS) at selected spectral points of the transparency windows centered around 4.0, 2.1 and 1.25 μm. The temperature dependence of the absorption continuum at 4.38 μm and 3.32 μm is measured in the 23-39 °C range. The self-continuum water vapor absorption is derived either from the baseline variation of spectra recorded for a series of pressure values over a small spectral interval or from baseline monitoring at fixed laser frequency, during pressure ramps. In order to avoid possible bias approaching the water saturation pressure, the maximum pressure value was limited to about 16 Torr, corresponding to a 75% humidity rate. After subtraction of the local water monomer lines contribution, self-continuum cross-sections, C_S, were determined with a few % accuracy from the pressure squared dependence of the spectra base line level. Together with our previous CRDS and OF-CEAS measurements in the 2.1 and 1.6 μm windows, the derived water vapor self-continuum provides a unique set of water vapor self-continuum cross-sections for a test of the MT_CKD model in four transparency windows. Although showing some important deviations of the absolute values (up to a factor of 4 at the center of the 2.1 μm window), our accurate measurements validate the overall frequency dependence of the MT_CKD2.8 model.

MJ10

PHOTOACOUSTIC SPECTROSCOPY OF PRESSURE- AND TEMPERATURE- DEPENDENCE IN THE O_2 A-BAND

MATTHEW J. CICH, *Jet Propulsion Laboratory, California Institute of Technology, Pasadena, CA, USA*; ELIZABETH M LUNNY, GAUTAM STROSCIO, *Division of Chemistry and Chemical Engineering, California Institute of Technology, Pasadena, CA, USA*; THINH QUOC BUI, *JILA, National Institute of Standards and Technology and Univ. of Colorado Department of Physics, University of Colorado, Boulder, Boulder, CO, USA*; PRIYANKA RUPASINGHE, *Physical Sciences, Cameron University, Lawton, OK, USA*; DANIEL HOGAN, *Department of Applied Physics, Stanford University, Stanford, CA, USA*; CAITLIN BRAY, *Department of Chemistry, Wesleyan University, Middletown, CT, USA*; DAVID A. LONG, JOSEPH T. HODGES, *Material Measurement Laboratory, National Institute of Standards and Technology, Gaithersburg, MD, USA*; TIMOTHY J. CRAWFORD, CHARLES MILLER, BRIAN DROUIN, *Jet Propulsion Laboratory, California Institute of Technology, Pasadena, CA, USA*; MITCHIO OKUMURA, *Division of Chemistry and Chemical Engineering, California Institute of Technology, Pasadena, CA, USA*.

NASA's Orbiting Carbon Observatory missions OCO-2 and OCO-3 require spectroscopic parameterization of the Oxygen A-Band absorption (757-775 nm) with unprecedented precision, to deliver space-based measurements of CO_2 column densities with an accuracy of better than 0.1%. Furthermore, with the long satellite-based pathlengths, the strongest A-Band lines are saturated. Accurate retrievals of O_2 column densities thus require precise modeling of the line shape, including the wings several linewidths from line center. The line shape model must go beyond the Voigt profile to include higher order effects such as Dicke narrowing, speed dependence, line mixing (LM), and collision-induced absorption (CIA). High precision laboratory data targeting these effects must be taken. Line mixing and collision induced absorption have proven to be especially problematic in satellite retrievals of O_2 column densities. LM and CIA are more prominent at lower temperatures and higher pressures. A temperature-stabilized photoacoustic spectrometer was therefore designed to study the temperature- and pressure-dependence of spectral line shapes at temperatures from 230-296 K and pressures up to 5 atm. Progress toward high resolution (2 MHz) measurements of the full A-Band will be presented. The observed lineshapes are analyzed with the Hartmann-Tran Profile (HTP), which incorporates LM and CIA , using the Labfit multispectrum fitting program, and the determination of LM and CIA effects will be presented.

MJ11 **4:52 – 5:07**

ULTRAVIOLET STUDY OF THE GAS PHASE HYDRATION OF METHYLGLYOXAL TO FORM THE GEMDIOL

JAY A KROLL, *Department of Chemistry and Biochemistry, University of Colorado, Boulder, CO, USA*; ANNE S. HANSEN, KRISTIAN H. MØLLER, *Department of Chemistry, University of Copenhagen, Copenhagen, Denmark*; JESSICA L AXSON, *Department of Environmental Health Sciences, University of Michigan, Ann Arbor, MI, USA*; HENRIK G. KJAERGAARD, *Department of Chemistry, University of Copenhagen, Copenhagen, Denmark*; VERONICA VAIDA, *Department of Chemistry and Biochemistry, University of Colorado, Boulder, CO, USA*.

Methylglyoxal is a known oxidation product of volatile organic compounds (VOCs) in Earth's atmosphere. While the gas phase chemistry of methylglyoxal is fairly well understood, its modeled concentration and role in the formation of secondary organic aerosol (SOA) continues to be controversial. The gas phase hydration of methylglyoxal to form a gemdiol has been shown to occur in infrared studies but has not been widely considered for water-restricted environments such as the atmosphere. However, this process may have important consequences for the atmospheric processing or VOCs. We have recorded UV spectroscopic measurements following the hydration of methylglyoxal and have compared these measurements to calculated spectra of the electronic transitions of methylglyoxal and methylglyoxal diol. We will report on these measurements and discuss the implications for understanding the atmospheric processing and fate of methylglyoxal and similar molecules

MJ12 **5:09 – 5:24**

DEVELOPMENT OF A QCL-BASED SPECTROMETER FOR SPECTROSCOPIC ANALYSIS OF BIOGENIC VOLATILE ORGANIC COMPOUNDS

MICHAEL CYRUS IRANPOUR, MINH NHAT TRAN, JACOB STEWART, *Department of Chemistry, Connecticut College, New London, CT, USA*.

Biogenic volatile organic compounds (BVOCs) are naturally occurring molecules that are emitted into the atmosphere by plants. BVOCs have an important role in atmospheric chemistry as they react readily with ozone, hydroxyl radicals, and nitric oxides to form aerosols and pollutants such as ozone in the troposphere. We are developing an IR spectrometer with the aim of measuring spectra of atmospheric samples of BVOCs to determine their concentrations. Using an external cavity quantum cascade laser (EC-QCL), we have acquired IR spectra of isoprene (C_5H_8) near 993 cm^{-1}. Isoprene represents an ideal target, as it is the simplest and most abundant BVOC. IR spectra of standard samples of isoprene were acquired in order to determine the detection limit of the spectrometer. We have also been working to improve the capabilities of the spectrometer by implementing wavelength modulation spectroscopy and increasing the path length through our samples by using a multipass cell. In this talk, we will present data from our initial measurements of the standard isoprene samples using a simple direct absorption setup as well as measurements using the improved spectrometer.

MK. Instrument/Technique Demonstration
Monday, June 19, 2017 – 1:45 PM
Room: 140 Burrill Hall

Chair: Thomas Giesen, University Kassel, Kassel, Germany

MK01   1:45 – 2:00

DUAL-COMB SPECTROSCOPY OF THE $\nu_1 + \nu_3$ BAND OF ACETYLENE: INTENSITY AND TRANSITION DIPOLE MOMENT

<u>KANA IWAKUNI</u>[a], *Department of Physics, Faculty of Science and Technology, Keio University, Yokohama, Japan*; SHO OKUBO, *National Metrology Institute of Japan (NMIJ), Ntional Institute of Advanced Industrial Science and Technology (AIST), Tsukuba, Japan*; KOICHI MT YAMADA, *Institute for Environmental Management Technology (EMTech), National Institute of Advanced Industrial Science and Technology (AIST), Tsukuba, Japan*; HAJIME INABA, ATSUSHI ONAE, *National Metrology Institute of Japan (NMIJ), Ntional Institute of Advanced Industrial Science and Technology (AIST), Tsukuba, Japan*; FENG-LEI HONG, *Department of Physics, Yokohama National University, Yokohama, Japan*; HIROYUKI SASADA, *Department of Physics, Faculty of Science and Technology, Keio University, Yokohama, Japan*.

The $\nu_1 + \nu_3$ vibration band of $^{12}C_2H_2$ is recorded with a homemade dual-comb spectrometer [b]. The spectral resolution and the accuracy of frequency determination are high, and the bandwidth is broad enough to take spectrum of the whole band in one shot. The last remarkable competence enables us to record all the spectral lines under constant experimental conditions. The linewidth and line strength of the P(26) to R(29) transitions are determined by fitting the line profile to Lambert-Beer's law with a Voigt function. In the course of analysis, we found the ortho-para dependence of the pressure-broadening coefficient [cd]. This time, we have determined the transition dipole moment of the $\nu_1 + \nu_3$ band. It is noted that the transition dipole moment determined from the ortho lines agrees with that from the para lines.

---

[a]present address; JILA, University of Colorado, Boulder, CO, USA

[b]S. Okubo *et al.*, Applied Physics Express <u>8</u>, 082402 (2015).

[c]K.Iwakuni *et al.*, 71th ISMS, WK15

[d]K. Iwakuni *et al.*, Physical Review Letters <u>117</u>, 143902 (2016).

MK02   2:02 – 2:17

HIGH-RESOLUTION DUAL-COMB SPECTROSCOPY WITH ULTRA-LOW NOISE FREQUENCY COMBS

WOLFGANG HÄNSEL, MICHELE GIUNTA, KATJA BEHA, , *Menlo Systems, GmbH, Martinsried, Germany*; <u>ADAM J. PERRY</u>, , *Menlo Systems, Inc., Newton, NJ, USA*; R. HOLZWARTH, , *Menlo Systems, GmbH, Martinsried, Germany*.

Dual-comb spectroscopy is a powerful tool for fast broad-band spectroscopy due to the parallel interrogation of thousands of spectral lines. Here we report on the spectroscopic analysis of acetylene vapor in a pressurized gas cell using two ultra-low noise frequency combs with a repetition rate around 250 MHz. Optical referencing to a high-finesse cavity yields a sub-Hertz stability of all individual comb lines (including the virtual comb lines between 0 Hz and the carrier) and permits one to pick a small difference of repetition rate for the two frequency combs on the order of 300 Hz, thus representing an optical spectrum of 100 THz (~ 3300 cm^{-1}) within half the free spectral range

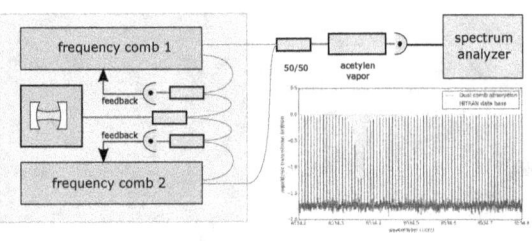

(125 MHz). The transmission signal is derived straight from a photodetector and recorded with a high-resolution spectrum analyzer or digitized with a computer-controlled AD converter. The figure to the right shows a schematic of the experimental setup which is all fiber-coupled with polarization-maintaining fiber except for the spectroscopic cell. The graph on the lower right reveals a portion of the recorded radio-frequency spectrum which has been scaled to the optical domain. The location of the measured absorption coincides well with data taken from the HITRAN data base. Due to the intrinsic linewidth of all contributing comb lines, each sampling point in the transmission graph corresponds to the probing at an optical frequency with sub-Hertz resolution. This resolution is maintained in coherent wavelength conversion processes such as difference-frequency generation (DFG), sum-frequency generation (SFG) or non-linear broadening (self-phase modulation), and is therefore easily transferred to a wide spectral range from the mid infrared up to the visible spectrum.

MK03

PROGRESS TOWARD INNOVATIONS IN CRYOGENIC ION CLUSTER SPECTROMETERS

CASEY J HOWDIESHELL, ETIENNE GARAND, *Department of Chemistry, University of Wisconsin–Madison, Madison, WI, USA.*

Cryogenic Ion Vibrational Spectroscopy (CIVS) is a useful technique that yields rich information about non-covalent interactions in various systems including catalytic complexes, small biologically relevant molecules, and solvent networks. Current instrumentation demands high production costs and large laboratory facilities. We have designed an affordable and compact instrument that is capable of current CIVS experiments. This setup utilizes an ion funnel and a Linear Trap Quadrupole (LTQ) which improves the ion density and allows for spectroscopic interrogation directly in the trap. Preliminary results and future innovations will be discussed.

MK04

CAVITY-ENHANCED SPECTROSCOPY OF MOLECULAR IONS IN THE MID-INFRARED WITH UP-CONVERSION DETECTION AND BREWSTER-PLATE SPOILERS

CHARLES R. MARKUS, JEFFERSON E. McCOLLUM, *Department of Chemistry, University of Illinois at Urbana-Champaign, Urbana, IL, USA*; JAMES NEIL HODGES, *Department of Chemistry and Biochemistry, Old Dominion University, Norfolk, VA, USA*; ADAM J. PERRY, , *Menlo Systems, Inc., Newton, NJ, USA*; BENJAMIN J. McCALL, *Departments of Chemistry and Astronomy, University of Illinois at Urbana-Champaign, Urbana, IL, USA.*

Molecular ions are challenging to study with conventional spectroscopic methods. Laboratory discharges produce ions in trace quantities which can be obscured by the abundant neutral molecules present. The technique Noise Immune Cavity Enhanced Optical Heterodyne Velocity Modulation Spectroscopy (NICE-OHVMS) overcomes these challenges by combining the ion-neutral discrimination of velocity modulation spectroscopy with the sensitivity of Noise-Immune Cavity-Enhanced Optical Heterodyne Molecular Spectroscopy (NICE-OHMS), and has been able to determine transition frequencies of molecular ions in the mid-infrared (mid-IR) with sub-MHz uncertainties when calibrated with an optical frequency comb[a]. However, the extent of these studies was limited by the presence of fringes due to parasitic etalons and the speed and noise characteristics of mid-IR detectors.

Recently, we have overcome these limitations by implementing up-conversion detection and dithered optics. We performed up-conversion using periodically poled lithium niobate to convert light from the mid-IR to the visible to be within the coverage of sensitive and fast silicon detectors while maintaining our heterodyne and velocity modulation signals. The parasitic etalons were removed by rapidly rotating CaF_2 windows with galvanometers, which is known as a Brewster-plate spoiler[b], which averaged out the fringes in detection. Together, these improved the sensitivity by more than an order of magnitude[c] and have enabled extended spectroscopic surveys of molecular ions in the mid-IR.

[a]J. N. Hodges, A. J. Perry, P. A. Jenkins II, B. M. Siller, and B. J. McCall, *J. Chem. Phys.* (2013), **139**, 164201.

[b]C. R. Webster, *J. Opt. Soc. Am. B* (1985), **2**, 1464.

[c]C. R. Markus, A. J. Perry, J. N. Hodges, and B. J. McCall, *Opt. Express* (2017), **25**, 3709–3721.

Intermission

MK05

CONTINUOUS-WAVE CAVITY RING-DOWN SPECTROSCOPY IN A PULSED UNIFORM SUPERSONIC FLOW

SHAMEEMAH THAWOOS, NICOLAS SUAS-DAVID, ARTHUR SUITS, *Department of Chemistry, University of Missouri, Columbia, MO, USA.*

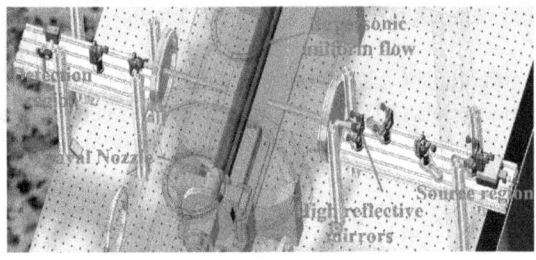

We introduce a new approach that couples a pulsed uniform supersonic flow with high sensitivity continuous wave cavity ringdown spectroscopy (UF-CRDS) operated in the near infrared (NIR).

This combination is related to the CRESU[a] technique developed in France and used for many years to study reaction kinetics at low temperature, and to the microwave based chirped-pulse uniform supersonic flow spectrometer (CPUF) developed in our group which has successfully demonstrated the use of pulsed uniform supersonic flow to probe reaction dynamics at temperatures as low as 22 K[b]. CRDS[c] operated with NIR permits access to the first overtones of C-H and O-H stretching/bending which, in combination with its extraordinary sensitivity opens new experiments complementary to the CPUF technique. The UF-CRDS apparatus (Figure) utilizes the pulsed uniform flow produced by means of a piezo-electric stack valve[d] in combination with a Laval nozzle. At present, two machined aluminum Laval nozzles designed for carrier gases Ar and He generate flows with a temperature of approximately 25 K and pressure around 0.15 mbar. This flow is probed by an external cavity diode laser in the NIR (1280-1380 nm). Laval nozzles designed using a newly developed MATLAB-based program will be used in the future. A detailed illustration of the novel UF-CRDS instrumentation and its performance will be presented along with future directions and applications.

[a] I. Sims, J. L. Queffelec, A. Defrance, C. Rebrion-Rowe, D. Travers, P. Bocherel, B. Rowe, I. W. Smith, J. Chem. Phys. 100, 4229-4241, (1994).

[b] C. Abeysekera, B. Joalland, N. Ariyasingha, L. N. Zack, I. R. Sims, R. W. Field, A. G. Suits, J. Phys. Chem. Lett. 6, 1599-1604, (2015).

[c] N. Suas-David, T. Vanfleteren, T. Földes, S. Kassi, R. Georges, M. Herman, J. Phys. Chem.A, 119, 10022-10034, (2015).

[d] C. Abeysekera, B. Joalland, Y. Shi, A. Kamasah, J. M. Oldham, A. G. Suits, Rev. Sci. Instrum. 85, 116107, (2014).

MK06

MIR AND FIR ANALYSIS OF INORGANIC SPECIES IN A SINGLE DATA ACQUISITION

PENG WANG, *Bruker Optics, Bruker Corporation, Billerica, USA*; SERGEY SHILOV, *Bruker, Bruker Optics, Billerica, USA.*

The extension of the mid IR towards the far IR spectral range below 400 cm^{-1} is of great interest for molecular vibrational analysis for inorganic and organometallic chemistry, for geological, pharmaceutical, and physical applications, polymorph screening and crystallinity analysis as well as for matrix isolation spectroscopy. In these cases, the additional far infrared region offers insight to low energy vibrations which are observable only there. This includes inorganic species, lattice vibrations or intermolecular vibrations in the ordered solid state. The spectral range of a FTIR spectrometer is defined by the major optical components such as the source, beamsplitter, and detector. The globar source covers a broad spectral range from 8000 to 20 cm^{-1}. However a bottle neck exists with respect to the beamsplitter and detector. To extend the spectral range further into the far IR and THz spectral ranges, one or more additional far IR beam splitters and detectors have been previously required. Two new optic components have been incorporated in a spectrometer to achieve coverage of both the mid and far infrared in a single scan: a wide range MIR-FIR beam splitter and the wide range DLaTGS detector that utilizes a diamond window. The use of a standard SiC IR source with these components yields a spectral range of 6000 down to 50 cm^{-1} in one step for all types of transmittance, reflectance and ATR measurements. Utilizing the external water cooled mercury arc high power lamp the spectral range can be ultimately extended down to 10 cm^{-1}. Examples of application will include emission in MIR-THz range, identification of pigments, additives in polymers, and polymorphism studies.

MK07

HIGH PRECISION 2.0 μm PHOTOACOUSTIC SPECTROMETER FOR DETERMINATION OF THE $^{13}CO_2/^{12}CO_2$ ISOTOPE RATIO

ZACHARY REED, *Chemical Sciences Division, National Institute of Standards and Technology, Gaithersburg, MD, USA*; JOSEPH T. HODGES, *Material Measurement Laboratory, National Institute of Standards and Technology, Gaithersburg, MD, USA.*

We have developed a portable photoacoustic spectrometer for high precision measurements of the $^{13}CO_2/^{12}CO_2$ isotope ratio and the absolute molar concentration of each isotope. The spectrometer extends on our previous work at 1.57 μm [1], and now employs two separate intensity modulated distributed feedback lasers and a fiber amplifier, operating in the 2.0 μm wavelength region. Each DFB is selected to probe individual spectrally isolated ro-vibrational transitions for $^{12}CO_2$ and $^{13}CO_2$. The spectrometer is actively temperature controlled, mitigating variations in the two spectral line intensities and the temperature dependent system response.

For measurements of ambient concentrations of carbon dioxide at nominally natural abundance in dry air, we demonstrate a measurement precision of 140 ppb for $^{12}CO_2$ with a 1 s averaging time and 10 ppb for $^{13}CO_2$ with a 60 s averaging time. Precision in $\delta 13C$ of better than 0.1 permil is demonstrated. The analyzer response is calibrated in terms of certified gas mixtures and compared to characterization by cavity ringdown spectroscopy. We also investigate how water vapor affects the photoacoustic signals by promoting collisional relaxation for each isotope.

[1] Z.D. Reed, B. Sperling, et al. App. Phys. B. 117, 645-657, 2014

MK08

OPTICAL DETECTION AND QUANTIFICATION OF RADIOCARBON DIOXIDE ($^{14}CO_2$) AT AND BELOW AMBIENT LEVELS

DAVID A. LONG, ADAM J. FLEISHER, QINGNAN LIU, JOSEPH T. HODGES, *Chemical Sciences Division, National Institute of Standards and Technology, Gaithersburg, MD, USA.*

Due to their age, fossil fuels and their byproducts are almost entirely depleted in radiocarbon (^{14}C). As a result, measurements of radiocarbon provide a unique tracer for determining the origin of products and emissions. Recent efforts at NIST have applied mid-infrared cavity ring-down spectroscopy to measurements of radiocarbon dioxide to allow for more rapid and less expensive measurements than are possible with traditional techniques such as accelerator mass spectrometry. I will discuss our present measurement detection limits and precision as well as discuss limiting noise sources and plans to further improve the instrument's stability and reproducibility.

MK09

USING WIDE SPECTRAL RANGE INFRARED SPECTROSCOPY TO OBTAIN BOTH SURFACE SPECIES AND CHANGES OF CATALYST ITSELF UNDER THE REACTION CONDITIONS

XUEFEI WENG, DING DING, HUAN LI, YANPING ZHENG, MINGSHU CHEN, *College of Chemistry and Chemical Engineering, Xiamen University, Xiamen, China.*

Fundamental understanding of catalysts under the reaction conditions is key for designing new catalysts, and improving catalysts and catalytic conversion processes. Such understanding can be achieved only by characterization of catalysts under the reaction conditions because catalyst structures and the mechanisms of catalytic reactions depend on the reaction environment. Raman spectroscopy is one of the few instrumental methods that in a single measurement can provide information about both solid catalysts and the molecules reacting on them. However, its sensitivities for the surface species and the surface changes under catalytic reaction are limited. Infrared spectroscopy is also a wide spectral range (6000-50 cm-1) technique that enables examination of the nature of molecular species, identification of solid phases. Unfortunately, most of the heterogeneous catalysts consist of oxides as the active components or as the supports, which strong IR adsorption (below 1200 cm-1) limits the in situ IR to measure only the surface species (4000 900 cm-1). In this presentation, we will present our new developments of in-situ infrared spectroscopies with a spectral range of 4000 400 cm-1, for both the reflection adsorption infrared spectroscopy (IRAS) and transparent infrared spectroscopy (FTIR, unpublished data), that are capable of measuring both the surface species and changes specific to the surface.

MK10   *Post-Deadline Abstract*   **4:35 – 4:50**

MEASUREMENTS OF ELECTRIC FIELD IN A NANOSECOND PULSE DISCHARGE BY 4-WAVE MIXING

EDMOND BARATTE, IGOR V. ADAMOVICH, MARIEN SIMENI SIMENI, KRAIG FREDERICKSON, *Department of Aerospace and Mechanical Engineering, The Ohio State University, Columbus, OH, USA.*

Picosecond four-wave mixing is used to measure temporally and Picosecond four-wave mixing is used to measure temporally and spatially resolved electric field in a nanosecond pulse dielectric discharge sustained in room air and in an atmospheric pressure hydrogen diffusion flame. Measurements of the electric field, and more precisely the reduced electric field (E/N) in the plasma is critical for determination rate coefficients of electron impact processes in the plasma, as well as for quantifying energy partition in the electric discharge among different molecular energy modes. The four-wave mixing measurements are performed using a collinear phase matching geometry, with nitrogen used as the probe species, at temporal resolution of about 2 ns . Absolute calibration is performed by measurement of a known electrostatic electric field. In the present experiments, the discharge is sustained between two stainless steel plate electrodes, each placed in a quartz sleeve, which greatly improves plasma uniformity. Our previous measurements of electric field in a nanosecond pulse dielectric barrier discharge by picosecond 4-wave mixing have been done in air at room temperature, in a discharge sustained between a razor edge high-voltage electrode and a plane grounded electrode (a quartz plate or a layer of distilled water). Electric field measurements in a flame, which is a high-temperature environment, are more challenging because the four-wave mixing signal is proportional to the to square root of the difference betwen the populations of N2 ground vibrational level (v=0) and first excited vibrational level (v=1). At high temperatures, the total number density is reduced, thus reducing absolute vibrational level populations of N2. Also, the signal is reduced further due to a wider distribution of N2 molecules over multiple rotational levels at higher temperatures, while the present four-wave mixing diagnostics is using spectrally narrow output of a ps laser and a high-pressure Raman cell, providing access only to a few N2 rotational levels. Because of this, the four-wave mixing signal in the flame is lower by more than an order of magnitude compared to the signal generated in room temperature air plasma. Preliminary experiments demonstrated four-wave mixing signal generated by the electric field in the flame, following ns pulse discharge breakdown. The electric field in the flame is estimated using four-wave mixing signal calibration vs. temperature in electrostatic electric field generated in heated air. Further measurements in the flame are underway.

MK11   *Post-Deadline Abstract*   **4:52 – 5:07**

N2 VIBRATIONAL TEMPERATURES AND OH NUMBER DENSITY MEASUREMENTS IN A NS PULSE DISCHARGE HYDROGEN-AIR PLASMAS

YICHEN HUNG, *Department of Chemistry, The Ohio State University, Columbus, OH, USA*; CAROLINE WINTERS, ELIJAH R JANS, KRAIG FREDERICKSON, IGOR V. ADAMOVICH, *Department of Aerospace and Mechanical Engineering, The Ohio State University, Columbus, OH, USA.*

This work presents time-resolved measurements of nitrogen vibrational temperature, translational-rotational temperature, and absolute OH number density in lean hydrogen-air mixtures excited in a diffuse filament nanosecond pulse discharge, at a pressure of 100 Torr and high specific energy loading. The main objective of these measurements is to study a possible effect of nitrogen vibrational excitation on low-temperature kinetics of HO2 and OH radicals. N2 vibrational temperature and gas temperature in the discharge and the afterglow are measured by ns broadband Coherent Anti-Stokes Scattering (CARS). Hydroxyl radical number density is measured by Laser Induced Fluorescence (LIF) calibrated by Rayleigh scattering. The results show that the discharge generates strong vibrational nonequilibrium in air and H2-air mixtures for delay times after the discharge pulse of up to 1 ms, with peak vibrational temperature of $T_v \approx 2000$ K at $T \approx 500$ K. Nitrogen vibrational temperature peaks ≈ 200 μs after the discharge pulse, before decreasing due to vibrational-translational relaxation by O atoms (on the time scale of a few hundred μs) and diffusion (on ms time scale). OH number density increases gradually after the discharge pulse, peaking at t 100-300 μs and decaying on a longer time scale, until t 1 ms. Both OH rise time and decay time decrease as H2 fraction in the mixture is increased from 1% to 5%. OH number density in a 1% H2-air mixture peaks at approximately the same time as vibrational temperature in air, suggesting that OH kinetics may be affected by N2 vibrational excitation. However, preliminary kinetic modeling calculations demonstrate that OH number density overshoot is controlled by known reactions of H and O radicals generated in the plasma, rather than by dissociation by HO2 radical in collisions with vibrationally excited N2 molecules, as has been suggested earlier. Additional measurements at higher specific energy loadings and kinetic modeling calculations are underway.

TA. Astronomy
Tuesday, June 20, 2017 – 8:30 AM
Room: 274 Medical Sciences Building

Chair: Leslie Looney, University of Illinois, Urbana, IL, USA

TA01   8:30–8:45

DISCOVERY OF ^{13}CCC in SgrB2(M)

THOMAS GIESEN, *Institute of Physics, University Kassel, Kassel, Germany*; BHASWATI MOOKERJEA, *Department of Astronomy & Astrophysics, Tata Institute of Fundamental Research, Mumbai, India*; JÜRGEN STUTZKI, *I. Physikalisches Institut, Universität zu Köln, Köln, Germany*; ALEXANDER A. BREIER, *Institute of Physics, University Kassel, Kassel, Germany*; THOMAS BUECHLING, *Institute of Physics, University of Kassel, Kassel, Germany*; GUIDO W FUCHS, *Physics Department, University of Kassel, Kassel, Germany.*

Small carbon chain molecules like linear C_3 are thought to play a crucial role in the formation of larger, complex molecules, including pre-biotic species. The formation pathways of organic molecules with carbon chains as backbones is by far not well understood. Studies of isotope fractionation have proven to be a useful tool of tracing chemical reaction pathways and to elucidate formation and destruction processes of interstellar molecules. Recent velocity-resolved observations in the far-infrared have resulted in the detection of C_3 ro-vibrational transitions in the warm envelopes of star-forming hot cores W31C, W49N and DR21(OH). Multiple far-infrared transitions of C_3 have also been detected towards the Galactic center molecular clouds SgrB2(M) and Sgr B2(N). Since C^+ is involved in an important step of the formation route of the C_3 molecule, it is likely that effects of isotopic fractionation of C^+ will manifest itself in the $^{12}C_3/^{13}CCC$ and $^{12}C_3/C^{13}CC$ ratios as well. Based on high resolution THz- laboratory measurements of C_3 and its ^{13}C-isotopologues conducted at the Kassel laboratories, we used the GREAT-receiver onboard SOFIA for a first ever detection of ^{13}CCC towards SgrB2(M). In this talk we present results and possible implications of the observation.

TA02   8:47–9:02

A SEARCH FOR THE HOCO RADICAL IN THE MASSIVE STAR-FORMING REGION Sgr B2(M)

TAKAHIRO OYAMA, MITSUNORI ARAKI, *Faculty of Science Division I, Tokyo University of Science, Shinjuku-ku, Tokyo, Japan*; SHURO TAKANO, *College of Engineering, Nihon University, Fukushima, Japan*; NOBUHIKO KUZE, *Faculty of Science and Technology, Sophia University, Tokyo, Japan*; YOSHIHIRO SUMIYOSHI, *Division of Pure and Applied Science, Faculty of Science and Technology, Gunma University, Maebashi, Japan*; KOICHI TSUKIYAMA, *Faculty of Science Division I, Tokyo University of Science, Shinjuku-ku, Tokyo, Japan*; YASUKI ENDO, *Department of Applied Chemistry, National Chiao Tung University, Hsinchu, Taiwan.*

Despite importance of the origin of life, long lasting challenges to detect the simplest amino acid glycine (H_2NCH_2COOH) in interstellar medium has not been successful. As a preliminary step toward search for glycine, detection of its precursor has received attention. It is considered that glycine is produced by the reaction of the HOCO radical and the aminomethyl radical(CH_2NH_2) on interstellar grain surface:

$$HOCO + CH_2NH_2 \rightarrow H_2NCH_2COOH. \quad (1)$$

HOCO is produced by the reaction of $OH + CO \rightarrow HOCO$ and/or $HCOOH \rightarrow HOCO + H$. However, HOCO and CH_2NH_2 have not been investigated in interstellar medium. Recently, we determined the accurate molecular constants of HOCO.[a] Thus, accurate rest frequencies were derived from the constants. In the present study, we carried out the observations of HOCO in the massive star-forming region Sgr B2(M), having variety of interstellar molecules, with Nobeyama 45 m radio telescope. Although HOCO could not be detected in Sgr B2(M), the upper limit of the column density was derived to be 9.0×10^{12} cm^{-2} via the spectrum in the 88 GHz region by the rotational diagram method. If the reaction (1) is a main process of the glycine production in this region, an extremely deep search is needed to detect glycine.

[a]T. Oyama *et al.*, *J. Chem. Phys.* **134**, 174303 (2011).

TA03   9:04–9:19

PRECISE DETERMINATION OF THE ISOTOPIC RATIOS OF HC$_3$N IN THE MASSIVE STAR-FORMING REGION Sgr B2(M)

TAKAHIRO OYAMA, MITSUNORI ARAKI, *Faculty of Science Division I, Tokyo University of Science, Shinjuku-ku, Tokyo, Japan*; SHURO TAKANO, *College of Engineering, Nihon University, Fukushima, Japan*; NOBUHIKO KUZE, *Faculty of Science and Technology, Sophia University, Tokyo, Japan*; YOSHIHIRO SUMIYOSHI, *Division of Pure and Applied Science, Faculty of Science and Technology, Gunma University, Maebashi, Japan*; KOICHI TSUKIYAMA, *Faculty of Science Division I, Tokyo University of Science, Shinjuku-ku, Tokyo, Japan*; YASUKI ENDO, *Department of Applied Chemistry, National Chiao Tung University, Hsinchu, Taiwan.*

Isotopic ratio is a critical parameter in understanding galactic chemical evolution. In addition, carbon isotopic ratio of an organic molecule reflects its formation mechanism. In the present study, we observed the simplest cyanopolyyne HC$_3$N and its isotopomers in the massive star-forming region Sgr B2(M) with Nobeyama 45 m radio telescope. The column density and the rotational temperature of HC$_3$N were determined to be 1.6×10^{15} cm^{-2} and 163 K, respectively. The ratios of the column densities for the ^{13}C isotopomers were derived to be [H^{13}CCCN]:[HC^{13}CCN]:[HCC^{13}CN] = 1:1.03(4):0.99(3), where the rotational temperature was fixed to that of HC$_3$N. The ratios are almost the same, suggesting no isotopic fractionation for the specific carbon atoms in HC$_3$N. Therefore, it is considered that the ^{13}C isotope exchange reactions do not contribute to make difference among the column densities of the three ^{13}C isotopomers in the relatively warm region of Sgr B2(M). In contrast, the reported ratios in TMC-1 and L1527 are 1:1.0(1):1.4(2)[a] and 1:1.01(2):1.35(3),[b] respectively, where the ratios show higher abundance of HCC^{13}CN.

We also observed the transitions in the vibrational excited states of HC$_3$N. The rotational temperature of 362 K in the ν_4, ν_5, ν_6 and ν_7 excited states was obviously different from that of the vibrational ground state.

[a]S. Takano *et al.*, *Astron. Astrophys.* **329**, 1156 (1998).
[b]M. Araki *et al.*, *ApJ* **833**, 291 (2016).

TA04   9:21–9:36

THE ^{12}C/^{13}C RATIO IN THE GALACTIC CENTER: IMPLICATIONS FOR GALACTIC CHEMICAL EVOLUTION AND ISOTOPE CHEMISTRY

DeWAYNE T HALFEN, *Department of Chemistry and Biochemistry, Department of Astronomy, The University of Arizona, Tucson, AZ, USA*; LUCY M. ZIURYS, *Department of Chemistry and Biochemistry; Department of Astronomy, Arizona Radio Observatory, University of Arizona, Tuscon, AZ, USA.*

Observations from a spectral-line survey of Sgr B2(N) of the ^{12}C and ^{13}C isotopologues of H$_2$CS, CH$_3$CCH, NH$_2$CHO, CH$_2$CHCN, and CH$_3$CH$_2$CN have been analyzed to more accurately establish the ^{12}C/^{13}C ratio in this cloud. The wide spectral coverage has enabled an accurate evaluation of the ^{12}C/^{13}C ratios in these low abundance molecules, based on numerous transitions. The lines typically exhibited two distinct velocity components at 64 and 73 km s^{-1}. The ^{12}C/^{13}C ratio was found to be in the range 18-24 for all 5 molecules, for optically thin transitions, with an average value of 20.5, and did not significantly vary between the two velocity components. The Galactic gradient has been revised to be ^{12}C/^{13}C = 6.08(0.48) D_{GC} + 15.7(2.9). Furthermore, the ^{12}C/^{13}C ratio did not change with substitution site on the molecule. Therefore, there appears to be very little chemical fractionation or isotope-selective photodissociation occurring in Sgr B2(N), and the ^{12}C/^{13}C ratios are a true reflection of the isotopic abundances generated by stellar nucleosynthesis.

Intermission

TA05

A STUDY OF THE c-C_3HD/c-C_3H_2 RATIO IN LOW-MASS STAR FORMING REGIONS.

JOHANNA CHANTZOS, SILVIA SPEZZANO, PAOLA CASELLI, ANA CHACON-TANARRO, *The Center for Astrochemical Studies, Max-Planck-Institut für extraterrestrische Physik, Garching, Germany.*

Deuterium fractionation increases significantly in cold ($T < 25$ K), dense ($n_H > 10^4$ cm^{-3}) molecular clouds, in which molecules like CO freeze out onto dust grains leading to an enhanced abundance of H_2D^+, D_2H^+ and D_3^+. c-C_3H_2 is formed and deuterated exclusively by gas-phase chemistry. This makes it to a very good indicator of gas-phase deuteration and therefore to an excellent tool to study the early phases of star formation.

We observed the c-C_3HD/c-C_3H_2 ratio toward 13 prestellar and 4 protostellar cores in the Taurus and Perseus Complex, respectively. In particular, the $3_{0,3} - 2_{1,2}$ and $2_{1,2} - 1_{0,1}$ transitions of the isotopologues c-C_3HD and c-$^{13}CC_2H_2$ were observed in all prestellar and protostellar cores with a very high S/N. In both samples a high deuteration factor was found. In the prestellar cores the c-C_3HD/c-C_3H_2 ratio varies between 5% and 13% while in protostellar cores is found to be 9%-23%.

I will present our results on the correlation between the deuterium fractionation of c-C_3H_2 and evolutionary indicators such as central density and dust temperature and compare them with the deuteration of N_2H^+ observed in the same sources.

TA06

DETECTIONS OF LONG CARBON CHAINS CH_3CCCCH, C_6H, LINEAR-C_6H_2 AND C_7H IN THE LOW-MASS STAR FORMING REGION L1527

MITSUNORI ARAKI, *Faculty of Science Division I, Tokyo University of Science, Shinjuku-ku, Tokyo, Japan*; SHURO TAKANO, *College of Engineering, Nihon University, Fukushima, Japan*; NAMI SAKAI, *RIKEN Center for Advanced Photonics, RIKEN, Wako, Japan*; SATOSHI YAMAMOTO, *Department of Physics and Research Center for the Early Universe, The University of Tokyo, Tokyo, Japan*; TAKAHIRO OYAMA, *Faculty of Science Division I, Tokyo University of Science, Tokyo , Japan*; NOBUHIKO KUZE, *Faculty of Science and Technology, Sophia University, Tokyo, Japan*; KOICHI TSUKIYAMA, *Faculty of Science Division I, Tokyo University of Science, Shinjuku-ku, Tokyo, Japan.*

Carbon chains in the warm carbon chain chemistry (WCCC) region has been searched in the 42–44 GHz region by using Green Bank 100 m telescope. Long carbon chains C_7H, C_6H, CH_3CCCCH, and linear-C_6H_2 and cyclic species C_3H and C_3H_2O have been detected in the low-mass star forming region L1527, performing the WCCC. C_7H was detected for the first time in molecular clouds. The column density of C_7H is derived to be 6.2×10^{10} cm^{-2} by using the detected $J = 24.5$–23.5 and 25.5–24.5 rotational lines. The $^2\Pi_{1/2}$ electronic state of C_6H, locating 21.6 K above the $^2\Pi_{3/2}$ electronic ground state, and the $K_a = 0$ line of the para species of linear-C_6H_2 were also detected firstly in molecular clouds. The column densities of the $^2\Pi_{1/2}$ and $^2\Pi_{3/2}$ states of C_6H in L1527 were derived to be 1.6×10^{11} and 1.1×10^{12} cm^{-2}, respectively. The total column density of linear-C_6H_2 is obtained to be 1.86×10^{11} cm^{-2}. While the abundance ratios of carbon chains in between L1527 and the starless dark cloud Taurus Molecular Cloud-1 Cyanopolyyne Peak (TMC-1 CP) have a trend of decrease by extension of carbon-chain length, column densities of CH_3CCCCH and C_6H are on the trend. However, the column densities of linear-C_6H_2, and C_7H are as abundant as those of TMC-1 CP in spite of long carbon chain, i.e., they are not on the trend. The abundances of linear-C_6H_2 and C_7H show that L1527 is rich for long carbon chains as well as TMC-1 CP.

TA07

POTENTIAL LINE STRUCTURE VARIABILITY IN DIB FEATURES OBSERVED IN PATHFINDER TRES SURVEY

CHARLES LAW, DAN MILISAVLJEVIC, , *Harvard-Smithsonian Center for Astrophysics, Cambridge, MA, USA*; KYLE N. CRABTREE, <u>SOMMER LYNN JOHANSEN</u>, *Department of Chemistry, The University of California, Davis, CA, USA.*

The Diffuse Interstellar Bands (DIBs) are hundreds of spectral lines observed in sightlines towards many stars in the optical and near-infrared. Although most of these transitions remain unassigned, four of them have recently been assigned to C_{60}^+ and C_{70}^+. In earlier observations of the visible spectrum of the extragalactic supernova SN 2012ap, we observed changes in the equivalent widths of DIBs on the timescale of its light curve, which indicated that some DIB carriers might exist closer to massive stars then previously believed. Motivated by these findings, we undertook a pathfinder survey of 17 massive stars with the Tillinghast Reflector Echelle Spectrograph at Fred L. Whipple Observatory in search of temporal variability in DIBs. In 3 of the 17 stars, we found possible evidence for variation in line substructure of DIBs $\lambda5797$ and $\lambda6614$. In this talk, we will discuss our efforts to model $\lambda5797$ toward MT-59 using contour simulations based on previously published spectral models from higher resolution observations. Although the SNR of this spectrum was only 5–15, our preliminary results suggest that the variations in molecular spectra over time might arise from changes in carrier temperature. These early results demonstrate the need for higher SNR spectra taken at multiple epochs to further explore potential temporal variability. If successful, time-variation could provide additional evidence to assist in identifying DIB carriers.

TA08

MODIFICATIONS OF THE RELATION BETWEEN COSMIC RAY IONIZATION RATE ζ AND H_3^+ COLUMN DENSITY IN THE CENTRAL MOLECULAR ZONE OF THE GALACTIC CENTER

<u>TAKESHI OKA</u>, *Department of Astronomy and Astrophysics and Department of Chemistry, The Enrico Fermi Institute, University of Chicago, Chicago, IL, USA.*

In deriving the simple formula, $\zeta L = 2k_e N(H_3^+)(n_C/n_H)_{SV} R/f(H_2)$, used to estimate cosmic ray H_2 ionization rate ζ from observed H_3^+ column density $N(H_3^+)$ in the Central Molecular Zone (CMZ) of the Galactic center (GC),[a] the following two effects were neglected: (1) the charge exchange reaction $H_2^+ + H \rightarrow H_2 + H^+$[b] which significantly reduces H_3^+ production rate if the fraction of molecular hydrogen $f(H_2)$ is much lower than 1, and (2) the production of electrons from ionization of H_2 and H which greatly increases the H_3^+ destruction rate if ζ is much higher than 10^{-15} s^{-1}. (Only electrons from VUV first ionization of C atoms had been considered). Recent more extensive analysis using the Meudon PDR code by Le Petit et al.[c] has indicated that these effects are not negligible in the CMZ.

While an extensive chemical model calculation is beyond the scope of our analysis, we have attempted to use our simple model considering only hydrogenic species and electrons to take these two effects into account. When (1) is introduced, the rate of H_3^+ production is approximated to be $\zeta n_H [f(H_2)]^2$,[d] which is ~ 3 times lower than the previous value for $f(H_2)$ = 0.6 reported by Le Petit et al.[c] When (2) is taken into account, the electron number density is approximated to be $n_e = n_C R + \zeta n_H/[2k_e n(H_3^+)]$ where the first and second term represents electrons from the C atoms and those from H_2 and H, respectively. The first term (in which R represents the increase of metallicity from the solar vicinity to the GC, $R \geq 3$) has the electron fraction $x_e = 5 \times 10^{-4}$ and the second term becomes significant at $\zeta \sim 10^{-15}$ s^{-1}. This introduces a non-linearity between ζ and $N(H_3^+)$ and the latter reaches a maximum at $\zeta \sim 10^{-14}$ s^{-1} and decreases as ζ increases further. Application of the results to the observed $N(H_3^+)$ will be discussed.

[a] Oka, T., Geballe, T. R., Goto, M., Usuda, T., McCall, B. J. 2005, ApJ, 632, 882

[b] Indriolo, N., McCall, B. J. 2012, ApJ, 745:91

[c] Le Petit, F., Ruaud, M., Bron, E., Godard, B., Roueff, E., Languignon, D., Le Bourlot, J. 2016, A&A, 585, A105

[d] Oka, T. 2013, Chem. Rev. 113, 8738

VIBRATIONAL SPECTROSCOPY OF He–O_2H^+ AND O_2H^+

HIROSHI KOHGUCHI, *Department of Chemistry, Hiroshima University, Hiroshima, Japan*; KOICHI MT YA-MADA, *EMTech, National Institute of Advanced Industrial Science and Technology (AIST) , Tsukuba, Japan*; PAVOL JUSKO, STEPHAN SCHLEMMER, OSKAR ASVANY, *I. Physikalisches Institut, Universität zu Köln, Köln, Germany.*

The elusive protonated oxygen, O_2H^+, has been characterized by vibrational action spectroscopy in a cryogenic 22-pole ion trap. On the one hand, the vibrational bands of the tagged He–O_2H^+ have been investigated, using a table-top OPO system for the known OH-stretch[a], whereas the FELIX[b] light source has been used to detect the hitherto unknown low-frequency O-O-H bend and O-O stretch. On the other hand, the untagged O_2H^+ has been detected for the first time by high-resolution rovibrational spectroscopy via its ν_1 OH stretch motion. 38 ro-vibrational fine structure transitions with partly resolved hyperfine satellites were measured (56 resolved lines in total). Spectroscopic parameters were determined by a fit to an asymmetric rotor model with a $^3A''$ electronic ground state. The band center is at 3016.73 cm^{-1}, which is in good agreement with experimental[a] and *ab initio*[c,d] predictions. Based on the spectroscopic parameters, the rotational spectrum is predicted, but not detected yet.

[a] S. A. Nizkorodov et al., Chem. Phys. Lett., 278, 26, 1997

[b] D. Oepts et al., Infrared Phys. Technol., 36, 297, 1995

[c] S. L. W. Weaver et al., Astrophys. J., 697, 601, 2009

[d] X. Huang and T. J. Lee, J. Chem. Phys., 129, 044312, 2008

TB. Mini-symposium: Multiple Potential Energy Surfaces

Tuesday, June 20, 2017 – 8:30 AM

Room: 116 Roger Adams Lab

Chair: Zhou Lin, Massachusetts Institute of Technology, Cambridge, MA, USA

TB01   *INVITED TALK*   8:30 – 9:00

MORE SPECTRA! A LOT MORE! BETTER TOO! NOW WHAT?

ROBERT W FIELD, *Department of Chemistry, MIT, Cambridge, MA, USA.*

I have been a card-carrying spectroscopist for 52 years. I began my career studying spectroscopic perturbations in CS and CO. I eventually graduated to vibrational polyads in acetylene and Multichannel Quantum Defect Theory (MQDT) models for Rydberg states of CaF. My experimental arsenal evolved from atomic resonance lamps to finicky cw dye lasers to user-friendly Nd:YAG pumped dye lasers, ending up with Chirped Pulse Millimeter Waves, non-finicky solid state cw lasers, and death-defying dreams about Stimulated Raman Adiabatic Passage (STIRAP). It has become possible to record an enormous quantity of unimaginably high quality spectra quickly. Increases by factors of 10^6 in spectral velocity have been claimed. Yet everything rests on assigning the spectrum. But the assignment game has changed. Instead of looking for patterns, we deal with meta-patterns. Our goal is to build a complex model that represents all of the energy levels and associates a multi-component eigenvector with each observed eigenstate. Eigenvectors can reveal what a molecule is thinking about doing when it grows up. Spectroscopy becomes a form of molecular psychoanalysis. A spectroscopist can observe the emergence and describe the mechanistic origin of new classes of large-amplitude intramolecular motions. This makes it possible to directly characterize things, such as transition states, which dogma has labeled "spectroscopically unobservable." Where is 21st century spectroscopy headed? I will discuss examples that include: spectroscopic perturbations of the S_2 $B^3\Sigma^-_u$ state, the SO_2 C state with its unequal SO bond-lengths, and the transition state for trans-cis isomerization in the S_1 state of acetylene.

TB02   9:04 – 9:19

TIME-RESOLVED MEASUREMENT OF THE C_2 $^1A\Pi u$ STATE POPULATION FOLLOWING PHOTODISSOCIATION OF THE S_1 STATE OF ACETYLENE USING FREQUENCY-MODULATION SPECTROSCOPY

ZHENHUI DU[a], JUN JIANG, ROBERT W FIELD, *Department of Chemistry, MIT, Cambridge, MA, USA.*

The excited-state population of the C_2 $^1A\Pi_u$ state produced in photolysis of S_1 acetylene was investigated. The pulsed UV laser (216.5 nm) excites acetylene into $J = 8$ e-symmetry level of the S_1 3^4 level, and subsequently dissociates the S_1 acetylene into C_2 fragments. A frequency-modulated near-infrared probe laser beam is used to detect the C_2 population in the $^1A\Pi_u$ state. The sensitivity and the fast response of the experimental setup has been verified by I_2 excited state measurements. The setup will be used to record the C_2 $A - X$ transitions, which are fitted with a Voigt function. The derived lineshape and line intensities will be analyzed, and we will use the information to calculate the A state populations of C_2 and map the populations with time-resolution following the photolysis.

---

[a]Department of Precision Instrument, Tianjin University, Tianjin, China

TB03

PRECISION MEASUREMENT OF THE ROVIBRATIONAL ENERGY-LEVEL STRUCTURE OF $^4\mathrm{He}_2^+$

LUCA SEMERIA, PAUL JANSEN, JOSEF A. AGNER, HANSJÜRG SCHMUTZ, FREDERIC MERKT, *Laboratorium für Physikalische Chemie, ETH Zurich, Zurich, Switzerland.*

He_2^+ is a three-electron system for which highly accurate *ab initio* calculations are possible. The latest calculations of the rovibrational energies of He_2^+ by Tung *et al.* [a] have a reported accuracy of 120 MHz, although they do not include relativistic and quantum electrodynamics (QED) effects.

We determined the rovibrational structure of $^4\mathrm{He}_2^+$ from measurements of the Rydberg spectrum of metastable $a\,^3\Sigma_u^+$ He_2 (He_2^* hereafter) and Rydberg-series extrapolation using multichannel quantum-defect-theory [b] [c]. He_2^* molecules are produced in supersonic beams with velocities tunable down to about 100 m/s by combining a cryogenic supersonic-beam source with a multistage Zeeman decelerator [d] [e]. They are then excited to high-np Rydberg states by single-photon excitation. In the experiments, we use a pulsed uv laser system, with a near Fourier-transform-limited bandwidth of 150 MHz. The Zeeman deceleration reduces the systematic uncertainty arising from a possible Doppler shift and greatly simplifies the spectral assignment because of its spin-rotational state selectivity [f].

Results will be presented on the rotational structure of the lowest three vibrational levels of He_2^+. The unprecedented accuracy that we have obtained for the $v^+ = 0$ rotational intervals of He_2^+ [g] enables the quantification of the relativistic and QED corrections by comparison with the results of Tung *et al.*[a]

[a] W.-C. Tung, M. Pavanello and L. Adamowicz, *J. Chem. Phys.*, 136, 104309, 2012.

[b] C. Jungen, *Elements of Quantum Defect Theory, in : Handbook of High-resolution Spectroscopy*, 2001.

[c] D. Sprecher, J. Liu, T. Krähenmann, M. Schäfer, and F. Merkt, *J. Chem. Phys.*, 140, 064304, 2014.

[d] A. W. Wiederkehr, S. D. Hogan, M. Andrist, H. Schmutz, B. Lambillotte, J. A. Agner, and F. Merkt., J. Chem. Phys., 135, 214202, 2011.

[e] M. Motsch, P. Jansen, J. A. Agner, H. Schmutz, and F. Merkt, *Phys. Rev. A*, 89, 043420, 2014.

[f] P. Jansen, L. Semeria, L. E. Hofer, S. Scheidegger, J. A. Agner, H. Schmutz, and F. Merkt. Phys. Rev. Lett., 115, 133202, 2015.

[g] L. Semeria, P. Jansen and F. Merkt, J. Chem. Phys., 145, 204301, 2016.

TB04

FORMATION OF H_2^+ AND ITS ISOTOPOMERS BY RADIATIVE ASSOCIATION: THE ROLE OF SHAPE AND FESHBACH RESONANCES

MAXIMILIAN BEYER, FREDERIC MERKT, *Laboratorium für Physikalische Chemie, ETH Zurich, Zurich, Switzerland.*

The recent observations [1,2] of shape and Feshbach resonances in the high-resolution photoelectron spectra of H_2, HD and D_2 in the vicinity of the dissociation thresholds of H_2^+, HD^+ and D_2^+ raise questions concerning their potential role in the formation of H_2^+ and its isotopomers in the early universe by radiative association, a topic of astrophysical interest [3]. Close-coupling calculations for the cross sections of the reactions

$$\mathrm{H}^+ + \mathrm{H} \to \mathrm{H}_2^+ + h\nu \tag{1}$$

$$\mathrm{H}^+ + \mathrm{D} \to \mathrm{HD}^+ + h\nu \tag{2}$$

$$\mathrm{D}^+ + \mathrm{H} \to \mathrm{HD}^+ + h\nu \tag{3}$$

$$\mathrm{D}^+ + \mathrm{D} \to \mathrm{D}_2^+ + h\nu, \tag{4}$$

will be presented which take into account nonadiabatic couplings involving rovibronic and hyperfine interactions, as well as relativistic and radiative corrections. The calculated energies and widths will be compared with the experimental results of Ref. [1,2] for H_2^+ and new data for HD^+ and D_2^+. The effect of the resonances on the radiative association rate coefficients will be discussed, also in comparison with earlier studies [4].

[1] M. Beyer and F. Merkt, *Phys. Rev. Lett.* **116**, 093001 (2016).

[2] M. Beyer and F. Merkt, *J. Mol. Spectrosc.* **330**, 147 (2016).

[3] Molecule formation in dust-poor environments, J. F. Babb and K. P. Kirby, in "The molecular astrophysics of stars and galaxies", T. W. Hartquist and D. A. Williams, eds., Oxford University Press, Oxford, 1998, pp. 11-34.

[4] D. E. Ramaker and J. M. Peek, *Phys. Rev. A* **13**, 58 (1976).

HOT BAND ANALYSIS AND KINETICS MEASUREMENTS FOR ETHYNYL RADICAL, C_2H, IN THE 1.49 μm REGION

ANH T. LE, GREGORY HALL, TREVOR SEARS[a], *Division of Chemistry, Department of Energy and Photon Sciences, Brookhaven National Laboratory, Upton, NY, USA.*

Ethynyl, C_2H, is an important intermediate in combustion processes and has been widely observed in interstellar space. Spectroscopically, it is of particular interest because it possesses three low-lying electronic surfaces: a ground $^2\Sigma^+$ state, and a low-lying $^2\Pi$ excited electronic state, which splits due to the Renner-Teller effect. Vibronic coupling among these states leads to a complicated, mixed-character, energy level structure. We have previously reported work[b] on three bands originating from the $\tilde{X}(0,0,0)\,^2\Sigma$ ground state to excited vibronic states: two $^2\Sigma - {}^2\Sigma$ transitions at 6696 and 7088 cm^{-1} and a $^2\Pi - {}^2\Sigma$ transition at 7108 cm^{-1}. In this work, the radicals were formed in a hot, non-thermal, population distribution by *u.v.* pulsed laser photolysis of a precursor. Kinetic measurements of the time-evolution of the ground state populations following collisional relaxation and reactive loss were also made, using some of the stronger rotational lines observed. Time-dependent signals in mixtures containing a variable concentration of precursor in argon suggested that vibronically hot C_2H radicals were less reactive than the relaxed, thermalized, radical. Two additional hot bands originating in states $\tilde{X}(0,1^1,0)\,^2\Pi$ and $\tilde{X}(0,2^0,0)\,^2\Sigma$, have now been identified in the same spectral region. In a new series of experiments, we have measured the kinetics of formation and decay of representative levels involving all the assigned transitions, i.e. originating in $\tilde{X}(0,v_2,0)$, with $v_2 = 0, 1$, and 2, in various concentrations of mixtures of precursor, inert gas and hydrogen. The new spectra also show greatly improved signal-to-noise ratio in comparison to our previous work, due to the use of a transient FM detection scheme, and additional spectral assignments seem likely. Both kinetics and spectroscopic results will be described in the talk.

Acknowledgments: Work at Brookhaven National Laboratory was carried out under Contract No. DE-SC0012704 with the U.S. Department of Energy, Office of Science, and supported by its Division of Chemical Sciences, Geosciences and Biosciences within the Office of Basic Energy Sciences.

[a]also: *Chemistry Department, Stony Brook University, Stony Brook, New York 11794*
[b]A. T. Le, G. E. Hall, T. J. Sears, *J. Chem. Phys.* **145** 074306, 2016

Intermission

PROBING THE STRUCTURES OF NEUTRAL B_{11} AND B_{12} USING HIGH-RESOLUTION PHOTOELECTRON IMAGING OF $B_{11}{}^-$ AND $B_{12}{}^-$

JOSEPH CZEKNER, LING FUNG CHEUNG, LAI-SHENG WANG, *Department of Chemistry, Brown University, Providence, RI, USA.*

We report high-resolution photoelectron imaging of $B_{11}{}^-$ and $B_{12}{}^-$ at 354.7 and 532.0 nm, respectively, resolving several low-frequency vibrational modes for neutral B_{11} and B_{12}. The vibrational information is highly valuable to verify the structures of the neutral clusters. Several isomers are considered, and vibrational frequencies are calculated for B_{11} and B_{12} using density functional theory. Comparisons between the experimental and calculated vibrational frequencies prove that the B_{11} neutral and anion both possess a perfectly planar C_{2v} structure. The $B_{12}{}^-$ anion is quasi-planar with C_s symmetry, while the experiment confirms that neutral B_{12} possesses C_{3v} symmetry. The high-resolution photoelectron spectra also allow the electron affinities of B_{11} and B_{12} to be measured more accurately as 3.401 and 2.221 eV, respectively. It is shown that high-resolution photoelectron imaging can be an effective method to determine structures of small neutral boron clusters, complementary to infrared spectroscopy.

TB07

VARIABLE MIXED ORBITAL CHARACTER IN THE PHOTOELECTRON ANGULAR DISTRIBUTION OF NO_2

BENJAMIN A LAWS, STEVEN J CAVANAGH, BRENTON R LEWIS, STEPHEN T GIBSON, *Research School of Physics and Engineering, Australian National University, Canberra, ACT, Australia.*

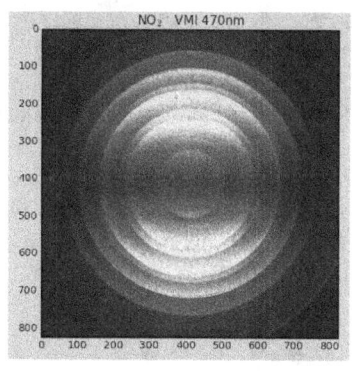

NO_2 a key component of photochemical smog and an important species in the Earth's atmosphere, is an example of a molecule which exhibits significant mixed orbital character in the HOMO. In photoelectron experiments the geometric properties of the parent anion orbital are reflected in the photoelectron angular distribution (PAD), an area of research that has benefited largely from the ability of velocity-map imaging (VMI) to simultaneously record both the energetic and angular information, with 100% collection efficiency.

Photoelectron spectra of NO_2^-, taken over a range of wavelengths (355nm-520nm) with the ANU's VMI spectrometer, reveal an anomalous jump in the anisotropy parameter near threshold. Consequently, the orbital behavior of NO_2^- appears to be quite different near threshold compared to detachment at higher photon energies. This surprising effect is due to the Wigner Threshold law, which causes p orbital character to dominate the photodetachment cross-section near threshold, before the mixed s/d orbital character becomes significant at higher electron kinetic energies.[a]

By extending recent work on binary character models[b] to form a more general expression, the variable mixed orbital character of NO_2^- is able to be described. This study provides the first multi-wavelength NO_2 anisotropy data, which is shown to be in decent agreement with much earlier zero-core model predictions[c] of the anisotropy parameter.

[a]K. J. Reed, A. H. Zimmerman, H. C. Andersen, and J. I. Brauman, *J. Chem. Phys.* **64**, 1368, (1976). doi:10.1063/1.432404

[b]D. Khuseynov, C. C. Blackstone, L. M. Culberson, and A. Sanov, *J. Chem. Phys.* **141**, 124312, (2014). doi:10.1063/1.4896241

[c]W. B. Clodius, R. M. Stehman, and S. B. Woo, *Phys. Rev. A.* **28**, 760, (1983). doi:10.1103/PhysRevA.28.760

[c]Research supported by the Australian Research Council Discovery Project Grant DP160102585

TB08

VIBRONIC COUPLING IN THE GROUND STATE OF VINYLIDENE $\tilde{X}^1A_1$ H_2CC

STEPHEN T GIBSON, BENJAMIN A LAWS, *Research School of Physics and Engineering, Australian National University, Canberra, ACT, Australia*; HUA GUO, *Chemistry, University of New Mexico, Albuquerque, NM, USA*; DANIEL NEUMARK, *Department of Chemistry, The University of California, Berkeley, CA, USA*; CARL LINEBERGER, *Department of Chemistry and Biochemistry, JILA - University of Colorado, Boulder, CO, USA*; ROBERT W FIELD, *Department of Chemistry, MIT, Cambridge, MA, USA.*

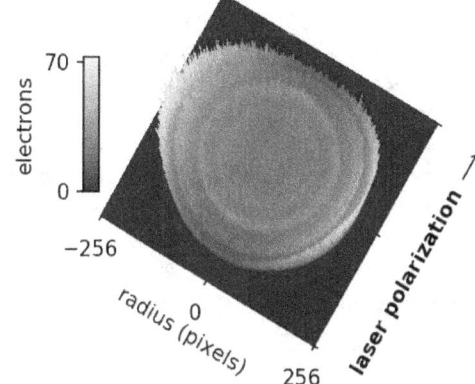

The nature of the isomeration process that turns vinylidene H_2CC to acetylene HCCH, requiring a 1,2-hydrogen atom shift across the molecule, is a long standing puzzle that has its origin in a 1989 photoelectron measurement of vinylidide (H_2CC^-)[a]. In recent years the photoelectron spectrum of vinylidide has been revisited, using improved experimental techniques, including velocity-map imaging for the detection of photoelectrons, low-temperature near-threshold methods (cryo-SEVI)[b], and sophisticated *ab inito* calculations[c]. The simple normal-mode structure, 1064 nm velocity-map image illustrated, is proving a challenge to decipher. However, the dramatic change in the photoelectron angular distribution of the inner-ring structure is characteristic of vibronic coupling[d]. The lowest electronic state with the correct symmetry, $\tilde{B}^1B_2$, is 4eV higher in energy.

[a]K. M. Ervin *et al. J. Chem. Phys.* **91** 5974 (1991).

[b]J. A. De Vine *et al. J. Am. Chem. Soc.* **138** 16417 (2016).

[c]L. Guo *et al. J. Phys. Chem.* **119** 8488 (2015).

[d]A. Weaver *et al. J. Chem. Phys.* **94** 1740 (1991).

Research supported by the Australian Research Council Discovery Project Grant DP160102585.

TB09   **11:20 – 11:35**

CONFORMATIONAL STUDY OF DIBENZYL ETHER

ALICIA O. HERNANDEZ-CASTILLO, CHAMARA ABEYSEKERA, DANIEL M. HEWETT, TIMOTHY S. ZWIER, *Department of Chemistry, Purdue University, West Lafayette, IN, USA.*

Understanding the initial stages of polycyclic aromatic hydrocarbon (PAH) aggregation, the onset of soot formation, is an important goal on the pathway to cleaner combustion processes. PAHs with short alkyl chains, present in fuel-rich combustion environments, can undergo reactions that will chemically link aromatic rings together. One such example of a linked diaryl compound is dibenzyl ether, C_6H_5-CH_2-O-CH_2-C_6H_5. The –CH_2-O-CH_2- linkage has a length and flexibility well-suited to forming a π-stacked conformation between the two phenyl rings. In this talk, we will explore the single-conformation spectroscopy of dibenzyl ether under jet-cooled conditions in the gas phase. Laser-induced fluorescence, chirped pulse Fourier transform microwave (8-18 GHz region), and single-conformation infrared spectroscopy in the alkyl CH stretch region were all carried out on the molecule, thereby interrogating its full array of electronic, vibrational and rotational degrees of freedom. This work is the first step in a broader study to determine the extent of π-stacking in linked aryl compounds as a function of linkage and PAH size.

TB10   **11:37 – 11:52**

THE EXOTIC EXCITED STATE BEHAVIOR OF 3-PHENYL-2-PROPYNENITRILE

KHADIJA M. JAWAD, CLAUDIA I VIQUEZ ROJAS, LYUDMILA V SLIPCHENKO, TIMOTHY S. ZWIER, *Department of Chemistry, Purdue University, West Lafayette, IN, USA.*

3-phenyl-2-propynenitrile (Ph-C$\equiv$C-C$\equiv$N) is of interest to the study of Titan's atmosphere as it is a likely product of the photochemical reaction between two known species in that environment: benzene and cyanoacetylene. The gas phase jet-cooled resonant two-photon ionization, laser induced fluorescence, and preliminary dispersed fluorescence spectra were previously reported without firm assignments due to the scarcity of totally symmetric vibrations and the prevalence of strong bands of b2 and b1 symmetry vibrations. These had called into question the identity and geometry of the excited state(s) involved in the transitions. We will here present the completed set of dispersed fluorescence data along with an analysis of the potential energy surfaces and vibronic coupling characteristic of the close-lying excited states in this intriguing molecule.

TB11   **11:54 – 12:09**

VIBRONIC EMISSION SPECTROSCOPY OF JET-COOLED BENZYL-TYPE RADICALS FROM CORONA DISCHARGE OF CHLORO-SUBSTITUTED O-XYLENE MOLECULES

YOUNG YOON, SANG LEE, *Department of Chemistry, Pusan National University, Pusan, Korea.*

Whereas benzyl radical, a prototypic aromatic free radical, has been the subject of numerous spectroscopic studies, chloro-substituted benzyl radicals have received less attention, due to the difficulties associated with production of radicals from precursors. The weak C-Cl bond can be easily dissociated in corona discharge of high voltage, leading to the formation of other benzyl-type radicals. During past years, we have concentrated the spectroscopy of chloro-substituted methylbenzyl radicals produced from corona discharge of precursor seeded in a large amount of helium carrier gas using a pinhole-type glass nozzle in a technique of corona excited supersonic expansion. From the analysis of the spectrum observed, we can easily distinguish the origin bands in the $D_1 \rightarrow D_0$ transition of the isomeric chloro-substituted methylbenzyl radicals with the additivity rule, [a][b] discovered from the analysis of a series of benzyl-type radicals. Also, the displacement of phenylic Cl by benzylic H was confirmed to be dependent on the distance between Cl and H atoms. The benzyl-type radicals produced in corona discharge from precursor were determined based on the bond dissociation energies and molecular structure of precursor molecules as well as the agreement of the observed with the calculated ones from Gaussian program, from which the 2-methyl-3-chlorobenzyl, 2-methyl-4-chlorobenzyl, 2-methyl-5-chlorobenzyl, and 2-methyl-6-chlorobenzyl radicals were newly identified.

[a]Y. W. Yoon, C. S. Huh, and S. K. Lee, *Chem. Phys. Lett.* **525-526**, 44-48 (2012).
[b]Y. W. Yoon and S. K. Lee, *J. Phys. Chem. A* **117**, 2485-2491 (2013).

TC. Structure determination

Tuesday, June 20, 2017 – 8:30 AM

Room: 1024 Chemistry Annex

Chair: Wolfgang Jäger, University of Alberta, Edmonton, AB, Canada

---

TC01 **8:30 – 8:45**

MICROWAVE SPECTRA OF THE TWO CONFORMERS OF PROPENE-3-d_1 AND A SEMIEXPERIMENTAL EQUILIBRIUM STRUCTURE OF PROPENE

NORMAN C. CRAIG, *Department of Chemistry and Biochemistry, Oberlin College, Oberlin, OH, USA*; J. DEMAISON, HEINZ DIETER RUDOLPH, *Section of Chemical Information Systems, Universität Ulm, Ulm, Germany*; RANIL M. GURUSINGHE, MICHAEL TUBERGEN, *Department of Chemistry and Biochemistry, Kent State University, Kent, OH, USA*; L. H. COUDERT, *Institut des Sciences Moléculaires d'Orsay, Université Paris-Sud, Orsay, France*; PETER SZALAY, *Institute of Chemistry, Eotvos University, Budapest, Hungary*; ATTILA CSÁSZÁR, *Research Group on Complex Chemical Systems, MTA-ELTE, Budapest, Hungary*.

FT microwave spectra have been observed and analyzed for the S (in-plane) and A (out-of-plane) conformers of propene-3-d_1 in the 10-22 GHz region. Both conformers display splittings due to deuterium quadrupole coupling; for the latter one only, a 19 MHz splitting due to internal rotation of the partially deuterated methyl group has been observed. In addition to rotational constants, the analysis yielded quadrupole coupling constants and parameters describing the tunneling splitting and its rotational dependence. Improved rotational constants for parent propene and the three $^{13}C_1$ species are recently available.[a] Use of vibration-rotation interaction constants computed at the MP2(FC)/cc-pVTZ level gave equilibrium rotational constants for these six species and for fourteen more deuterium isotopologues with diminished accuracy from early literature data. A semiexperimental equilibrium structure, r_e^{SE}, has been determined for propene by fitting fourteen structural parameters to the equilibrium rotational constants. The new r_e^{SE} structure compares well with an ab initio equilibrium structure computed with the all-electron CCSD(T)/cc-pV(Q,T)Z model and with a structure obtained using the mixed regression method with predicates and equilibrium rotational constants.

[a]N. C. Craig, P. Groner, A. R. Conrad, R. Gurusinghe, M. J. Tubergen J. Mol. Spectrosc. 248, 1-6 (2016).

TC02 **8:47 – 9:02**

CONFORMATIONAL STUDIES OF 1-OCTYNE FROM ROTATIONAL SPECTROSCOPY

MARK P. MATURO, DANIEL A. OBENCHAIN, ROBERT MELCHREIT, *Department of Chemistry, Wesleyan University, Middletown, CT, USA*; S. A. COOKE, *Natural and Social Science, Purchase College SUNY, Purchase, NY, USA*; STEWART E. NOVICK, *Department of Chemistry, Wesleyan University, Middletown, CT, USA*.

Alkanes of the form $CH_3(CH_2)_nCH_3$ generally favor ground state geometries that have co-planar carbon atoms. In this study, we have looked at a long chain hydrocarbon with a terminal carbon-carbon triple bond, viz., 1-octyne. Guided by the results of the 1-hexyne studies,[a] three possible low energy conformers were studied which we reference as anti-anti (AA, straight chain), anti-gauche (AG, terminal methyl group is gauche), and gauche-anti (GA, ethyl group is gauche). An initial broadband chirp-pulse was performed between 7-13 GHz and a total of sixty-eight transitions were fit. Additional measurements on a Balle Flygare cavity instrument yielded an additional seventy-three lines belonging to three of the conformers. Transitions for all 8 of the singly substituted ^{13}C isotopologues, in natural abundance, have also been observed for the AA conformer. *Ab-initio* optimizations at the MP2/6-311++g(2d,2p) level of theory and basis set for these three conformers will be compared to experimental rotational constants. Structure determinations of the AA conformer will also be discussed.

[a]Atticks, K.; Bohn, R. K.; Michaels H. H. Int'l J. of Quantum Chem. 2001, 85, 514-519; Utzat, K.; Bohn, R. K.; Michaels H. H. J. Mol. Struct. 2007, 841, 22-27

TC03   9:04 – 9:19

HIGH-RESOLUTION ROTATIONAL SPECTROSCOPY OF A MOLECULAR ROTARY MOTOR

SERGIO R DOMINGOS, *CoCoMol, Max-Planck-Institut für Struktur und Dynamik der Materie, Hamburg, Germany*; ARJEN CNOSSEN, *Stratingh Institute for Chemistry and Zernike Institute for Advanced Materials, University of Groningen, Groningen, The Netherlands*; CRISTOBAL PEREZ, *CoCoMol, Max-Planck-Institut für Struktur und Dynamik der Materie, Hamburg, Germany*; WYBREN JAN BUMA, *Van' t Hoff Institute for Molecular Sciences, University of Amsterdam, Amsterdam, Netherlands*; WESLEY R BROWNE, BEN L FERINGA, *Stratingh Institute for Chemistry and Zernike Institute for Advanced Materials, University of Groningen, Groningen, The Netherlands*; MELANIE SCHNELL, *CoCoMol, Max-Planck-Institut für Struktur und Dynamik der Materie, Hamburg, Germany*.

To develop synthetic molecular motors and machinery that can mimic their biological counterparts has become a stimulating quest in modern synthetic chemistry. Gas phase studies of these simpler synthetic model systems provide the necessary isolated conditions that facilitate the elucidation of their structural intricacies. We report the first high-resolution rotational study of a synthetic molecular rotary motor based on chiral overcrowded alkenes[a] ($C_{27}H_{20}$) using chirp-pulsed Fourier transform microwave spectroscopy[b]. Rotational constants and quartic centrifugal distortion constants were determined based on a fit using more than two hundred rotational transitions spanning $5 \leq J \leq 21$ in the 2−4 GHz frequency range. Despite the lack of polar groups, the rotor's asymmetry produces strong a− and b−type rotational transitions arising from a single predominant conformer. Evidence for fragmentation of the rotor allows for unambiguous identification of the isolated rotor components. The experimental spectroscopic parameters of the rotor are compared and discussed against current high-level *ab initio* and density functional theory methods.

[a] Vicario et al. Chem. Commun., 5910-5912 (2005)
[b] Brown et al. Rev. Sci. Instrum., 79, 053103 (2008)

TC04   9:21 – 9:36

THE MOLECULAR STRUCTURE OF MONOFLUOROBENZALDEHYDES

ISSIAH BYEN LOZADA, WENHAO SUN, JENNIFER VAN WIJNGAARDEN, *Department of Chemistry, University of Manitoba, Winnipeg, MB, Canada*.

The pure rotational spectra of 2- and 3-fluorobenzaldehyde have been investigated using a chirped pulse Fourier transform microwave (FTMW) spectrometer in the range of 8-18 GHz and a Balle-Flygare FTMW spectrometer in the range of 4-26 GHz. As in a previous study of monofluorobenzaldehydes,[a] only transitions due to a single planar conformer were observed for 2-fluorobenzaldehyde (O-trans) whereas two planar conformers (O-trans and O-cis) of 3-fluorobenzaldehydes were confirmed. Transitions due to the seven unique ^{13}C isotopologues of each of the three molecules have been observed for the first time. Their rotational constants were used to derive the effective ground state (r_0) and substitution (r_s) structures. The results compare favourably with the equilibrium (r_e) geometries which were determined following geometry optimization at the MP2/aug-cc-pVTZ level of theory.

[a] José L. Alonso and Rosa M. Villamañán, J. Chem. Soc., Faraday Trans. 2, 1989, 85(2), 137-149

TC05   9:38 – 9:53

CHIRPED-PULSE FOURIER TRANSFORM MICROWAVE SPECTROSCOPY OF THE 2,3-DIFLUOROPYRIDINE-CARBON DIOXIDE COMPLEX

SYDNEY A GASTER, CAMERON M FUNDERBURK, GORDON G BROWN, *Department of Science and Mathematics, Coker College, Hartsville, SC, USA*.

The pure rotational spectrum of the 2,3-difluoropyridine-CO_2 complex was measured on a chirped-pulsed Fourier transform microwave (CP-FTMW) spectrometer in the 3 − 18 GHz frequency range. The spectrum was analyzed to find the spectroscopic constants of the complex, including the quadrupole coupling constants and the centrifugal distortion constants. The spectrum of the 2,3-difluoropyridine-$^{13}CO_2$ complex was also measured and analyzed. Experimental constants were compared to the results of *ab initio* calculations.

Intermission

TC06 10:12 – 10:27

THE IMPORTANCE OF A GOOD FIT: THE MICROWAVE SPECTRA AND MOLECULAR STRUCTURES OF *TRANS*-1,2-DIFLUOROETHYLENE-HYDROGEN CHLORIDE AND *CIS*-1,2-DIFLUOROETHYLENE-HYDROGEN CHLORIDE

HELEN O. LEUNG, MARK D. MARSHALL, LEONARD H. YOON, *Chemistry Department, Amherst College, Amherst, MA, USA.*

Previously studied complexes of hydrogen chloride with fluoroethylenes demonstrate that the secondary interaction between the chlorine atom of the HCl and a hydrogen of the ethylene occurs with the sterically accessible *cis* H-atom in vinyl fluoride rather than the electrostatically favorable geminal hydrogen. However, with 1,1,2-trifluoroethylene the opposite occurs and electrostatics is favored over sterics. The two hydrogen atoms in *trans*-1,2-difluoroethylene are electrostatically equivalent and each offer the possibility of interacting in a geminal or in a *cis* fashion. Thus, the observed structure of *trans*-1,2-difluoroethylene-HCl, with a secondary interaction to the *cis* H-atom, is consistent with the favorable steric interactions associated with this configuration. On the other hand, in *cis*-1,2-difluoroethylene only electrostatically-equivalent geminal H-atoms are present. We find that rather than adopting the sterically unfavorable arrangement found in 1,1,2-trifluoroethylene-HCl, the hydrogen chloride in this complex instead forms a bifurcated hydrogen bond with the two F-atoms, and there is no secondary interaction involving the chlorine atom.

TC07 10:29 – 10:44

TAKING THE NEXT STEP WITH HALOGENATED OLEFINS: MICROWAVE SPECTROSCOPY AND MOLECULAR STRUCTURES OF TETRAFLUORO- AND CHLORO-TRIFLUORO PROPENES AND THEIR COMPLEXES WITH THE ARGON ATOM

MARK D. MARSHALL, HELEN O. LEUNG, MILES A. WRONKOVICH, MEGAN E TRACY, LABONI HOQUE, ALLISON M RANDY-COFIE, ALINA K DAO, *Chemistry Department, Amherst College, Amherst, MA, USA.*

The determination of the structures of heterodimers of haloethylenes with protic acids has provided a wealth of information and a few surprises concerning intermolecular forces and the sometimes cooperative and sometimes competing effects of electrostatic, steric, and dispersion forces. In seeking to apply this knowledge to larger systems with a wider variety of possible interactions and binding sites, we extend the carbon chain by one atom via the addition of a trifluoromethyl moeity. As a first step the microwave rotational spectra of the halopropene monomers, 2,3,3,3-tetrafluoropropene, 2-chloro-3,3,3-trifluoropropene, (*E*)-1-chloro-3,3,3-trifluoropropene, and (*Z*)-1-chloro-3,3,3-trifluoropropene, and their complexes with the argon atom are obtained and analyzed to obtain molecular structures.

TC08 10:46 – 11:01

GERMANIUM DICARBIDE: EVIDENCE FOR A T–SHAPED GROUND STATE STRUCTURE

OLIVER ZINGSHEIM, *I. Physikalisches Institut, Universität zu Köln, Köln, Germany*; MARIE-ALINE MARTIN-DRUMEL, *CNRS, Institut des Sciences Moléculaires d'Orsay, Orsay, France*; SVEN THORWIRTH, STEPHAN SCHLEMMER, *I. Physikalisches Institut, Universität zu Köln, Köln, Germany*; CARL A GOTTLIEB, *Radio and Geoastronomy Division, Harvard-Smithsonian Center for Astrophysics, Cambridge, MA, USA*; JÜRGEN GAUSS, *Institut für Physikalische Chemie, Universität Mainz, Mainz, Germany*; MICHAEL C McCARTHY, *Atomic and Molecular Physics, Harvard-Smithsonian Center for Astrophysics, Cambridge, MA, USA.*

The preferred equilibrium structure of germanium dicarbide (GeC_2) has been an open question for decades: while high-level quantum chemical calculations predict an L-shaped ground state structure, the very flat potential energy surface of the species prevents a T-shaped structure from being entirely ruled out[1]. By recording for the first time the rotational spectrum of GeC_2 using sensitive microwave and millimeter techniques, we establish that the molecule adopts a vibrationally-averaged T-shaped structure in the ground state. From isotopic substitution of 14 isotopologues, a precise r_0 structure has been derived. This structural work should serve as an important benchmark for future calculations.

[1] Sari et al., *J. Chem. Phys.* **117** 10008 (2002)

TC09 11:03 – 11:18

CARBON CHAINS CONTAINING GROUP IV ELEMENTS: ROTATIONAL DETECTION OF GeC_4 AND GeC_5

MICHAEL C McCARTHY, *Atomic and Molecular Physics, Harvard-Smithsonian Center for Astrophysics, Cambridge, MA, USA*; MARIE-ALINE MARTIN-DRUMEL, *CNRS, Institut des Sciences Moleculaires d'Orsay, Orsay, France*; SVEN THORWIRTH, *I. Physikalisches Institut, Universität zu Köln, Köln, Germany.*

Following the recent discovery of T-shaped GeC_2 by chirped-pulse FT microwave spectroscopy, evidence has been found for two longer carbon chains, GeC_4 and GeC_5, guided by high-level quantum chemical calculations of their molecular structure. Like their isovalent Si-bearing counterparts, those with an even number of carbon atoms are predicted to possess $^1\Sigma$ ground states, while odd-numbered carbon chains have low-lying $^3\Sigma$ linear isomers; all are predicted to be highly polar. With the exception of ^{73}Ge, rotational lines of the other four Ge isotopic species have been observed between 6 and 18 GHz. From these measurements, the Ge-C bond length has been determined to high precision, and can be compared to that found in other Ge species, such as GeC [1] and GeC_3Ge [2] studied previously at rotational resolution. Somewhat surprisingly, the spectrum of GeC_5 very closely resembles that of $^1\Sigma$ molecule, presumably owing to the very large spin-orbit constant of atomic Ge, which is manifest as an equally large spin-spin constant in the chain. A comparison between the production of SiC_n and GeC_n chains by laser ablation, including the absence of those with $n = 3$, will be given.

[1] C. R. Brazier and J. I. Ruiz, *J. Mol. Spectrosc.*, **270**, 26-32 (2011).
[2] S. Thorwirth *et al.*, *J. Phys. Chem. A*, **120**, 254-259 (2016).

TC10 11:20 – 11:35

MICROWAVE SPECTRA OF Ar···AgI AND H_2O···AgI PRODUCED BY LASER ABLATION

JOHN C MULLANEY, CHRIS MEDCRAFT, NICK WALKER, *School of Chemistry, Newcastle University, Newcastle-upon-Tyne, United Kingdom*; ANTHONY LEGON, *School of Chemistry, University of Bristol, Bristol, United Kingdom.*

Complexes of argon and water with silver iodide have been formed in the gas phase by laser ablation of a silver iodide rod and studied using a chirped-pulse Fourier transform microwave spectrometer. Ar···AgI was characterized by its rotational spectrum and *ab initio* calculations carried out at the CCSD(T)(F12c)/cc-pVTZ-F12 explicitly correlated level of theory. The molecule was shown to be linear in the ground state, with atoms in the order shown. The Ar···Ag and Ag-I bond lengths, r_0(Ar···Ag) = 2.6759 Å and r_0(Ag-I) = 2.5356 Å, were determined. Other factors such as the dissociation energy, the intermolecular quadratic stretching force constant and the change in ionicity of AgI upon forming the complex were also determined and will be discussed with comparison to the series Ar···AgX (X = F, Cl, Br and I). Data of the H_2O···AgI complex will also be presented with isotopic studies ongoing.

TD. Large amplitude motions, internal rotation

Tuesday, June 20, 2017 – 8:30 AM

Room: B102 Chemical and Life Sciences

Chair: V. Ilyushin, Institute of Radio Astronomy of NASU, Kharkiv, Ukraine

---

TD01   8:30 – 8:45

THE MICROWAVE SPECTROSCOPY STUDY OF 1,2-DIMETHOXYETHANE

WEIXING LI, ANNALISA VIGORITO, CAMILLA CALABRESE, LUCA EVANGELISTI, *Dipartimento di Chimica G. Ciamician, Università di Bologna, Bologna, Italy*; LAURA B. FAVERO, *Istituto per lo Studio dei Materiali Nanostrutturati, Consiglio Nazionale delle Ricerche (ISMN-CNR), Bologna, Italy*; ASSIMO MARIS, SONIA MELANDRI, *Dipartimento di Chimica G. Ciamician, Università di Bologna, Bologna, Italy*.

With Pulsed-Jet Fourier Transform MicroWave (PJ-FTMW) spectroscopy and Stark modulated Free Jet Millimeter-Wave absorption (FJ-AMMW) spectroscopy, the rotational spectra of two conformers of 1,2-Dimethoxyethane were identified and characterized. Besides the normal species, the spectra of all the mono-substituted ^{13}C isotopologues in natural abundance were also measured. By fitting the rotational transitions split by the methyl internal rotations using both XIAM and ERHAM programs, the spectroscopic parameters were obtained and compared. The rotational constants indicated the conformers to be TGT and TGG', respectively. With the rotational constants of the normal and ^{13}C species, the coordinates of the substituted carbon atoms could be calculated with Kraitchmann's equations. The carbon-frameworks further confirmed the assignment of the two conformations. The V_3 barriers of the two methyl groups' internal rotations were also experimentally determined.

TD02   8:47 – 9:02

AB INITIO EFFECTIVE ROVIBRATIONAL HAMILTONIANS FOR NON-RIGID MOLECULES VIA CURVILINEAR VMP2

BRYAN CHANGALA, *JILA, National Institute of Standards and Technology and Univ. of Colorado Department of Physics, University of Colorado, Boulder, CO, USA*; JOSHUA H BARABAN, *Department of Chemistry, University of Colorado, Boulder, CO, USA*.

Accurate predictions of spectroscopic constants for non-rigid molecules are particularly challenging for *ab initio* theory. For all but the smallest systems, "brute force" diagonalization of the full rovibrational Hamiltonian is computationally prohibitive, leaving us at the mercy of perturbative approaches. However, standard perturbative techniques, such as second order vibrational perturbation theory (VPT2), are based on the approximation that a molecule makes small amplitude vibrations about a well defined equilibrium structure. Such assumptions are physically inappropriate for non-rigid systems. In this talk, we will describe extensions to curvilinear vibrational Møller-Plesset perturbation theory (VMP2) that account for rotational and rovibrational effects in the molecular Hamiltonian. Through several examples, we will show that this approach provides predictions to nearly microwave accuracy of molecular constants including rotational and centrifugal distortion parameters, Coriolis coupling constants, and anharmonic vibrational and tunneling frequencies.

ANOMALOUS CENTRIFUGAL DISTORTION IN NH_2

MARIE-ALINE MARTIN-DRUMEL, OLIVIER PIRALI, <u>L. H. COUDERT</u>, *Institut des Sciences Moléculaires d'Orsay, Université Paris-Sud, Orsay, France.*

The NH_2 radical spectrum, first observed by Herzberg and Ramsay,[a] is dominated by a strong Renner-Teller effect[b] giving rise to two electronic states: the bent $X\,^2B_1$ ground state and the quasi-linear $A\,^2A_1$ excited state. The NH_2 radical has been the subject of numerous high-resolution investigations and its electronic and ro-vibrational transitions[c] have been measured. Using synchrotron radiation, new rotational transitions have been recently recorded and a value of the rotational quantum number N as large as 26 could be reached.[d] In the $X\,^2B_1$ ground state, the NH_2 radical behaves like a triatomic molecule displaying spin-rotation splittings. Due to the lightness of the molecule, a strong coupling between the overall rotation and the bending mode arises whose effects increase with N and lead to the anomalous centrifugal distortion evidenced in the new measurements.[d]

In this talk the Bending-Rotation approach[e] developed to account for the anomalous centrifugal distortion of the water molecule is modified to include spin-rotation coupling and applied to the fitting of high-resolution data pertaining to the ground electronic state of NH_2. A preliminary line position analysis of the available data[c,d] allowed us to account for 1681 transitions with a unitless standard deviation of 1.2. New transitions could also be assigned in the spectrum recorded by Martin-Drumel *et al.*[d] In the talk, the results obtained with the new theoretical approach will be compared to those retrieved with a Watson-type Hamiltonian and the effects of the vibronic coupling between the ground $X\,^2B_1$ and the excited $A\,^2A_1$ electronic state will be discussed.

[a] Herzberg and Ramsay, *J. Chem. Phys.* **20** (1952) 347

[b] Dressler and Ramsay, *Phil. Trans. R. Soc. A* **25** (1959) 553

[c] Hadj Bachir, Huet, Destombes, and Vervloet, *J. Molec. Spectrosc.* **193** (1999) 326; McKellar, Vervloet, Burkholder, and Howard, *J. Molec. Spectrosc.* **142** (1990) 319; Morino and Kawaguchi, *J. Molec. Spectrosc.* **182** (1997) 428

[d] Martin-Drumel, Pirali, and Vervloet, *J. Phys. Chem. A* **118** (2014) 1331

[e] Coudert, *J. Molec. Spectrosc.* **165** (1994) 406

THE EFFECT OF TORSION - VIBRATION COUPLINGS ON THE ν_9 AND ν_1 BANDS IN THE $CH_3OO\cdot$ RADICAL

<u>MENG HUANG</u>, TERRY A. MILLER, *Department of Chemistry and Biochemistry, The Ohio State University, Columbus, OH, USA*; ANNE B McCOY, *Department of Chemistry, University of Washington, Seattle, WA, USA.*

There are three CH stretch modes in the $CH_3OO\cdot$, namely the totally symmetric stretch, ν_2, the out-of-phase symmetric stretch, ν_1, and the antisymmetric stretch, ν_9. However, only two strong partially rotationally resolved vibrational transitions are observed in the CH stretch region of the infrared spectrum of $CH_3OO\cdot$. Moreover, the Q-branches of both bands are significantly broader than the simulations using an asymmetric rigid rotor model. Previously, the rotational contour of the ν_2 band in the experimental spectrum has been simulated by considering the fundamental as well as sequence band transitions involving torsionally excited levels populated at room temperature[a]. Using a four dimension Hamiltonian that involves the three CH stretches and the CH_3 torsion, the torsional sequence bands of the ν_2 stretch were calculated to be slightly shifted from the origin band as a result of the couplings between the CH stretches and CH_3 torsion which are particularly large due to an accidental degeneracy between the combination level involving the ν_2 stretch and n quanta in the CH_3 torsion and the combination level involving the ν_1 or ν_9 stretch and $n-1$ quanta in the CH_3 torsion. In this study we focus on the part of $CH_3OO\cdot$ vibrational spectrum containing the ν_1 and ν_9 bands. Along with the 4-dimensional Hamiltonian which was used to simulate the spectra in the ν_2 region, we analyze the effect of torsion-stretch couplings on the ν_1 and ν_9 bands based on the model developed by Hougen[b] to describe methanol. To account for the accidental degeneracies in CH_3OO, we extend the previous model to include the combination levels involving the ν_2 stretch and the CH_3 torsion. Unlike the torsional sequence bands and fundamental of ν_2, the tunneling splittings of the torsional sequence bands and fundamentals of ν_1 and ν_9 have different signs, which results in the weak intensity of ν_1 fundamental and broadening in the rotational contour of the torsional sequence bands of ν_9. The simulation of the ν_9 and ν_1 bands including the torsional sequence bands using the parameters which are consistent with the ones used in the simulation of the ν_2 band is in good agreement with the experimental spectrum.

[a] K.-H. Hsu, Y.-H. Huang, Y.-P. Lee, M. Huang, T. A. Miller and A. B. McCoy *J. Phys. Chem. A*, 2016, 120 (27), 4827

[b] J.-T. Hougen *J. Mol. Spec.*, 2001, 207, 60

TD05 9:38 – 9:53

ROVIBRATIONAL QUANTUM DYNAMICS OF THE METHANE-WATER DIMER

JÁNOS SARKA, ATTILA CSÁSZÁR, *Complex Chemical Systems Research Group, MTA-ELTE, Budapest, Hungary*; EDIT MÁTYUS, *Institute of Chemistry, Eötvös Loránd University, Budapest, Hungary.*

The challenging quantum dynamical description of the $CH_4 \cdot H_2O$ complex has been solved variationally[a] to provide theoretical explanation and assignment to the high-resolution spectroscopic measurements of the methane-water dimer carried out some twenty years ago.[b] The computational results are in excellent agreement with the reported experimental transitions and the experimentally observed reversed rovibrational sequences, i.e., formally negative rotational excitation energies, are also obtained in the computations. In order to better understand the origin of these peculiar features in the energy-level spectrum, we studied[c] all four possible combinations of the light and heavy isotopologues of methane and water and analyzed their rovibrational states using two limiting model systems: the rigidly rotating (RR) molecule and the coupled rotor (CR) system corresponding to the coupling of the two rotating monomers.

All rovibrational quantum dynamical computations[a,c] were carried out with rigid monomers and $J = 0,1,2$ total angular momentum quantum numbers using the fourth-age quantum chemical code GENIUSH[d,e] and two different methane-water potential energy surfaces (PES).[f,g] The numerical and formal analysis of the wave functions give insight into a fascinating complex world worth for further theoretical and experimental inquiries.

[a] J. Sarka, A. G. Császár, S. C. Althorpe, D. J. Wales and E. Mátyus, Phys. Chem. Chem. Phys. 18, 22816 (2016).

[b] L. Dore, R. C. Cohen, C. A. Schmuttenmaer, K. L. Busarow, M. J. Elrod, J. G. Loeser and R. J. Saykally, J. Chem. Phys. 100, 863 (1994).

[c] J. Sarka, A. G. Császár and E. Mátyus, Phys. Chem. Chem. Phys. accepted for publication (2017).

[d] E. Mátyus, G. Czakó and A. G. Császár, J. Chem. Phys. 130, 134112 (2009).

[e] C. Fábri, E. Mátyus and A. G. Császár, J. Chem. Phys. 134, 074105 (2011).

[f] O. Akin-Ojo and K. Szalewicz, J. Chem. Phys. 123, 134311 (2005).

[g] C. Qu, R. Conte, P. L. Houston and J. M. Bowman, Phys. Chem. Chem. Phys. 17, 8172 (2015).

Intermission

TD06 10:12 – 10:27

A COMBINED GIGAHERTZ AND TERAHERTZ SYNCHROTRON-BASED FOURIER TRANSFORM INFRARED SPECTROSCOPIC INVESTIGATION OF ORTHO-D-PHENOL

SIEGHARD ALBERT, ZIQIU CHEN, CSABA FÁBRI, ROBERT PRENTNER, MARTIN QUACK, DANIEL ZINDEL, *Laboratory of Physical Chemistry, ETH Zurich, Zürich, Switzerland.*

Tunneling switching is a fundamental phenomenon of interest in molecular quantum dynamics including also chiral molecules and parity violation.[a,b,c] Deuterated phenols have been identified as prototypical achrial candidates.[d] We report the high resolution spectroscopic investigation of the ortho-D-phenol in the GHz and THz ranges following our recent discovery of tunneling switching in its isotopomer meta-D-phenol.[e] Here we report new results on ortho-D-phenol. The pure rotational spectra were recorded in the range of 72-117 GHz and assigned to the syn- and anti- structures in the ground and the first excited torsional states. Specific torsional states were assigned based on a comparison of experimental rotational constants with the quasiadiabatic channel reaction path Hamiltonian (RPH) calculations. The torsional fundamental at $308\ cm^{-1}$ and the first hot band at $275\ cm^{-1}$ were subsequently assigned. The analyses of pure rotational and rovibrational spectra shall be discussed in detail in relation to possible tunneling switching.

[a] M. Quack , *Fundamental Symmetries and Symmetry Violations from High-resolution Spectroscopy, Handbook of High Resolution Spectroscopy*, M. Quack and F. Merkt eds.,John Wiley & Sons Ltd, Chichester, New York, 2001, vol. 1, ch. 18, pp. 659-722.

[b] R. Prentner, M. Quack, J. Stohner and M. Willeke, *J. Phys. Chem. A* **119**, 12805-12822 (2015).

[c] S. Albert, Z. Chen, C. Fábri, R. Prentner M. Quack and D. Zindel, paper at this meeting.

[d] S. Albert, Ph. Lerch, R. Prentner and M. Quack, *Angew. Chem. Int. Ed.* **52**, 346-349 (2013).

[e] S. Albert, Z. Chen, C. Fábri,P. Lerch, R. Prentner and M. Quack, *Mol.Phys.* **114**, 2751-2768 (2016) and *71st International Symposium on Molecular Spectroscopy*, Urbana-Champaign, USA, June 20-24, Talk FE04 (2016).

TD07

ANALYSIS OF THE ν_6 ASYMMETRIC NO STRETCH BAND OF NITROMETHANE

MAHESH B. DAWADI, LOU DEGLIUMBERTO, <u>DAVID S. PERRY</u>, *Department of Chemistry, The University of Akron, Akron, OH, USA*; HOWARD METTEE, *Department of Chemistry, Youngstown State University, Youngstown, OH, USA*; ROBERT L. SAMS, *Chemical Physics, Pacific Northwest National Laboratory, Richland, WA, USA.*

The b-type band near 1583 cm^{-1} has been assigned for $m \leq 3$, $K_a'' \leq 10$, $J'' \leq 20$. The ground state combination differences derived from these assigned levels were fit with the RAM36 program with an RMS deviation of 0.0006 cm^{-1}. The upper state levels are split into multiplets by perturbations. A subset of the available upper state combination differences for $m = 0$, $K_a' \leq 7$, $J' \leq 10$ were fit with the same program, but with rather poorer precision (0.01 cm^{-1}) than for the ground state.

TD08

TORSIONAL, VIBRATIONAL AND VIBRATION-TORSIONAL LEVELS IN THE S_1 AND GROUND CATIONIC D_0^+ STATES OF PARA-FLUOROTOLUENE

<u>ADRIAN M. GARDNER</u>, WILLIAM DUNCAN TUTTLE, LAURA E. WHALLEY, ANDREW CLAYDON, JOSEPH H. CARTER, TIMOTHY G. WRIGHT, *School of Chemistry, University of Nottingham, Nottingham, United Kingdom.*

The S_1 electronic state and ground state of the cation of *para*-fluorotoluene (*p*FT) have been investigated using resonance-enhanced multiphoton ionization (REMPI) spectroscopy and zero-kinetic-energy (ZEKE) spectroscopy.[a] Here we focus on the low wavenumber region where a number of "pure" torsional, fundamental vibrational and vibration-torsional levels are expected; assignments of observed transitions are discussed, which are compared to results of published work on toluene (methylbenzene) from the Lawrance group.[b] The similarity in the activity observed in the excitation spectrum of the two molecules is striking.

[a] A. M. Gardner, W. D. Tuttle, L. Whalley, A. Claydon, J. H. Carter and T. G. Wright, *J. Chem. Phys.*, **145**, 124307 (2016).
[b] J. R. Gascooke, E. A. Virgo, and W. D. Lawrance *J. Chem. Phys.*, **143**, 044313 (2015).

TD09

MOLECULAR SYMMETRY ANALYSIS OF LOW-ENERGY TORSIONAL AND VIBRATIONAL STATES IN THE S_0 AND S_1 STATES OF p-XYLENE TO INTERPRET THE REMPI SPECTRUM

<u>PETER GRONER</u>, *Department of Chemistry, University of Missouri - Kansas City, Kansas City, MO, USA*; ADRIAN M. GARDNER, WILLIAM DUNCAN TUTTLE, TIMOTHY G. WRIGHT, *School of Chemistry, University of Nottingham, Nottingham, United Kingdom.*

The electronic transition $S_1 \leftarrow S_0$ of p-xylene (pXyl) has been observed by REMPI spectroscopy.[a] Its analysis required a detailed investigation of the molecular symmetry of pXyl whose methyl groups are almost free internal rotors. The molecular symmetry group of pXyl has 72 operators.[b] This group, called $[33]D_{2h}$, is isomorphic to G_{36}(EM),[c] the double group for ethane and dimethyl acetylene even though it is NOT a double group for pXyl. Loosely speaking, the group symbol, $[33]D_{2h}$, indicates that is for a molecule with two threefold rotors on a molecular frame with D_{2h} point group symmetry. The transformation properties of the (i) free internal rotor basis functions for the torsional coordinates, (ii) the asymmetric rotor (Wang) basis functions for the Eulerian angles, (iii) nuclear spin functions, (iv) potential function, and (v) transitions dipole moment functions were determined. The forms of the torsional potential in the S_0 and S_1 states and the dependence of the first order torsional splittings on the potential coefficients have been obtained.

[a] AM Gardner, WD Tuttle, P. Groner, TG Wright, J. Chem. Phys., submitted Dec 2016
[b] P Groner, JR Durig, J. Chem. Phys., 66 (1977) 1856
[c] PR Bunker, P Jensen, Molecular Symmetry and Spectroscopy (1998, NRC Research Press, Ottawa, 2nd ed.)

TD10   **11:20 – 11:35**

TORSIONAL, VIBRATIONAL AND VIBRATION-TORSIONAL LEVELS IN THE S_1 AND GROUND CATIONIC D_0^+ STATES OF PARA-XYLENE

<u>ADRIAN M. GARDNER</u>, WILLIAM DUNCAN TUTTLE, *School of Chemistry, University of Nottingham, Nottingham, United Kingdom*; PETER GRONER, *Department of Chemistry, University of Missouri - Kansas City, Kansas City, MO, USA*; TIMOTHY G. WRIGHT, *School of Chemistry, University of Nottingham, Nottingham, United Kingdom.*

Insight gained from examining the "pure" torsional, vibrational and vibration-torsional (vibtor) levels of the single rotor molecules: toluene (methylbenzene)[a] and *para*-fluorotoluene (*p*FT),[b] is applied to the double rotor *para*-xylene (*p*-dimethylbenzene) molecule .[c] Resonance-enhanced multiphoton ionization (REMPI) spectroscopy and zero-kinetic-energy (ZEKE) spectroscopy are employed in order to investigate the S_1 and ground cationic states of para-xylene. Observed transitions are assigned in the full molecular symmetry group (G_{72}) for the first time.

[a] J. R. Gascooke, E. A. Virgo, and W. D. Lawrance, *J. Chem. Phys.*, **143**, 044313 (2015).

[b] A. M. Gardner, W. D. Tuttle, L. Whalley, A. Claydon, J. H. Carter and T. G. Wright, *J. Chem. Phys.*, **145**, 124307 (2016).

[c] A. M. Gardner, W. D. Tuttle, P. Groner and T. G. Wright, *J. Chem. Phys.*, (2017, in press).

TE. Fundamental interest

Tuesday, June 20, 2017 – 8:30 AM

Room: 161 Noyes Laboratory

Chair: Josh Vura-Weis, University of Illinois at Urbana-Champaign, Urbana, IL, USA

TE01   8:30–8:45

COVALENT AND NONCOVALENT INTERACTIONS BETWEEN BORON AND ARGON: AN INFRARED PHOTODIS-SOCIATION SPECTROSCOPIC STUDY OF ARGON-BORON OXIDE CATION COMPLEXES

JIAYE JIN, WEI LI, GUANJUN WANG, MINGFEI ZHOU, *Fudan University, Department of Chemistry, Shanghai, China.*

Although a wide range of compounds of the heavy rare-gas elements are experimentally known, very few chemically bound molecules have been experimentally observed for the lighter noble gases. Here we report a combined infrared photodissociation spectroscopic and theoretical study on a series of argon-boron oxide cation complexes prepared via a laser vaporization supersonic ion source in the gas phase. Infrared spectroscopic combined with state-of-the-art quantum chemical calculations indicate that the $[ArB_3O_{4,5}]^+$, $[ArB_4O_{5-7}]^+$ and $[ArB_5O_7]^+$ cation complexes have planar structures each involving an aromatic boroxol ring and an argon-boron covalent bond formed between the in-plane 2p atomic orbitals of Ar and boron. In contrast, the $[ArB_3O_4]^+$ cation complex is characterized to be a weakly bound complex with a BO chain structure.

TE02   8:47–9:02

OBSERVATION OF QUANTUM BEATING IN Rb AT 2.1 THz AND 18.2 THz: LONG-RANGE Rb*-Rb INTERACTIONS.

WILLIAM GOLDSHLAG, BRIAN J RICCONI, J. GARY EDEN, *Department of Electrical and Computer Engineering, University of Illinois at Urbana-Champaign, Urbana, IL, USA.*

The interaction of Rb 7s $^2S_{1/2}$, 5d $^2D_{3/2,5/2}$ and 5p $^2P_{3/2}$ atoms with the background species at long range (100-1000Å) has been observed by pump-probe ultrafast laser spectroscopy. Parametric four-wave mixing in Rb vapor with pairs of 50-70 fs pulses produces coherent Rb 6P-5S emission at 420 nm that is modulated by Rb quantum beating. The two dominant beating frequencies are 18.2 THz and 2.07 THz, corresponding to quantum beating between 7S and 5D states and to the $(5D-5P_{3/2})-(5P_{3/2}-5S)$ defect, respectively. Analysis of Rabi oscillations in these pump-probe experiments allows for the mean interaction energy at long range to be determined. The figure shows Fourier transform spectra of representative Rabi oscillation waveforms.

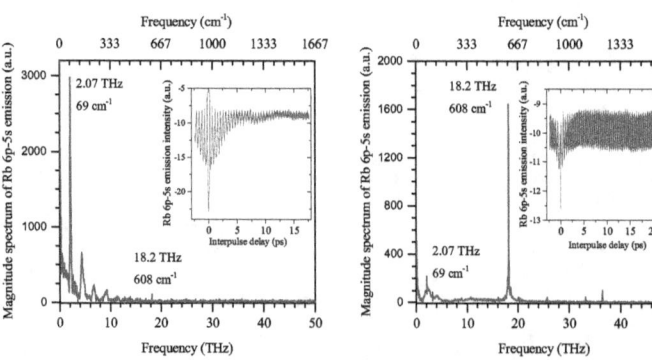

The waveform and spectrum at left illustrate quantum beating in Rb at 2.1 THz. The spectrum at right is dominated by the 18.2 THz frequency component generated by 7S-5D beating in Rb. Insets show respective temporal behaviors of the 6P-5S line near the coherent transient (zero interpulse delay).

TE03   9:04 – 9:19

ROTATIONAL SPECTRUM OF SACCHARINE

ELENA R. ALONSO, SANTIAGO MATA, JOSÉ L. ALONSO, *Grupo de Espectroscopia Molecular, Lab. de Espectroscopia y Bioespectroscopia, Unidad Asociada CSIC, Universidad de Valladolid, Valladolid, Spain.*

A significant step forward in the structure-activity relationships of sweeteners was the assignment of the AH-B moiety in sweeteners by Shallenberger and Acree [a,b]. They proposed that all sweeteners contain an AH-B moiety, known as glucophore, in which A and B are electronegative atoms separated by a distance between 2.5 to 4 Å. H is a hydrogen atom attached to one of the electronegative atom by a covalent bond. For saccharine, one of the oldest artificial sweeteners widely used in food and drinks, two possible B moieties exist ,the carbonyl oxygen atom and the sulfoxide oxygen atom although there is a consensus of opinion among scientists over the assignment of AH-B moieties to HN-SO. In the present work, the solid of saccharine (m.p. 220ºC) has been vaporized by laser ablation (LA) and its rotational spectrum has been analyzed by broadband CP-FTMW and narrowband MB-FTMW Fourier transform microwave techniques. The detailed structural information extracted from the rotational constants and ^{14}N nuclear quadrupole coupling constants provided enough information to ascribe the glucophore's AH and B sites of saccharine.

[a]R. S. Shallenberger, T. E. Acree. Nature 216, 480-482 Nov 1967.
[b]R. S. Shallenberger. Taste Chemistry; Blackie Academic & Professional, London, (1993).

TE04   9:21 – 9:36

MICROWAVE OBSERVATION OF THE O_2-CONTAINING COMPLEX, O_2-HCl

FRANK E MARSHALL, NICOLE MOON, THOMAS D. PERSINGER, RICHARD DAWES, G. S. GRUBBS II, *Department of Chemistry, Missouri University of Science and Technology, Rolla, MO, USA.*

In the realm of small-molecule van der Waals interactions, there exists much experimental and theoretical data for most fundamental atmospheric components. For complexes containing O_2, however, there is actually very little experimental data. This is most likely due to the spin complications brought about by the $^3\Sigma$ state of the molecule. In this talk, the authors will detail the first known measurement of the complex O_2-HCl along with experimental and theoretical analyses of the complex. Previously measured O_2-HF[a] analysis have been used as a guide and this talk will outline similarities and differences in the two species.

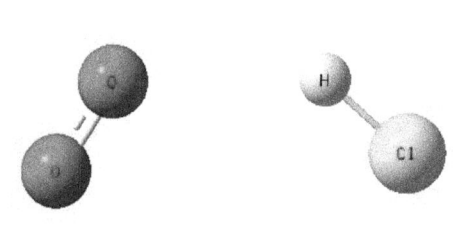

[a]S. Wu, G. Sedo, E. M. Grumstrup, and K. R. Leopold, *J. Chem. Phys.*, **127** (2007) 204315.

TE05   9:38 – 9:53

SUPERRADIANCE IN A STRONGLY-COUPLED MULTI-LEVEL SYSTEM

STEPHEN COY, DAVID GRIMES, TIMOTHY J BARNUM, ROBERT W FIELD, *Department of Chemistry, MIT, Cambridge, MA, USA.*

We have observed superradiance in barium Rydberg levels in a multi-level system involving emission from a directly populated level and another level coupled by cascade. A number of interesting effects are observed. These include pi phase discontinuities in the emission associated with the transition between emitting superradiant or subradiant modes, frequency shifts that vary across the superradiant output pulse, and intensity competition between emission on direct and cascade-populated transitions. Our experimental barium level structure will be described, and the observed interactions between the competing levels will be shown.

TE06   9:55 – 10:10

ABSTRACT WITHDRAWN

Intermission

TE07 10:29 – 10:44

DETECTION OF THE MW TRANSITION BETWEEN ORTHO AND PARA STATES

HIDETO KANAMORI, ZEINAB TAFTI DEHGHANI, ASAO MIZOGUCHI, *Department of Physics, Tokyo Institute of Technology, Tokyo, Japan*; YASUKI ENDO, *Department of Applied Chemistry, National Chiao Tung University, Hsinchu, Taiwan.*

Thorough the detailed analysis of the hyperfine resolved rotational transitions[ab], we have been pointed out that there exists not a little interaction between *ortho* and *para* states in the molecular Hamiltonian of S_2Cl_2. Using the *ortho-para* mixed molecular wavefunctions derived from the Hamiltonian, we calculated the transition moment and frequency of the *ortho-para* forbidden transitions in the cm- and mm-wave region, and picked up some promising candidate transitions for the spectroscopic detection.

In the experiment, the S_2Cl_2 vapor with Ar buffer gas in a supersonic jet condition was used with FTMW spectrometer at National Chiao Tung University. As a result, seven hyperfine resolved rotational transitions in the cm-wave region were detected as the *ortho-para* transition at the predicted frequency within the experimental error range. The observed intensity was 10^{-3} smaller than that of an allowed transition, which is also consistent with the prediction. This is the first time the electric dipole transition between *ortho* and *para* states has been detected in a free isolated molecule.

[a] A. Mizoguchi, S. Ota, H. Kanamori, Y. Sumiyoshi, and Y. Endo, J. Mol. Spectrosc, 250, 86 (2008)
[b] Z. T. Dehghani, S. Ota, A. Mizoguchi and H. Kanamori, J. Phys. Chem. A, 117(39), 10041, (2013)

TE08 10:46 – 11:01

NEAR-INFRARED SPECTROSCOPY OF SMALL PROTONATED WATER CLUSTERS

J. PHILIPP WAGNER, *Department of Chemistry, University of Georgia, Athens, GA, USA*; DAVID C McDON-ALD, *Chemistry, University of Georgia, Athens, GA, USA*; ANNE B McCOY, *Department of Chemistry, University of Washington, Seattle, WA, USA*; MICHAEL A DUNCAN, *Department of Chemistry, University of Georgia, Athens, GA, USA.*

Small protonated water clusters and their argon tagged analogues of the general formula $H^+(H_2O)_n Ar_m$ have been generated in a pulsed electric discharge source. Clusters containing $n = 1 - 8$ water molecules were mass-selected and their absorptions in the near-infrared were probed with a tunable Nd: YAG pumped OPA/OPA laser system in the region from $4850-7350$ cm^{-1}. A doublet corresponding to overtones of the free O−H stretches of the external waters was observed around 7200 cm^{-1} that was continuously decreasing in intensity with increasing cluster size. Broad, mostly featureless absorptions were found around 5300 cm^{-1} associated with stretch/bend combinations and with the hydrogen bonded waters in the core of the clusters. Vibrational assignments were substantiated by comparison to anharmonic frequency computations via second-order vibrational perturbation theory (VPT2) at the MP2/aug-cc-pVTZ level of theory.

TE09 11:03 – 11:18

NON COVALENT INTERACTIONS IN LARGE DIAMONDOID DIMERS IN THE GAS PHASE - A MICROWAVE STUDY

CRISTOBAL PEREZ, *CoCoMol, Max-Planck-Institut für Struktur und Dynamik der Materie, Hamburg, Germany*; MARINA SEKUTOR, ANDREY A. FOKIN, *Institute for Organic Chemistry, Justus Liebig University of Giessen, Giessen, Germany*; SEBASTIAN BLOMEYER, YURY V. VISHNEVSKIY, NORBERT W. MITZEL, *Lehrstuhl für Anorganische Chemie und Strukturchemie, Institut für Anorganische Chemie, Bielefeld, Germany*; PETER R. SCHREINER, *Institute for Organic Chemistry, Justus Liebig University of Giessen, Giessen, Germany*; MELANIE SCHNELL, *MPSD, Max Planck Institute for the Structure and Dynamics of Matter, Hamburg, Germany*.

Accurate structure determination of large molecules still represents an ambitious challenge. Interesting benchmark systems for structure determination are large diamondoid dimers, whose structures are governed by strong intramolecular interactions. Recently, diamondoid dimers with unusually long central C–C bonds (up to 1.71 Å) were synthesized. This long central C-C bond was rationalized by numerous CH$\cdots$HC-type dispersion attractions between the two halves of the molecule. The thermodynamic stabilization of molecules equipped with bulky groups has provided a conceptually new rationale, since until then it had been assumed that such molecules are highly unstable. We performed a broadband CP-FTMW spectroscopy study in the 2-8 GHz frequency range on oxygen-substituted diamondoid dimers ($C_{26}H_{34}O_2$, 28 heavy atoms) as well as diadamantyl ether to provide further insight into their structures. The experimental data are compared with results from quantum-chemical calculations and gas-phase electron diffraction. For the ether, we even obtained ^{13}C and ^{18}O isotopologues to generate the full heavy-atom substitution structure.

TE10 11:20 – 11:35

VIBRATIONAL STARK EFFECT TO PROBE THE ELECTRIC-DOUBLE LAYER OF THE IONIC LIQUID-METAL ELECTRODES

NATALIA GARCIA REY, ALEXANDER KNIGHT MOORE, SHUICHI TOYOUCHI, DANA DLOTT, *Department of Chemistry, University of Illinois at Urbana-Champaign, Urbana, IL, USA*.

Vibrational sum frequency generation (VSFG) spectroscopy is used to study the effect of room temperature ionic liquids (RTILs) *in situ* at the electrical double layer (EDL). RTILs have been recognized as electrolytes without solvent for applications in batteries, supercapacitors and electrodeposition[1]. The molecular response of the RTIL in the EDL affects the performance of these devices. We use the vibrational Stark effect on CO as a probe to detect the changes in the electric field affected by the RTIL across the EDL on metal electrodes. The Stark effect is a shift in the frequency in response to an externally applied electric field and also influenced by the surrounding electrolyte and electrode[2]. The CO Stark shift is monitored by the CO-VSFG spectra on Pt or Ag in a range of different imidazolium-based RTILs electrolytes, where their composition is tuned by exchanging the anion, the cation or the imidazolium functional group. We study the free induction decay (FID)[3] of the CO to monitor how the RTIL structure and composition affect the vibrational relaxation of the CO. Combining the CO vibrational Stark effect and the FID allow us to understand how the RTIL electrochemical response, molecular orientation response and collective relaxation affect the potential drop of the electric field across the EDL, and, in turn, how determines the electrical capacitance or reactivity of the electrolyte/electrode interface.

[1] Fedorov, M. V.; Kornyshev, A. A., Ionic Liquids at Electrified Interfaces. Chem. Rev. 2014, 114, 2978-3036.

[2] (a) Lambert, D. K., Vibrational Stark Effect of Adsorbates at Electrochemical Interfaces. Electrochim. Acta 1996, 41, 623-630. (b) Oklejas, V.; Sjostrom, C.; Harris, J. M., SERS Detection of the Vibrational Stark Effect from Nitrile-Terminated SAMs to Probe Electric Fields in the Diffuse Double-Layer. J. Am. Chem. Soc. 2002, 124, 2408-2409.

[3] Symonds, J. P. R.; Arnolds, H.; Zhang, V. L.; Fukutani, K.; King, D. A.,Broadband Femtosecond Sum-Frequency Spectroscopy of CO on Ru{1010} in the Frequency and Time Domains. J. Chem. Phys. 2004, 120, 7158-7164.

TE11 11:37 – 11:52

IN-SITU GENERATED GRAPHENE AS THE CATALYTIC SITE FOR VISIBLE-LIGHT MEDIATED ETHYLENE EPOXIDATION ON AG NANOCATALYSTS

XUEQIANG ALEX ZHANG, *Chemistry, University of Illinois at Urbana-Champaign, URBANA-CHAMPAIGN, IL, USA*; PRASHANT JAIN, *Department of Chemistry, University of Illinois at Urbana-Champaign, Urbana, IL, USA.*

Despite the harsh conditions for chemical conversion, ethylene oxide produced from ethylene epoxidation on Ag-based heterogeneous catalyst constitutes one of the largest volume chemicals in chemical industry. Recently, photocatalytic epoxidation of ethylene over plasmonic Ag nanoparticles enables the chemical conversion under significantly decreased temperature and ambient pressure conditions. Yet a detailed understanding of the photocatalytic process at the reactant/catalyst interface is under debate. Surface enhanced Raman spectroscopy (SERS) is a powerful vibrational spectroscopy technique that enables the localized detection of rare and/or transient chemical species with high sensitivity under in situ and ambient conditions. Using SERS, we are able to monitor at individual sites of an Ag nanocatalyst the visible-light-mediated adsorption and epoxidation of ethylene. From detected intermediates, we find that the primary step in the photoepoxidation is the transient formation of graphene catalyzed by the Ag surface. Density functional theory (DFT) simulations that model the observed SERS spectra suggest that the defective edge sites of the graphene formed on Ag constitute the active site for C_2H_4 adsorption and epoxidation. Further studies with pre-formed graphene/Ag catalyst composites confirm the indispensable role of graphene in visible-light-mediated ethylene epoxidation. Carbon is often thought to be either an innocent support or a poison for metallic catalysts; however our studies reveal a surprising role for crystalline carbon layers as potential co-catalysts.

TF. Mini-symposium: ALMA's Molecular View
Tuesday, June 20, 2017 – 1:45 PM
Room: 274 Medical Sciences Building

Chair: Susanne Aalto, Chalmers University of Technology, Onsala, Sweden

---

TF01   ***INVITED TALK***   1:45 – 2:15

PHYSICOCHEMICAL PROCESSES ON ICE DUST TOWARDS DEUTERIUM ENRICHMENT

NAOKI WATANABE, *Institute of Low Temperature Science, Hokkaido University, Sapporo, JAPAN.*

Water and some organic molecules were found to be deuterium enriched toward various astronomical targets. Understanding the deuterium-fractionation process pertains directly to know how and when molecules are created. Although gas phase chemistry is certainly important for deuterium enrichment, the role of physicochemical processes on the dust surfaces should be also considered. In fact, the extreme deuterium enrichment of formaldehyde and methanol requires the dust grain-surface process. In this context, we have performed a series of experiments on the formation of deuterated species of water and simple organic molecules. From the results of these experiments and related works, I will discuss the key processes for the deuterium enrichment on dust. For deuterium chemistry, another important issue is the ortho-to-para ratio (OPR) of H_2, which is closely related to the formation of H_2D^+ and thus the deuterium fractionation of molecules in the gas phase. Because the radiative nuclear spin conversion of H_2 is forbidden, the ortho-para conversion is very slow in the gas phase. In contrast, it was not obvious how the nuclear spins behave on cosmic dust. Therefore, it is desirable to understand how the OPR of H_2 is determined on the dust surfaces. We have tackled this issue experimentally. Using experimental techniques of molecular beam, photostimulated-desorption, and resonance-enhanced multiphoton ionization, we measured the OPRs of H_2 photodesorbed from amorphous solid water at around 10 K, which is an ice dust analogue. It was first demonstrated that the rate of spin conversion from ortho to para drastically increases from 2.4×10^{-4} to 1.7×10^{-3} s^{-1} within the very narrow temperature window of 9.2 to16 K. The observed strong temperature cannot be explained by solely state-mixing models ever proposed but by the energy dissipation model via two phonon process. I will present our recent experiments regarding this.

TF02   2:19 – 2:34

CO IN PROTOSTARS (COPS): *HERSCHEL*-SPIRE SPECTROSCOPY OF EMBEDDED PROTOSTARS

YAO-LUN YANG, *Department of Astronomy, The University of Texas at Austin, Austin, TX, USA*; JOEL D. GREEN, *Office of Public Outreach, Space Telescope Science Institute, Baltimore, MD, USA*; NEAL J EVANS II, *Department of Astronomy, The University of Texas at Austin, Austin, TX, USA.*

Protostars form from cold dense cores dominated by molecular gas and dust, showing excess continuum and rich spectra beyond 100 μm that are best observed by *Herschel Space Observatory*. Molecular emission reveals the properties of the surrounding gas and the underlying physical processes that govern the early stage of star formation. The CO in Protostars (COPS) *Herschel* program observes 27 embedded protostars with SPIRE, including several dominant molecular species, such as CO, ^{13}CO, H_2O, and HCO^+. The COPS dataset covers a unique wavelength range, allowing us to investigate the early stage of star formation across a large sample of sources. We detect CO rotational lines from $J_{up} = 4$ to 36, ^{13}CO lines from $J_{up} = 5$ to 10, and six H_2O lines, along with [N II] and [C I]. We have created an uniformly calibrated dataset with the data from Dust, Ice, and Gas In Time (DIGIT) *Herschel* Key Program and archival photometry, in which we characterize each source by its spectral energy distribution and evolutionary class. With an automatic line fitting pipeline, we detect 323 lines from 25 sources from which we successfully extracted 1D spectra, and 3068 lines from 27 sources observed in all spatial pixels of SPIRE. We analyze the correlations of the line strengths of every line pair from all lines detected with two methods from ASURV package, Spearman's ρ, which test whether the line strengths relation can be described by a monotonic function, and the Kendall z-value, which quantifies the similarity of the ordering of the line strengths of two lines. The distribution of correlations shows a systematic tendency coinciding with the wavelength coverages of the instruments, suggesting that the correlations should only be compared within the lines observed by each module. Within each module, the correlations of two CO line pairs show high correlations, which decrease as the difference of the upper J-level of the two CO lines increases. The smooth gradients of the distribution of correlations hint that the temperature and density of CO gas are continuously varying throughout the embedding envelope. If all CO gas in the envelope shares a same temperature or density, the correlations would be strong for two CO lines originating from two very different J-levels. We find no obvious clustering in the distribution of correlations, while a group of CO lines could have shown particularly strong correlations if their properties were dominated by a same physical process.

TF03

THE CO TRANSITION FROM DIFFUSE MOLECULAR GAS TO DENSE CLOUDS

JOHNATHAN S RICE, STEVEN FEDERMAN, *Physics and Astronomy, University of Toledo, Toledo, OH, USA.*

The atomic to molecular transitions occurring in diffuse interstellar gas surrounding molecular clouds are affected by the local physical conditions (density and temperature) and the radiation field penetrating the material. Our optical observations of CH, CH^+, and CN absorption from McDonald Observatory and the European Southern Observatory are useful tracers of this gas and provide the velocity structure needed for analyzing lower resolution ultraviolet observations of CO and H_2 absorption from Far Ultraviolet Spectroscopic Explorer. We explore the changing environment between diffuse and dense gas by using the column densities and excitation temperatures from CO and H_2 to determine the gas density. The resulting gas densities from this method are compared to densities inferred from other methods such as C_2 and CN chemistry. The densities allow us to interpret the trends from the combined set of tracers. Groupings of sight lines, such as those toward h and χ Persei or Chameleon provide a chance for further characterization of the environment. The Chameleon region in particular helps illuminate CO-dark gas, which is not associated with emission from H I at 21 cm or from CO at 2.6 mm. Expanding this analysis to include emission data from the GOT C+ survey allows the further characterization of neutral diffuse gas, including CO-dark gas.

TF04

HIGH RESOLUTION ROTATIONAL SPECTROSCOPY OF HCSSH: A CS_2 PROXY IN THE ISM

DOMENICO PRUDENZANO, JACOB LAAS, *The Center for Astrochemical Studies, Max-Planck-Institut für extraterrestrische Physik, Garching, Germany*; MARIA ELISABETTA PALUMBO, *Catania Astrophysical Observatory, INAF - Osservatorio Astrofisico di Catania, Catania, Italy*; PAOLA CASELLI, *The Center for Astrochemical Studies, Max-Planck-Institut für extraterrestrische Physik, Garching, Germany.*

In the last few decades sulfur bearing molecules have become a relevant topic in astrochemistry. The observed overall abundances of these compounds in the dense gas and around young stellar objects is indeed not in agreement with the estimated cosmic abundance of sulfur (Tieftrunk et al. 1994; Palumbo et al. 1997). Many studies point to polysulphanes and sulphur polymers, mainly S_8, as possible sulfur reservoirs, which from solid phase might be released into gas phase as simpler sulfur compounds, e.g. in shocked or hot environments (Wakelam et al. 2004; Laas, in prep.). Laboratory studies on dust and ice analogues indicate CS_2 as a potential decomposition product of the sulfur residue (Jiménez-Escobar et al. 2014 and references therein). Nevertheless, this species is not detectable by radio-telescopes due to lack of permanent dipole moment. Dithioformic acid (HCSSH), a possible byproduct of interstellar CS_2, may thus serve as a proxy for this non-polar S-bearing molecule. Millimeter and sub-millimeter spectra have been recorded and analyzed for the trans and cis conformers of HCSSH, up to 478 GHz. We employed the frequency modulation sub-millimeter absorption spectrometer recently developed at the Center for Astrochemical Studies (CAS) in Garching. HCSSH was produced by a glow discharge mixture of CS_2 and H_2 diluted in Ar. Accurate rest frequencies, which might serve as guidance for astronomical searches have been obtained thanks to our recent experiment. In particular trans-HCSSH, the lowest-energy conformer, is the best candidate for a potential detection.

TF05

THE ASTROPHYSICAL WEEDS: ROTATIONAL TRANSITIONS IN EXCITED VIBRATIONAL STATES

JOSÉ L. ALONSO, LUCIE KOLESNIKOVÁ, ELENA R. ALONSO, SANTIAGO MATA, *Grupo de Espectroscopia Molecular, Lab. de Espectroscopia y Bioespectroscopia, Unidad Asociada CSIC, Universidad de Valladolid, Valladolid, Spain.*

The number of unidentified lines in the millimeter and submillimeter wave surveys of the interstellar medium has grown rapidly. The major contributions are due to rotational transitions in excited vibrational states of a relatively few molecules that are called the astrophysical weeds [a],[b]. The necessary data to deal with spectral lines from astrophysical weeds species can be obtained from detailed laboratory rotational measurements in the microwave and millimeter wave region. A general procedure is being used at Valladolid combining different time and/or frequency domain spectroscopic tools of varying importance for providing the precise set of spectroscopic constants that could be used to search for this species in the ISM. This is illustrated in the present contribution through its application to several significant examples.

---

[a]Fortman, S. M., Medvedev, I. R., Neese, C.F., & De Lucia, F.C. 2010, ApJ,725, 1682

[b]Rotational Spectra in 29 Vibrationally Excited States of Interstellar Aminoacetonitrile, L. Kolesniková, E. R. Alonso, S. Mata, and J. L. Alonso, The Astrophysical Journal Supplement Series 2017, (in press).

Intermission

TF06   **3:44 – 3:59**

TIME-SENSITIVE CHEMICAL TRACERS WITHIN SHOCKED ASTROPHYSICAL SOURCES

ANDREW M BURKHARDT, *Department of Astronomy, The University of Virginia, Charlottesville, VA, USA*; CHRISTOPHER N SHINGLEDECKER, ROMANE LE GAL, *Department of Chemistry, The University of Virginia, Charlottesville, VA, USA*; BRETT A. McGUIRE, *NAASC, National Radio Astronomy Observatory, Charlottesville, VA, USA*; ANTHONY REMIJAN, *ALMA, National Radio Astronomy Observatory, Charlottesville, VA, USA*; ERIC HERBST, *Department of Chemistry, The University of Virginia, Charlottesville, VA, USA.*

In regions of star formation, astrophysical shocks have been found to be both common and influential on the molecular make-up of the surrounding material. The formation of complex organic molecules (COMs) in the interstellar medium relies on interplay between gas-phase and grain-surface physical and chemical processes, including shock-induced non-thermal desorption of COMs formed on the grain. We will utilize the gas-grain chemical network model NAUTILUS, coupled with the inclusion of a high-temperature network and a parametric shock model, with the inclusion of collisional dust heating and sputtering processes from gas-phase particles, in order to study the effects shocks have on the chemical complexity of a prototypical shocked outflow over a range of chemically-relevant timescales. Here, we will present the effectiveness of using chemical tracers to determine the presence of a shock long after it has passed, as well as the prominence of post-shock gas-phase formation and destruction of shock-tracing molecules.

TF07   **4:01 – 4:16**

SIO OUTFLOWS AS TRACERS OF MASSIVE STAR FORMATION IN INFRARED DARK CLOUDS

MENGYAO LIU, *Physical Chemistry, University of Florida, Gainesville, FL, USA.*

We present ALMA Cycle 2 observations of SiO(5-4) outflows towards 30 Infrared Dark Cloud (IRDC) clumps, which are spatially resolved down to $\sim$0.05pc. Out of the 30 clumps observed, we have detected SiO emission in 20 clumps. We discuss the association of SiO with mm continuum and FIR emission, and fit the SEDs of potential protostellar sources with radiative transfer models based on the Turbulent Core Model. In 6 of the 20 clumps the SiO emission is stronger than the 10 sigma noise level and appears to trace outflows being driven by protostellar sources that are also revealed as nearby mm continuum peaks. We locate the dense protostellar cores associated with the outflows in position-velocity space utilizing dense gas tracers DCN(3-2), DCO$^+$(3-2) and C^{18}O(2-1). The different morphology and kinematics of the outflows indicate different core structures, accretion histories and ambient cloud environments. The mass and energetics of the outflows indicate that these 6 protostars are in a relatively early evolutionary stage and some may eventually become massive stars.

TF08 4:18–4:33

QUANTUM-CHEMICAL CALCULATIONS REVEALING THE EFFECTS OF MAGNETIC FIELDS ON METHANOL

BOY LANKHAAR, *Onsala Space Observatory, Chalmers University of Technology, Onsala, Sweden*; AD VAN DER AVOIRD, *Institute for Molecules and Materials (IMM), Radboud University Nijmegen, Nijmegen, Netherlands*; WOUTER H.T. VLEMMINGS, *Onsala Space Observatory, Chalmers University of Technology, Onsala, Sweden*; GERRIT GROENENBOOM, *Institute for Molecules and Materials (IMM), Radboud University Nijmegen, Nijmegen, Netherlands*; HUIB JAN VAN LANGEVELDE, , *Joint Institute for VLBI in Europe, Dwingeloo, Netherlands*; GABRIELE SURCIS, *Istituto Nazionale di Astrofisica, Osservatorio Astronomico di Cagliari, Selargius, Italy.*

Maser observations of both linear and circular emission have provided unique information on the magnetic field in the densest regions of star forming regions. While linear polarization observations provide morphological constraints, the magnetic field strength determination is done by comparing the Zeeman-induced velocity shifts between left- and right-circularly polarized emission of molecular maser species. Soon, full-polarization observations with be possible with ALMA, making magnetic field measurements with unprecedented spatial resolution possible. In particular, methanol is of special interest as it is the most abundant maser species and its different transitions probe unique areas of high-mass proto-stellar disks and outflows. However, its exact Zeeman-parameters are unknown. Experimental efforts to determine the Zeeman-parameters have failed. Here we present quantum chemical calculations to the Zeeman-parameters of methanol, along with calculations to the hyperfine structure, which are also necessary to interpret the Zeeman effect in methanol. We present the proper treatment of the torsional motion in computing hyperfine and Zeeman effects. Our results on the hyperfine structure show good agreement with recent experimental data. We find that the Zeeman-effect in methanol is non-linear and comment on its applicability in astronomical magnetic field studies. We give an outlook on rigorously treating non-linear Zeeman effects in radiative transfer modeling of maser-species interacting with a magnetic field.

TF09 4:35–4:50

ZEEMAN EFFECT IN SULFUR MONOXIDE: A PROBE TO OBSERVE MAGNETIC FIELDS IN STAR FORMING REGIONS?

GABRIELE CAZZOLI, *Dep. Chemistry 'Giacomo Ciamician', University of Bologna, Bologna, Italy*; VALERIO LATTANZI, *The Center for Astrochemical Studies, Max-Planck-Institut für extraterrestrische Physik, Garching, Germany*; SONIA CORIANI, *Department of Chemistry, Technical University of Denmark, Kgs. Lyngby, Denmark*; JÜRGEN GAUSS, *Institut für Physikalische Chemie, Universität Mainz, Mainz, Germany*; CLAUDIO CODELLA, *Arcetri Observatory, INAF, Florence, Italy*; ANDRÉS ASENSIO RAMOS, *Instituto de Astrofísica de Canarias, Instituto de Astrofísica de Canarias, La Laguna, Spain*; JOSE CERNICHARO, *Molecular Astrophysics, ICMM, Madrid, Spain*; CRISTINA PUZZARINI, *Dep. Chemistry 'Giacomo Ciamician', University of Bologna, Bologna, Italy.*

Magnetic fields play a fundamental role in star formation processes and the best method to evaluate their intensity is is to measure the Zeeman effect of atomic and molecular lines. However, a direct measurement of the Zeeman spectral pattern from interstellar molecular species is challenging due to the high sensitivity and high spectral resolution required. So far, the Zeeman effect has been detected unambiguously in star forming regions for very few non-masing species, such as OH and CN. We decided to investigate the ability of sulfur monoxide (SO), which is one of the most abundant species in star forming regions, for probing the intensity of magnetic fields via Zeeman effect. The Zeeman effect for several rotational transitions of SO in the (sub-)mm spectral regions has been investigated by using a frequency-modulated, computer-controlled spectrometer, and by applying a magnetic field parallel to the radiation source. To support the experimental determination of the g factors of SO, a systematic quantum-chemical investigation of these parameters for both SO and O_2 has been carried out. An effective experimental-computational strategy for providing accurate g factors as well as for identifying the rotational transitions showing the strongest Zeeman effect has been presented. Our investigation supports SO as a good candidate for probing magnetic fields in high-density star forming regions.

TF10 4:52 – 5:07

MAPPING MAGNETIC FIELDS IN MOLECULAR CLOUDS WITH THE CN ZEEMAN EFFECT

RICHARD CRUTCHER, *Department of Astronomy, University of Illinois at Urbana-Champaign, Urbana, IL, USA.*

The role of magnetic fields in star formation remains controversial. Observations of the Zeeman effect provide the only available technique for directly measuring the strengths of magnetic fields in molecular clouds. We have mapped the Zeeman effect toward the massive star forming complex W3OH in the CN N=2-1 transition at 226 GHz with both the IRAM 30-m telescope and the CARMA array and have combined these data to produce a fully spatially sampled map of the magnetic field along the line of sight, with approximately 4 arcsec resolution. These are both the first CN Zeeman maps and the first detections of the Zeeman effect in the CN N=2-1 transition. We will present this map and discuss the astrophysical implications. This work may be considered to be a pathfinder for future similar ALMA observations, which have the potential to advance considerably our understanding of the role of magnetic fields in the star formation process.

TF11 5:09 – 5:24

LAYING THE GROUNDWORK FOR FUTURE ALMA DIRECT MAGNETIC FIELD DETECTION IN PROTOSTELLAR ENVIRONMENTS

ERIN GUILFOIL COX, ROBERT J HARRIS, LESLIE LOONEY, DOMINIQUE M. SEGURA-COX, RICHARD CRUTCHER, *Department of Astronomy, University of Illinois at Urbana-Champaign, Urbana, IL, USA*; ZHI-YUN LI, *Department of Astronomy, The University of Virginia, Charlottesville, VA, USA*; JOHN TOBIN, *Homer L Dodge Department of Physics and Astronomy, University of Oklahoma, Norman, OK, USA*; IAN STEPHENS, *Radio and Geoastronomy Division, Harvard-Smithsonian Center for Astrophysics, Cambridge, MA, USA*; GILES NOVAK, *Physics and Astronomy, Northwestern University, Evanston, IL, USA*; MANUEL FERNANDEZ-LOPEZ, *Instituto Argentino de Radioastronomía, Centro Científico Tecnológico La Plata, Villa Elisa, Argentina.*

Magnetic fields are a crucial element of the star formation process on many scales, from controlling jet and outflow formation on large scales, determining the structure of any protostellar disk, to modulating the accretion rate onto the central protostar. Both the three-dimensional structure and the field strength are important in determining the outcome of star formation. Unfortunately, the method most commonly used to infer magnetic field structure – linearly polarized dust continuum emission – is limited to the plane-of-sky field structure, and gives no reliable information on field strength. Alternatively, observations of the Zeeman effect in transitions of paramagnetic molecules, especially CN, are one of the best prospects for making such measurements due to the molecules' high Zeeman coefficients. In particular, these observations have been used in determining field strengths on cloud-size scales. However, CN and other paramagnetic molecules have, to our knowledge, never been observed in the envelopes/disks of Class 0 protostars at $\sim$arcsecond resolution, due both to sensitivity and resolution limits of previous generations of millimeter-wave interferometers. Because field strengths near the protostar are so important to understand the star formation process, we have conducted a snapshot ALMA Band 3 (3 mm / 113 GHz) survey of the 10 brightest Class 0 protostars in the Perseus, Taurus, and ρ Ophiuchus molecular clouds in the regions surrounding five transitions of four paramagnetic species, including CN, SO, C_2S, and C_4H. We present this survey – the principle goal of which was to assess the brightness of the lines within $\sim$ 1000 AU of the protostar – and assess the likelihood of using ALMA observations of the Zeeman effect to determine protostellar magnetic field strength.

TG. Mini-symposium: Multiple Potential Energy Surfaces

Tuesday, June 20, 2017 – 1:45 PM

Room: 116 Roger Adams Lab

Chair: John Parkhill, The University of Notre Dame, Notre Dame, IN, US

TG01 1:45 – 2:00

ROAMING ISOMERIZATION OF PHOTOEXCITED HALOGENATED ALKANES IN THE GAS AND LIQUID PHASES
^a

VENIAMIN A. BORIN, SERGEY M. MATVEEV, DARYA S. BUDKINA, CHRISTOPHER M. HICKS, ANDREY S. MERESHCHENKO, EVGENIIA V. BUTAEVA, VASILY V. VOROBYEV, ALEXANDER N TARNOVSKY, *Department of Chemistry and Center for Photochemical Sciences, Bowling Green State University, Bowling Green, OH, USA.*

Recent experimental and computational gas-phase studies have brought to light a new type of unimolecular decomposition called a "roaming mechanism. It has only been observed in the gas phase, and whether it also occurs in solution is an intriguing question. Using ultrafast transient absorption spectroscopy, we report direct isomerization of $CHBr_3$, BBr_3, and PBr_3 geminal tribromides in solution within the first 100 fs after S_1-excitation. The gas-phase conditions do not affect the earliest course of similar isomerization of $CHBr_3$. High-level ab initio simulations on XBr_3 (X = B, P, and CH) suggest that isomerization is governed by an energetically and dynamically accessible S_1/S_0 conical intersection and can be best described as a roaming-mediated pathway. Following the initial relaxation from the Franck-Condon point, "wandering" of the central atoms and migration of Br atom starts on a planar region of the S_1 surface, and in the vicinity of the conical intersection (40 fs) the XBr_2 and Br fragments become separated to ≥ 3 Å. After passage through the conical intersection, the partially dissociated bromine atom slips off the XBr_2 bisector plane, and forms the Br–Br bond of the BrXBr–Br isomer (60 fs). We give examples of similar roaming isomerization in several other di- and polyhalogenated alkanes.

^aThe authors gratefully acknowledges grants from the National Science Foundation (CHE-0847707, 0923360, and 1626420) which supported this work.

TG02 2:02 – 2:17

NONRADIATIVE DECAY ROUTE OF CINNAMATE DERIVATIVES STUDIED BY FREQUENCY AND TIME DOMAIN LASER SPECTROSCOPY IN THE GAS PHASE, MATRIX ISOLATION FTIR SPECTROSCOPY AND QUANTUM CHEMICAL CALCULATIONS

TAKAYUKI EBATA, *Department of Chemistry, Graduate School of Science, Hiroshima University, Higashihiroshima City, Japan.*

The nonraddiative dececy route involving trans → cis photo-isomerization from the S_1 ($\pi\pi^*$) state has been investigated for several trans-cinnamate derivatives, which are known as sunscreen reagents. We examined two types of substitution effects.

One is structural isomer such as ortho-, meta-, and para-hydroxy-methylcinnmate (o-, m-, p-HMC). The S_1 lifetime of p-HMC is less than 8 ps at zero-point level, and it undergoes rapid $S_1 \rightarrow {}^1n\pi^* \rightarrow T_1$ decay via multiple conical intersections. Finally, the trans → cis isomerization proceeds in the T_1 state. On the other hand, both o- and m-HMC show very slow decay. Their S_1 lifetimes are in the order of 100 ps even at the excess energy of 2000-3000 cm^{-1}.

The other is the effect of the complexity of ester group in para-subsitituted species, such as para-methoxy-methyl, -ethyl and -2ethylhexyl cinnamate (p-MMC, p-MEC, p-M2EHC). p-MMC and p-MEC show sharp $S_0 \rightarrow S_1$ ($\pi\pi^*$) vibronic bands, while p-M2EHC shows only broad structureless feature even under the jet-cooled condition. In addition, we found that the $S_0 \rightarrow {}^1n\pi^*$ absorption appears at 1000 cm^{-1} below the $S_0 \rightarrow S_1$ ($\pi\pi^*$) transition in p-MEC and p-M2EHC, but not in p-MMC. Thus, the complexity of the ester group is very important for the appearance of the $^1n\pi^*$ state.

TG03 2:19 – 2:34

PHOTOISOMERIZATION DYNAMICS OF AZOBENZENE DERIVATIVES IN SOLUTION USING BROADBAND ULTRAFAST SPECTROSCOPY

ABDELQADER JAMHAWI, *Department of Chemistry, University of Louisville, Louisville, KY, USA*; ISHAN FURSULE, QUNFEI ZHOU, BRAD J. BARRON, MATTHEW J. BECK, *Department of Chemical and Materials Engineering, University of Kentucky, Lexington, KY, USA*; JINJUN LIU, *Department of Chemistry, University of Louisville, Louisville, KY, USA*.

Photochromic molecular switches such as azobenzene, stilbene, and their derivatives have a wide range of potential applications, among which are drug delivery, sensory, data storage, and optoelectronics. Compared to stilbene, azobenzene can isomerize under more constrained conditions like condensed matrices or within polymer chains because of different possible reaction channels. Thanks to such unique properties, azobenzene derivatives can be used to build photoactive monomer units for ring-opening metathesis polymerization (ROMP). Unraveling the details of the isomerization mechanisms of this family of molecular switches is therefore of vital importance, especially for designing task-specific ROMPs. In this work, we employ femtosecond pump-broadband-probe transient absorption (TA) spectroscopy to investigate the ultrafast isomerization dynamics of 4,4'-diaminoazobenzene (ABn), and 4,4'-dibutenylamidoazobenzene (AB22) in solution. Following an nπ* excitation, ground-state bleaching signal of *cis-* and *trans-* isomers reveals a product formation rates slower than that of the parent compound. Furthermore, excited-state absorption features observed within the detected spectral range indicate vibrational energy relaxation processes on the sub-picosecond to a few picoseconds time scale prior to population decay to ground electronic states of the two conformers via a conical intersection. Interpretation of the observed photo-induced dynamics on the basis of quantum chemical calculations will also be presented.

TG04 2:36 – 2:51

TIME-RESOLVED SIGNATURES ACROSS THE INTRAMOLECULAR RESPONSE IN SUBSTITUTED CYANINE DYES

MUATH NAIRAT, MORGAN WEBB, MICHAEL ESCH, VADIM V. LOZOVOY, BENJAMIN G LEVINE, MARCOS DANTUS, *Department of Chemistry, Michigan State University, East Lansing, MI, USA*.

The optically populated excited state wave packet propagates along multidimensional intramolecular coordinates soon after photoexcitation. This action occurs alongside an intermolecular response from the surrounding solvent. Disentangling the multidimensional convoluted signal enables the possibility to separate and understand the initial intramolecular relaxation pathways over the excited state potential energy surface. Here we track the initial excited state dynamics by measuring the fluorescence yield from the first excited state as a function of time delay between two color femtosecond pulses for several cyanine dyes, having different electronic configurations. We find that when the high frequency pulse precedes the low frequency one and for timescales up to 200 fs, the excited state can be depleted through stimulated emission with efficiency that is dependent on the molecular electronic structure. A similar observation at even shorter times was made by scanning the chirp (frequencies ordering) of a femtosecond pulse. These changes reflect the rate at which the nuclear coordinates of the excited state leave the Franck-Condon (FC) region and progress towards achieving equilibrium. Through functional group substitution, we explore these dynamic changes as a function of dipolar change following photoexcitation. We show that with proper knowledge and control over the phase of the excitation pulses, we can extract the relative energy relaxation rates through which the early intramolecular modes are populated at the FC geometry soon after excitation

TG05   2:53 – 3:08

CAN INTERNAL CONVERSION BE CONTROLLED BY MODE-SPECIFIC VIBRATIONAL EXCITATION IN POLY-ATOMIC MOLECULES

MICHAEL EPSHTEIN, ALEXANDER PORTNOV, <u>ILANA BAR</u>, *Physics, Ben-Gurion University of the Negev, Beer-Sheva, Israel.*

Nonadiabatic processes, dominated by dynamic passage of reactive fluxes through conical intersections (CIs) are considered to be appealing means for manipulating reaction paths. One approach that is considered to be effective in controlling the course of dissociation processes is the selective excitation of vibrational modes containing a considerable component of motion. Here, we have chosen to study the predissociation of the model test molecule, methylamine and its deuterated isotopologues, excited to well-characterized quantum states on the first excited electronic state, S_1, by following the N-H(D) bond fission dynamics through sensitive H(D) photofragment probing. The branching ratios between slow and fast H(D) photofragments, the internal energies of their counter radical photofragments and the anisotropy parameters for fast H photofragments, confirm correlated anomalies for predissociation initiated from specific rovibronic states, reflecting the existence of a dynamic resonance in each molecule. This resonance strongly depends on the energy of the initially excited rovibronic states, the evolving vibrational mode on the repulsive S_1 part during N-H(D) bond elongation, and the manipulated passage through the CI that leads to radicals excited with C-N-H(D) bending and preferential perpendicular bond breaking, relative to the photolyzing laser polarization, in molecules containing the NH_2 group. The indicated resonance plays an important role in the bifurcation dynamics at the CI and can be foreseen to exist in other photoinitiated processes and to control their outcome.

Intermission

TG06   3:27 – 3:42

THE MOLECULAR GEOMETRIC PHASE AND LIGHT-INDUCED CONICAL INTERSECTIONS

<u>EMIL J ZAK</u>, *Department of Physics and Astronomy, University College London, Gower Street, London WC1E 6BT, United Kingdom.*

Potential energy surfaces for electronic states of molecules in strong electromagnetic fields can be described in the dressed-state formalism, which introduces light-induced potentials. A light-induced conical intersection (LICI) [1] appears when two electronic states intersect due to the presence of an external electric field and when the dipole coupling between the field and the molecule vanishes. There are several aspects of quantum dynamics near LICIs, which still require a thorough investigation. How do non-adiabatic effects manifest themselves in polyatomic molecules in strong electromagnetic fields? Are the natural conical-intersections (NCI) and the light-induced conical intersections identical in nature? Do topological effects (Berry phase) [2] influence the nuclear dynamics around NCIs and LICIs? To answer these questions, a computer code for time-propagation of the ro-vibronic wavefunction on multiple coupled potential energy surfaces has been developed. The time-independent zero-order basis is taken from the DUO suite [3], which solves the full ro-vibronic Schrödinger equation for diatomic molecules. Non-adiabatic nuclear dynamics near LICIs will be presented on the examples of NaH and CaF molecules, with a perspective for extension to polyatomics.

[1] G. J. Halász, A Vibók, M. Šindelka, N. Moiseyev, L. S. Cederbaum, 2011 J. Phys. B: At. Mol. Opt. Phys. 44 175102

[2] C. Wittig, Phys. Chem. Chem. Phys., 2012, 14, 6409–6432

[3] S. N. Yurchenko, L. Lodi, J. Tennyson, A. V. Stolyarov, Comput. Phys. Commun., 202, 262, 2016

TG07   **3:44 – 3:59**

SEMI-CLASSICAL DYNAMICS STUDIES OF THE PHOTODISSOCIATION OF ICN⁻ and BrCN⁻

<u>BERNICE OPOKU-AGYEMAN</u>, *Department of Chemistry and Biochemistry, The Ohio State University, Columbus, OH, USA*; ANNE B McCOY, *Department of Chemistry, University of Washington, Seattle, WA, USA.*

We present the results of surface hopping studies of the photodissociation of ICN⁻ and BrCN⁻ following UV and visible excitations to states that can dissociate to $X^- + CN$ or $X^* + CN^-$ (X = I or Br). Based on previous quantum dynamics studies carried out on the anions, the electronic states that are accessed by the initial UV or visible excitation were found to be coupled, and both photoproducts were observed in the dissociation process. The calculated branching ratios indicated stronger non-adiabatic interactions between the adiabatic electronic states in BrCN⁻ than in ICN⁻.[abc] In this work, we employ Tully's surface hopping algorithm[d] implemented in classical dynamics simulations to investigate the fragmentation processes of these anions. We calculate the branching ratios and the partitioning of the energies among the various degrees of freedom during the dynamics. The results of the surface hopping algorithm are then compared to the reported quantum dynamics results after which the surface hopping approach can then be applied to investigate the dynamics of argon clusters of the ICN⁻ and BrCN⁻. In addition to comparison between the quantum and classical dynamics calculations, preliminary results for the semi-classical calculations on the ICN⁻ show that during the dissociation process, some of the anions live longer than others. Once the anions are solvated, we expect the presence of the argon atoms to stabilize the complexes as reported in previous work, resulting in longer lifetimes. Since experimental studies carried out on the solvated anions show the existence of recombined products even for clusters as small as ICN⁻(Ar) or ICN⁻(Ar)₂, when the clusters are initially excited with visible or UV radiations, respectively,[bc] the observations of these long-lived anions provide a step towards understanding the dynamics processes that lead to the recombined products in the clusters.

[a]B. Opoku-Agyeman, A. S. Case, J. H. Lehman, W. Carl Lineberger and A. B. McCoy, *J. Chem Phys.* **141**, 084305 (2014).

[b]A. S. Case, A. B. McCoy, W. C. Lineberger, *J. Phys. Chem. A*, **117**(50), 13310(2013).

[c]A. S. Case, E. M. Miller, J. P. Martin, Y. J. Lu, L. Sheps, A. B. McCoy, W. C. Lineberger, *Angew. Chem., Int. Ed.* **51**(11), 2651 (2012).

[d]J. C. Tully, *J. Chem Phys.* **93**, 1061 (1990)

TG08   **4:01 – 4:16**

ULTRAFAST MOLECULAR PHOTODISSOCIATION DYNAMICS STUDIED BY FEMTOSECOND PHOTOELECTRON-PHOTOION COINCIDENCE SPECTROSCOPY

<u>BERNHARD THALER</u>, PASCAL HEIM, WOLFGANG E. ERNST, MARKUS KOCH, *Institute of Experimental Physics, Graz University of Technology, Graz, Austria.*

To completely characterize photodissociation mechanisms with time-resolved spectroscopy, it is essential to obtain unequivocal experimental information about the fragmentation dynamics induced by the laser pulse. We apply time-resolved photoelectron-photoion coincidence (PEPICO) detection in combination with different excitation schemes to obtain a mechanistic picture of the fragmentation process. For gas phase acetone molecules excited to high lying Rydberg states we are able to disentangle different ionization channels and investigate the fragmentation behavior of each channel separately. In particular, the high differentiability of PEPICO allows to distinguish channels where fragmentation proceeds after ionization from channels with fragmentation in the neutral.

We show that excited Rydberg state population undergoes internal conversion due to coupling to valence states, which takes place within (150 ± 30) fs. The corresponding non-adiabatic, ultrafast relaxation dynamics to lower lying states causes conversion of electronic to vibrational energy and is found to play a crucial role in the fragmentation process (see figure 1). By studying the influence of photon energy, pulse duration, chirp and intensity

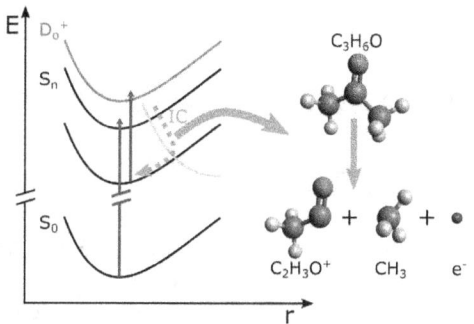

Figure 1: Schematic conversion process of electronic to vibrational energy leading to fragmentation in the ion.

of the laser pulses, we are able to determine the energy-threshold that is required for fragmentation, as well as corresponding fragmentation ratios. Surprisingly, for excitation with pulses possessing a strong negative chirp we observe significantly reduced fragmentation, indicating different internal conversion pathways and the associated intramolecular vibrational redistribution.

TG09 4:18 – 4:33

Cl-LOSS DYNAMICS OF VINYL CHLORIDE CATION IN B STATE: ROLE OF C STATE

X ZHOU, *Hefei National Laboratory for Physical Science at Microscale, University of Science and Technology of China, Hefei, China.*

Dissociative photoionization of vinyl chloride (C2H3Cl) in the 11.0 14.2 eV photon energy range was investigated with the method of threshold photoelectron photoion coincidence (TPEPICO) velocity map imaging. Three electronic states, A2A', B2A" and C2A', of C2H3Cl+ cation were prepared and their dissociation dynamics were investigated respectively. A unique fragment ion, C2H3+, was observed within the present excitation energy range. From the TPEPICO 3-dimensional time-sliced velocity map images of C2H3+, kinetic energy release distributions (KERD) and anisotropy parameters in dissociation of internal energy-selected C2H3Cl+ cation were obtained. At 13.14 eV, the total KERD showed a bimodal distribution consisting of a Boltzmann and a Gaussian-type component, indicating competing statistical and non-statistical dissociation mechanisms. An additional component of Gaussian-type was found in KERD at 13.65 eV with a center located at lower kinetic energy. With the aid of re-calculated Cl-loss potential energy curves with time-dependent density functional theory, the overall dissociative photoionization mechanisms of C2H3Cl+ cation in the B2A" and C2A' states are proposed. The inconsistence of the previous conclusions on dissociation mechanism of C2H3Cl+ is stated.

TG10 4:35 – 4:50

VIBRATIONAL MODE-SPECIFIC AUTODETACHMENT AND COUPLING OF CH2CN-

JUSTIN LYLE, *Chemistry, Washington University in Saint Louis, Saint Louis, MO, USA*; RICHARD MABBS, *Department of Chemistry, Washington University, St. Louis, MO, USA.*

The Cyanomethyl Anion, CH_2CN-, and neutral radical have been studied extensively, with several findings of autodetachment about the totally symmetric transition, as well as high resolution experiments revealing symmetrically forbidden and weak vibrational features. We report photoelectron spectra using the Velocity-Mapped Imaging Technique in 1-2 cm^{-1} increments over a range of 13460 to 15384 cm^{-1} that has not been previously examined. These spectra include excitation of the ground state cyanomethyl anion into the direct detachment thresholds of previously reported vibrational modes for the neutral radical. Significant variations from Franck-Condon behavior were observed in the branching ratios for resolved vibrational features for excitation in the vicinity of the thresholds involving the ν_3 and ν_5 modes. These are consistent with autodetachment from rovibrational levels of a dipole bound state acting as a resonance in the detachment continuum. The autodetachment channels involve single changes in vibrational quantum number, consistent with the vibrational propensity rule but in some cases reveal relaxation to a different vibrational mode indicating coupling between the modes and/or a breakdown of the normal mode approximation.

TG11 4:52 – 5:07

INFLUENCE OF SPIN-ORBIT QUENCHING ON THE SOLVATION OF INDIUM IN HELIUM DROPLETS

RALF MEYER, JOHANN V. POTOTSCHNIG, WOLFGANG E. ERNST, ANDREAS W. HAUSER, *Institute of Experimental Physics, Graz University of Technology, Graz, Austria.*

Recent experimental interest of the collaborating group of M. Koch on the dynamics of electronic excitations of indium in helium droplets triggered a series of computational studies on the group 13 elements Al, Ga and In and their indecisive behavior between wetting and non wetting when placed onto superfluid helium droplets. We employ a combination of multi-configurational self consistent field calculations (MCSCF) and multireference configuration interaction (MRCI) to calculate the diatomic potentials. Particularly interesting is the case of indium with an Ancilotto parameter[a] λ close to the threshold value of 1.9.

As shown by Reho et al.[b] the spin-orbit splitting of metal atoms solvated in helium droplets is subject to a quenching effect. This can drastically change the solvation behavior. In this work we extend the approach presented by Reho et al. to include distance dependent spin-orbit coupling. The resulting potential surfaces are used to calculate the solvation energy of the ground state and the first excited state with orbital-free helium density functional theory.

[a]F. Ancilotto, P. B. Lerner and M. W. Cole, *Journal of Low Temperature Physics*, 1995, **101**, 1123-1146
[b]J. H. Reho, U. Merker, M. R. Radcliff, K. K. Lehmann and G. Scoles, *The Journal of Physical Chemistry A*, 2000, **104**, 3620-3626

$n \rightarrow \pi^*$ NON-COVALENT INTERACTION IS WEAK BUT STRONG IN ACTION

SANTOSH KUMAR SINGH, ALOKE DAS, *Department of Chemistry, Indian Institute of Science Education and Research, Pune, Maharshtra, India.*

$n \rightarrow \pi^*$ interaction is a newly discovered non-covalent interaction which involves delocalization of lone pair (n) electrons of an electronegative atom into π^* orbital of a carbonyl group or an aromatic ring. It is widely observed in materials, biomolecules (protein, DNA, RNA), amino acids, neurotransmitter and drugs. However, due to its weak strength and counterintuitive nature its existence is debatable. Such weak interactions are often masked by solvent effects in condense phase or physiological conditions thereby, making it difficult to prove the presence of such weak interactions. Therefore, we have used isolated gas phase spectroscopy in combination with quantum chemical calculations to study $n \rightarrow \pi^*$ interaction in several molecules where, our molecular systems are free from solvent effects or any external forces. Herein I will be discussing two of the molecular systems (phenyl formate and salicin) where, we have observed the significance of $n \rightarrow \pi^*$ interaction in determining the conformational specificity of the molecules. We have proved the existence of $n \rightarrow \pi^*$ interaction for the first time through IR spectroscopy by probing the carbonyl stretching frequency of phenyl formate. Our study is further pursued on a drug named salicin where, we have observed that its conformational preferences is ruled by $n \rightarrow \pi^*$ interaction even though a strong hydrogen bonding interaction is present in the molecule. Our results show that $n \rightarrow \pi^*$ interaction, in spite of its weak strength, should not be overlooked as it existence can play an important role in governing the structures of molecules like other strong non-covalent interactions do.

TH. Instrument/Technique Demonstration

Tuesday, June 20, 2017 – 1:45 PM

Room: 1024 Chemistry Annex

Chair: Christopher F. Neese, The Ohio State University, Columbus, OH, USA

TH01   1:45 – 2:00

A NEW 2.0-6.0 GHz CHIRPED PULSE FOURIER TRANSFORM MICROWAVE SPECTROMETER: INSTRUMENTAL
ANALYSIS AND INITIAL MOLECULAR RESULTS

NATHAN A SEIFERT, JAVIX THOMAS, WOLFGANG JÄGER, YUNJIE XU, *Department of Chemistry, University of Alberta, Edmonton, AB, Canada.*

Low frequency microwave spectroscopy (< 10 GHz) is ideal for studies of large molecular systems including higher order molecular complexes. The cold rotational temperature of a pulsed jet makes detections in this region highly attractive for these larger molecular systems with small rotational constants. Here, we report on the construction and initial benchmarking results for a new 2.0-6.0 GHz CP-FTMW spectrometer, similar in design to the 2.0-8.0 GHz spectrometer designed in Brooks Pate's group at the University of Virginia[a], that takes advantage of numerous improvements in solid-state microwave devices and high-speed digitizers.

In addition to details and analysis of the new instrumental design, comparisons to the previous generation 7.5-18.0 GHz spectrometer at the University of Alberta will be presented using the microwave spectrum of methyl lactate as a benchmark. Finally, initial results for several novel molecular systems studied using this new spectrometer, including the tetramer of 2-fluoroethanol, will be presented.

[a]C. Perez, S. Lobsiger, N. A. Seifert, D. P. Zaleski, B. Temelso, G. C. Shields, Z. Kisiel, B. H. Pate, Chem. Phys. Lett., **2013**, *571*, 1-15.

TH02   2:02 – 2:17

AN 18-26 GHz SEGMENTED CHIRPED PULSE FOURIER TRANSFORM MICROWAVE SPECTROMETER FOR ASTROCHEMICAL APPLICATIONS

AMANDA STEBER, *The Centre for Ultrafast Imaging (CUI), Universität Hamburg, Hamburg, Germany*; MARIYAM FATIMA, CRISTOBAL PEREZ, MELANIE SCHNELL, *CoCoMol, Max-Planck-Institut für Struktur und Dynamik der Materie, Hamburg, Germany.*

In the past decade, astrochemistry has seen an increase in interest. As higher throughput and increased resolution radio astronomy facilities come online, faster laboratory instrumentation that directly covers the frequency ranges of these facilities is needed. The 18-26 GHz region is of interest astronomically as many cold organic molecules have their peak intensity in this region. We present here a new segmented chirped pulse Fourier transform microwave (CP-FTMW) spectrometer operating between 18-26 GHz. Using state-of-the-art digital electronics and the segmented approach[1], this design has the potential to be faster and cheaper than the previously presented broadband design. Characterization of the instrument using OCS will be presented, along with a comparison to the previously built and optimized 18-26 CP-FTMW built at the University of Virginia. It will be coupled with a discharge nozzle[2], and its applications to astrochemistry will be explored in this talk.

[1] Neill, J.L., Harris, B.J., Steber, A.L., Douglass, K.O., Plusquellic, D.F., Pate, B.H. *Opt. Express*, 21, 19743-19749, **2013**.

[2] McCarthy, M.C., Chen, W., Travers, M.J., Thaddeus, P. *Astrophys. J. Suppl. Ser.*, 129, 611-623, **2000**.

TH03   **2:19 – 2:34**

A HIGHLY-INTEGRATED SUPERSONIC-JET FOURIER TRANSFORM MICROWAVE SPECTROMETER

QIAN GOU, <u>GANG FENG</u>, *School of Chemistry and Chemical Engineering, Chongqing University, Chongqing, China*; JENS-UWE GRABOW, *Institut für Physikalische Chemie und Elektrochemie, Gottfried-Wilhelm-Leibniz-Universität, Hannover, Germany.*

A highly integrated supersonic-jet Fourier-transform microwave spectrometer of coaxially oriented beam-resonator arrangement (COBRA) type, covering 2-20GHz, has been recently built at Chongqing University, China. Built up almost entirely in an NI PXIe chassis, we take the advantage of the NI PXIe-5451 Dual-channel arbitrary waveform generator and the PXIe-5654 RF signal generator to create a spectrometer with wobbling capacity for fast resonator tuning. Based on the I/Q modulation, associate with PXI control and sequence boards built at the Leibniz Universitat Hannover, the design of the spectrometer is much simpler and very compact. The Fabry–Pérot resonator is semi confocal with a spherical reflector of 630 mm diameter and a radius of 900 mm curvature and one circulator plate reflector of 630 mm diameter. The vacuum is effectuated by a three-stage mechanical (two-stage rotary vane and roots booster) pump at the fore line of a DN630 ISO-F 20000 L/s oil-diffusion pump. The supersonic-jet expansion is pulsed by a general valve Series 9 solenoid valve which is controlled by a general valve IOTA one driver governed by the experiment-sequence generation. First molecular examples to illustrate the performance of the new setup will include OCS and CF_3CHFCl.

TH04   **2:36 – 2:51**

LARGE OLIGOMERS STABILIZED BY WHB NETWORKS: PENTAMERS OF DIFLUOROMETHANE AND ITS WATER CLUSTERS

<u>EMILIO J. COCINERO</u>, ICIAR URIARTE, *Physical Chemistry Department, Universidad del País Vasco (UPV/EHU), Bilbao, Spain*; LUCA EVANGELISTI, CAMILLA CALABRESE, *Dep. Chemistry 'Giacomo Ciamician', University of Bologna, Bologna, Italy*; GIACOMO PRAMPOLINI, *Istituto di Chimica dei Composti OrganoMetallici (ICCOM-CNR), UOS di Pisa, Consiglio Nazionale delle Ricerche, Pisa, Italy*; IVO CACELLI, *Dipartimento di Chimica e Chimica Industriale, Università di Pisa, Pisa, Italy*; BROOKS PATE, *Department of Chemistry, The University of Virginia, Charlottesville, VA, USA.*

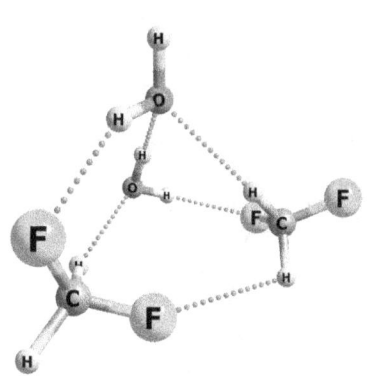

Microwave spectroscopy has been restricted to the investigation of small molecules in the last years. However, with the advent of FTMW and CP-FTMW spectroscopies coupled with laser vaporization techniques it has turned into a very competitive methodology in the studies of moderate-size molecules. In particular, the studies of relatively large molecular aggregates[a,b] are very interesting, being a bridge between microsystems and molecular bulk.

Here, we present the study of two pentarmers of difluoromethane $(CH_2F_2)_5$ and the water clusters $(CH_2F_2)_1\cdots(H_2O)_2$, $(CH_2F_2)_2\cdots(H_2O)_1$ and $(CH_2F_2)_2\cdots(H_2O)_2$ stabilized by weak hydrogen bonds networks (O-H$\cdots$F, C-H$\cdots$F and C-H$\cdots$O interactions). The experiments were carried out in the CP-FTMW spectrometers of Bilbao (Spain)[c] and Virginia (USA). In addition, the experimental work was supported by theoretical calculations. The force fields were specifically parameterized for reproduce others oligomers where WHB interactions play a crucial role.

[a]T. Tang, Y. Xu, A. R. W. McKellar and W. Jäger, *Science* **29**, 297, 2002.
[b]C. Pérez, M. T. Muckle, D. P. Zaleski, N. A. Seifert, B. Temelso, G. H. Shields, Z. Kisiel and B. H. Pate, *Science* **336**, 897, 2012.
[c]I. Uriarte, C. Pérez, E. Caballero-Mancebo, F. J. Basterretxea, A. Lesarri, J. A. Fernández and E. J. Cocinero *Chem. Eur. J.* in press, 2017.

TH05 2:53 – 3:08

MEASURING CONFORMATIONAL ENERGY DIFFERENCES USING PULSED-JET MICROWAVE SPECTROSCOPY

CAMERON M FUNDERBURK, SYDNEY A GASTER, TIFFANY R TAYLOR, GORDON G BROWN, *Department of Science and Mathematics, Coker College, Hartsville, SC, USA.*

The conformational energy differences of various chemicals have been measured using chirped-pulse Fourier transform microwave (CP-FTMW) spectroscopy. The hypothesis is that the relative intensities measured in a pulsed-jet instrument are proportional to the conformer populations present before the expansion occurs. Therefore, by measuring the relative intensities in a CP-FTMW spectrum, we aim to determine the relative energy difference between conformers. Experimentally, pulsed-jet CP-FTMW data will be compared to energy differences reported in the literature and to room-temperature CP-FTMW data acquired at Coker College. Results from *ab initio* calculations will also be used for comparison.

TH06 3:10 – 3:25

METHOXYETHANOL, ETHOXYETHANOL, AND SPECTRAL COMPLEXITY

J. H. WESTERFIELD, ERIKA RIFFE, MARIA PHILLIPS, ERIKA JOHNSON, STEVEN SHIPMAN, *Department of Chemistry, New College of Florida, Sarasota, FL, USA.*

Over the last few years, we have been working to improve the AUTOFIT program[a] and extend it to work on more complex spectra, especially spectra collected near room temperature. In this talk, we will discuss the problem of spectral complexity and the challenges it poses for moving to increasingly complicated systems. This will be highlighted by the cases of methoxyethanol, in which AUTOFIT was able to easily extract contributions from the ground state and four vibrationally excited states, and ethoxyethanol, in which AUTOFIT had difficulty identifying more than the ground vibrational state without the assistance of additional double resonance measurements.

[a]Seifert, N.A., Finneran, I.A., Perez, C., Zaleski, D.P., Neill, J.L., Steber, A.L., Suenram, R.D., Lesarri, A., Shipman, S.T., Pate, B.H., J. Mol. Spec. 312, 13-21 (2015)

Intermission

TH07 3:44 – 3:59

LABORATORY HETERODYNE SPECTROMETERS OPERATING AT 100 AND 300 GHZ

JAKOB MAßEN, NADINE WEHRES, *I. Physikalisches Institut, Universität zu Köln, Köln, Germany*; MARIUS HERMANNS, *I. Physikalisches Institut, University of Cologne, Cologne, Germany*; FRANK LEWEN, BETTINA HEYNE, *I. Physikalisches Institut, Universität zu Köln, Köln, Germany*; CHRISTIAN ENDRES, *The Center for Astrochemical Studies, Max-Planck-Institut für extraterrestrische Physik, Garching, Germany*; URS GRAF, NETTY HONINGH, STEPHAN SCHLEMMER, *I. Physikalisches Institut, Universität zu Köln, Köln, Germany.*

Two new laboratory heterodyne emission spectrometers are presented that are currently used for high-resolution rotational spectroscopy of complex organic molecules. The room temperature heterodyne receiver operating between 80-110 GHz, as well as the SIS heterodyne receiver operating between 270-370 GHz allow access to two very important frequency regimes, coinciding with Bands 3 and 7 of the ALMA (Atacama Large Millimeter Array) telescope. Taking advantage of recent progresses in the field of mm/submm technology, we build these two spectrometers using an XFFFTS (eXtended Fast Fourier Transform Spectrometer) for spectral acquisition. The instantaneous bandwidth is 2.5 GHz in a single sideband, spread over 32768 channels. Thus, the spectral resolution is about 76 kHz per channel and thus comparable to high resolution spectra from telescopes. Both receivers are operated in double sideband mode resulting in a total instantaneous bandwidth of 5 GHz. The system performances, in particular the noise temperatures and stabilities are presented. Proof-of-concept is demonstrated by showing spectra of methyl cyanide obtained with both spectrometers. While the transition frequencies for this molecule are very well known, intensities of those transitions can also be determined with high accuracy using our new instruments. This additional information shall be exploited in future measurements to improve spectral predictions for astronomical observations. Other future prospects concern the study of more complex organic species, such as ethyl cyanide. These aspects of the new instruments as well as limitations of the two distinct receivers will be discussed.

TH08 4:01 – 4:16

COMPLEX MOLECULES IN THE LABORATORY - A COMPARISON OF CHRIPED PULSE AND EMISSION SPEC-
TROSCOPY

MARIUS HERMANNS, NADINE WEHRES, JAKOB MAßEN, STEPHAN SCHLEMMER, *I. Physikalisches
Institut, Universität zu Köln, Köln, Germany.*

Detecting molecules of astrophysical interest in the interstellar medium strongly relies on precise spectroscopic data
from the laboratory. In recent years, the advancement of the chirped-pulse technique has added many more options available
to choose from. The Cologne emission spectrometer is an additional path to molecular spectroscopy. It allows to record
instantaneously broad band spectra with calibrated intensities. Here we present a comparison of both methods: The Cologne
chirped-pulse spectrometer as well as the Cologne emission spectrometer both cover the frequency range of 75-110 GHz,
consistent with the ALMA Band 3 receivers. High sensitive heterodyne receivers with very low noise temperature amplifiers
are used with a typical bandwidth of 2.5 GHz in a single sideband. Additionally the chirped-pulse spectrometer contains a
high power amplifier of 200 mW for the excitation of molecules. Room temperature spectra of methyl cyanide and comparison
of key features, such as measurement time, sensitivity, limitations and commonalities are shown in respect to identification of
complex molecules of astrophysical importance. In addition, future developments for both setups will be discussed.

TH09 4:18 – 4:33

A 530-590 GHZ SCHOTTKY HETERODYNE RECEIVER FOR HIGH-RESOLUTION MOLECULAR SPECTROSCOPY
WITH LILLE'S FAST-SCAN FULLY SOLID-STATE DDS SPECTROMETER

A. PIENKINA, L. MARGULÈS, R. A. MOTIYENKO, *Laboratoire PhLAM, UMR 8523 CNRS - Université
Lille 1, Villeneuve d'Ascq, France*; MARTINA C. WIEDNER, ALAIN MAESTRINI, FABIEN DEFRANCE,
*LERMA, Observatoire de Paris, PSL Research University, CNRS, Sorbonne Universités, UPMC Univ. Paris 06,
Paris, France.*

Laboratory spectroscopy, especially at THz and mm-wave ranges require the advances in instrumentation techniques
to provide high resolution of the recorded spectra with precise frequency measurement that facilitates the mathematical
treatment. We report the first implementation of a Schottky heterodyne receiver, operating at room temperature and covering
the range between 530 and 590 GHz, for molecular laboratory spectroscopy.

A 530-590 GHz non-cryogenic Schottky solid-state receiver[a] was designed at LERMA, Observatoire de Paris and fabricated in partnership with LPN- CNRS (Laboratoire de Photonique et de Nanostructures), and was initially developed for
ESA Jupiter Icy Moons Explorer (JUICE), intended to observe Jupiter and its icy moon atmospheres. It is based on a sub-
harmonic Schottky diode mixer, designed and fabricated at LERMA-LPN, pumped by a Local Oscillator (LO), consisting of a
frequency Amplifier/Multiplier chains (AMCs) from RPG (Radiometer Physics GmBh). The performance of the receiver was
demonstrated by absorption spectroscopy of CH_3CH_2CN with Lille's fast-scan DDS spectrometer. A series of test measurements showed the receiver's good sensitivity, stability and frequency accuracy comparable to those of 4K QMC bolometers,
thus making room-temperature Schottky receiver a competitive alternative to 4K QMC bolometers to laboratory spectroscopy
applications. We will present the first results with such a combination of a compact room temperature Schottky heterodyne
receiver and a fast-scan DDS spectrometer.

This work was funded by the French ANR under the Contract No. ANR-13-BS05-0008-02 IMOLABS.

[a]J. Treuttel, L. Gatilova, A. Maestrini *et al.*, 2016, *IEEE Trans. Terahertz Science and Tech.*, **6**, 148-155.

TH10 4:35–4:50

MILLIMETER WAVE SPECTROSCOPY IN A SEMI-CONFOCAL FABRY-PEROT CAVITY

BRIAN DROUIN, ADRIAN TANG, THEODORE J RECK, DEACON J NEMCHICK, MATTHEW J. CICH, TIMOTHY J. CRAWFORD, *Jet Propulsion Laboratory, California Institute of Technology, Pasadena, CA, USA*; ALEXANDER W RAYMOND, *Applied Physics, Harvard University, Cambridge, MA, USA*; M.-C. FRANK CHANG, ROD M. KIM, *Electrical Engineering, University of California - Los Angeles, Los Angeles, CA, USA*.

A new generation of CMOS circuits operating at 89-104 GHz with improved output power and pulse switch isolation have enhanced the performance of the miniaturized pulsed-echo Fourier transform spectrometer under development for planetary exploration at the Jet Propulsion laboratory. Additional progress has been made by creating a waveguide-fed structure for the novel planar coupler design. This structure has enabled characterization of each component in the system and enabled spectroscopy to be done with conventional millimeter hardware that enables (1) direct comparisons to the CMOS components, (2) enhanced bandwidth of 74-109 GHz, and (3) amplification of the transmitter prior to cavity injection. We have now demonstrated the technique with room temperature detections on multiple species including N_2O, OCS, CH_3CN, CH_3OH, CH_3NH_2, CH_3CHO, CH_3Cl, HDO, D_2O, CH_3CH_2CN and CH_3CH_2OH. Of particular interest to spectroscopic work in the millimeter range is the ongoing incorporation of a $\Delta\Sigma$ radio-frequency source into the millimeter-wave lock-loop - this has improved the phase-noise of the tunable CMOS transceiver to better than the room-temperature Doppler limit and provides a promising source for general use that may replace the high end microwave synthesizers. We are in the process of building a functional interface to the various subsystems. We will present a trade-space study to determine the optimal operating conditions of the pulse-echo system.

TH11 4:52–5:07

DETERMINING THE CONCENTRATIONS AND TEMPERATURES OF PRODUCTS IN A CF_4/CHF_3/N_2 PLASMA VIA SUBMILLIMETER ABSORPTION SPECTROSCOPY

YASER H. HELAL, CHRISTOPHER F. NEESE, FRANK C. DE LUCIA, *Department of Physics, The Ohio State University, Columbus, OH, USA*; PAUL R. EWING, , *Applied Materials, Austin, TX, USA*; ANKUR AGARWAL, BARRY CRAVER, PHILLIP J. STOUT, MICHAEL D. ARMACOST, , *Applied Materials, Sunnyvale, CA, USA*.

Plasmas used for the manufacturing of semiconductor devices are similar in pressure and temperature to those used in the laboratory for the study of astrophysical species in the submillimeter (SMM) spectral region. The methods and technology developed in the SMM for these laboratory studies are directly applicable for diagnostic measurements in the semiconductor manufacturing industry. Many of the molecular neutrals, radicals, and ions present in processing plasmas have been studied and their spectra have been cataloged or are in the literature. In this work, a continuous wave, intensity calibrated SMM absorption spectrometer was developed as a remote sensor of gas and plasma species. A major advantage of intensity calibrated rotational absorption spectroscopy is its ability to determine absolute concentrations and temperatures of plasma species from first principles without altering the plasma environment. An important part of this work was the design of the optical components which couple $500 - 750$ GHz radiation through a commercial inductively coupled plasma chamber. The measurement of transmission spectra was simultaneously fit for background and absorption signal. The measured absorption was used to calculate absolute densities and temperatures of polar species. Measurements for CHF_3, CF_2, FCN, HCN, and CN made in a CF_4/CHF_3/N_2 plasma will be presented. Temperature equilibrium among species will be shown and the common temperature is leveraged to obtain accurate density measurements for simultaneously observed species. The densities and temperatures of plasma species are studied as a function of plasma parameters, including flow rate, pressure, and discharge power.

TI. Large amplitude motions, internal rotation

Tuesday, June 20, 2017 – 1:45 PM

Room: B102 Chemical and Life Sciences

Chair: Peter Groner, University of Missouri, Kansas City, MO, USA

TI01   **1:45 – 2:00**

BROADBAND FTMW SPECTROSCOPY OF 2-METHYLIMIDAZOLE AND COMPLEXES WITH WATER AND ARGON

<u>CHRIS MEDCRAFT</u>, *School of Chemistry, Newcastle University, Newcastle-upon-Tyne, United Kingdom*; JULIANE HEITKÄMPER, *Institute of Physical Chemistry, Karlsruhe Institute of Technology, Karlsruhe, Germany*; JOHN C MULLANEY, NICK WALKER, *School of Chemistry, Newcastle University, Newcastle-upon-Tyne, United Kingdom.*

The rotational spectrum of 2-methylimidazole has been measured using laser ablation chirped-pulse Fourier transform microwave spectroscopy from 2-18.5 GHz. 2-methylimidazole was laser vaporised then entrained within an argon buffer gas undergoing supersonic expansion allowing for efficient rotational cooling. Carbon-13 and nitrogen-15 isotopologues were measured in natural abundance and substitution coordinates have been determined. The barrier to internal rotation of the methyl group was found to be 122.697(20) cm^{-1}. Nuclear quadropole coupling constants for the two nitrogen nuclei were determined via a rigid rotor fit of the A internal rotor state. Complexes with water and argon were also observed and fit in a similar way.

TI02   *Post-Deadline Abstract - Original Abstract Withdrawn*   **2:02 – 2:17**

CONNECTION BETWEEN THE SU(3) ALGEBRAIC MODEL AND CONFIGURATION SPACE FOR BENDING MODES OF LINEAR MOLECULES: APPLICATION TO ACETYLENE

<u>LEMUS RENATO</u>, ESTEZEZ-FREGOZO MARÍA DEL MAR, *Estructura de la Materia, Instituto de ciencias Nucleares, Mexico City, Mexico.*

An approach to connect the $su(3)$ dynamical group- used to describe the bending modes of linear molecules- with configuration space is discussed. The $SU(3)$ group may be seen as a consequence of adding a scalar boson to the $SU(2)$ space of two degenerate harmonic oscillators. The resulting $SU(3)$ group becomes the dynamical group for the bending degrees of freedom of linear molecules, but the connection to configuration space is not obvious. This work aims at providing this connection. Our approach is based on the basis of establishing a mapping between the algebraic and configuration states. An arbitrary operator in configuration space is then expanded in terms of generators of the dynamical algebra. The coefficients are determined through a minimization procedure and given in terms of matrix elements defined in configuration space. As an application we consider the vibrational description of the bending modes of the acetylene molecule, where the force constants are estimated in the framework of the $U(3) \times U(3)$ model.

TI03   2:19 – 2:34

MICROWAVE AND FIR SPECTROSCOPY OF DIMETHYLSULFIDE IN THE GROUND, FIRST AND SECOND EXCITED TORSIONAL STATES

V. ILYUSHIN, IULIIA ARMIEIEVA, OLGA DOROVSKAYA, MYKOLA POGREBNYAK, IGOR KRAPIVIN, E. A. ALEKSEEV, *Radiospectrometry Department, Institute of Radio Astronomy of NASU, Kharkov, Ukraine*; L. MARGULÈS, R. A. MOTIYENKO, *Laboratoire PhLAM, UMR 8523 CNRS - Université Lille 1, Villeneuve d'Ascq, France*; F. KWABIA TCHANA, *CNRS, Universités Paris Est Créteil et Paris Diderot, LISA, Créteil, France*; ATEF JABRI, *Department of Chemistry, MONARIS, CNRS, UMR 8233, Sorbonne Universités, UPMC Univ Paris 06, Paris, France*; LAURENT MANCERON, *AILES beam line, Synchrotron Soleil, Gif-sur-Yvette, France*; SIGURD BAUERECKER, CHRISTOF MAUL, *Institut für Physikalische und Theoretische Chemie, Technische Universität Braunschweig, Braunschweig, Germany*.

A new study [a] of the dimethylsulfide $((CH_3)_2 S)$ spectrum is reported. The new measurements have been carried out using the Kharkiv spectrometer in the Institute of Radio Astronomy of NASU (Ukraine) and using the Lille spectrometer in the PhLAM laboratory (France). The new millimeter and submillimeter wave measurements cover the frequency range from 49 GHz to 660 GHz. The rotational transitions belonging to the three lowest torsional states of the molecule as well as the new assignments in the FIR torsional band (AILES beamline of the synchrotron SOLEIL) and the microwave data available in the literature [b] have been analyzed using recently developed model for the molecules with two equivalent methyl rotors and C_{2v} symmetry at equilibrium (PAM_C2v_2tops program)[c]. In the talk the details of this new study will be discussed.

[a]This work was done under support of the Volkswagen foundation. The assistance of Science and Technology Center in Ukraine is acknowledged (STCU partner project P686).
[b]A. Jabri, V. Van, H. V. L. Nguyen, H. Mouhib, F. Kwabia Tchana , L. Manceron , W. Stahl, I. Kleiner, A&A 589, A127 (2016).
[c]Ilyushin V. V., Hougen J. T. J. Mol. Spectrosc. 289 (2013) 41-49.

TI04   2:36 – 2:51

ADVANCES IN GLOBAL MODELLING OF METHYL MERCAPTAN $CH_3{}^{32}SH$ TORSION-ROTATION SPECTRUM

V. ILYUSHIN, IULIIA ARMIEIEVA, *Radiospectrometry Department, Institute of Radio Astronomy of NASU, Kharkov, Ukraine*; OLENA ZAKHARENKO, HOLGER S. P. MÜLLER, FRANK LEWEN, STEPHAN SCHLEMMER, *I. Physikalisches Institut, Universität zu Köln, Köln, Germany*; LI-HONG XU, RONALD M. LEES, *Department of Physics, University of New Brunswick, Saint John, NB, Canada*.

A progress in the analysis of the methyl mercaptan $CH_3 SH$ spectrum in its ground, first and second excited torsional states is reported [a]. The available in the literature data [b] were reanalyzed using RAM36 code [c] with the main improvement achieved for the root mean square deviation of the microwave data. The updated Hamiltonian model was applied to the further assignments of the methyl mercaptan spectrum using the records obtained in the THz region in the previous study[b]. Also a new measurement campaign in subTHz frequency range is planned for the nearest future using a set of spectrometers available in Köln. In the talk the details of this new study will be discussed.

[a]This work was done under support of the Volkswagen foundation. The assistance of Science and Technology Center in Ukraine is acknowledged (STCU partner project P686).
[b]L.-H. Xu, R. M. Lees, G. T. Crabbe, J. A. Myshrall, H. S. P. Müller, C. P. Endres, O. Baum, F. Lewen, S. Schlemmer, K. M. Menten, and B. E. Billinghurst J. Chem. Phys. 137, 104313 (2012).
[c]V. Ilyushin, Z. Kisiel, L. Pszczółkowski, H. Mäder, J. T. Hougen // J. Mol. Spectrosc. Vol. 259, pp. 26-38 (2010).

Intermission

TI05

THE MICROWAVE SPECTROSCOPY OF CD_3SH

<u>KAORI KOBAYASHI</u>, SHOZO TSUNEKAWA, *Department of Physics, University of Toyama, Toyama, Japan*; NOBUKIMI OHASHI, , *Kanazawa University, Kanazawa, Japan.*

Methyl mercaptan (CH_3SH) is a sulfur-containing interstellar molecule, first identified in Sgr B2. [a] Although the abundance of methyl mercaptan was not too high compared with methanol, the identification of CD_3OH towards the IRAS 16293-2422 stimulated our microwave study of CD_3SH. Deuterium fractionation may be possible for the methyl mercaptan molecule. There are many laboratory microwave spectroscopic results on the normal species but the studies on this isotopolog is quite limited. We have observed the spectra of CD_3SH in the 12-240 GHz region without gap. The pure rotational transitions in the ground and first methyl torsional excited states were assigned. We applied pseudo-PAM Hamiltonian to analyze these transitions. We will report the status of the analysis.

[a]R. A. Linke, M. A. Frerking, and P. Thaddeus, *Astrophys. J.* **234**, L139 (1979).

TI06

QUANTUM CHEMICAL CALCULATIONS OF TORSIONALLY MEDIATED HYPERFINE SPLITTINGS IN STATES OF E SYMMETRY OF ACETALDEHYDE (CH_3CHO)

<u>LI-HONG XU</u>, ELIAS M. REID, BRADLEY GUISLAIN, *Department of Physics, University of New Brunswick, Saint John, NB, Canada*; JON T. HOUGEN, *Sensor Science Division, National Institute of Standards and Technology, Gaithersburg, MD, USA*; E. A. ALEKSEEV, *Quantum Radiophysics Department, Kharkiv National University and Institute of Radioastronomy of NASU, Kharkov, Ukraine*; IGOR KRAPIVIN, *Radiospectrometry Department, Institute of Radio Astronomy of NASU, Kharkov, Ukraine.*

Hyperfine splittings in methanol have been revisited in three recent publications. (i) Coudert et al. [JCP 143 (2015) 044304] published an analysis of splittings observed in the low-J range. They calculated 32 spin-rotation, 32 spin-spin, and 16 spin-torsion hyperfine constants using the ACES2 package. Three of these constants were adjusted to fit hyperfine patterns for 12 transitions. (ii) Three present authors and collaborators [JCP 145 (2016) 024307] analyzed medium to high-J experimental Lamb-dip measurements in methanol and presented a theoretical spin-rotation explanation that was based on torsionally mediated spin-rotation hyperfine operators. These contain, in addition to the usual nuclear spin and overall rotational operators, factors in the torsional angle α of the form $e^{\pm in\alpha}$. Such operators have non-zero matrix elements between the two components of a torsion-rotation ^{tr}E state, but have zero matrix elements within a ^{tr}A state. More than 55 hyperfine splittings were successfully fitted using three parameters and the fitted values agree well with ab initio values obtained in (i). (iii) Lankhaar et al. [JCP 145 (2016) 244301] published a reanalysis of the data set from (i), using CFOUR recalculated hyperfine constants based on their rederivation of the relevant expressions. They explain why their choice of fixed and floated parameters leads to numerical values for all parameters that seem to be more physical than those in (i). The results in (ii) raise the question of whether large torsionally-mediated spin-rotation splittings will occur in other methyl-rotor-containing molecules. This abstract presents ab initio calculations of torsionally mediated hyperfine splittings in the E states of acetaldehyde using the same three operators as in (ii) and spin-rotation constants computed by Gaussian09. We explored the first 13 K states for J from 10 to 40 and $\nu_t = 0$, 1, and 2. Our calculations indicate that hyperfine splittings in CH_3CHO are just below current measurement capability. This conclusion is confirmed by available experimental measurements.

TI07

MICROWAVE SPECTROSCOPY OF 2-PENTANONE

MAIKE ANDRESEN, *Institute for Physical Chemistry, RWTH Aachen University, Aachen, Germany*; HA VINH
LAM NGUYEN, ISABELLE KLEINER, *Laboratoire Interuniversitaire des Systèmes Atmosphériques (LISA),
CNRS et Universités Paris Est et Paris Diderot, Créteil, France*; WOLFGANG STAHL, *Institute for Physical
Chemistry, RWTH Aachen University, Aachen, Germany.*

Methyl propyl ketone (MPK) or 2-Pentanone is known to be an alarm pheroromone released by the mandibular glands of
the bees. It is a highly volatile compound. This molecule was studied by a combination of quantum chemical calculations and
microwave spectroscopy in order to get informations about the lowest energy conformers and their structures.The rotational
spectrum of 2-pentanone was measured using the molecular beam Fourier transform microwave spectrometer in Aachen
operating between 2 and 26.5 GHz. Ab initio calculations determine 4 conformers but only two of them are observed in our
jet-beam conditions.The lowest conformer has a C_1 structure and its spectrum shows internal rotation splittings arising from
two methyl groups. The internal splittings of 305 transitions for this conformer were analyzed using the XIAM code [a]. It led
to the determination of the values for the barrier heights hindering the internal rotation of two methyl groups of 239 cm^{-1}
and 980 cm^{-1} respectively. The next energy conformer has a C_s structure and the analysis of the internal splittings of 134
transitions using the XIAM code and the BELGI code [b] led to the determination of internal rotation barrier height of 186
cm^{-1}. Comparisons of quantum chemistry and experimental results will be discussed.

[a]H. Hartwig, H. Dreizler, Z. Naturforsch. 51a, 923 (1996).

[b]J. T. Hougen, I. Kleiner and M. Godefroid, J. Mol. Spectrosc., 163, 559-586 (1994).

TI08

COMPETITION BETWEEN TWO LARGE-AMPLITUDE MOTION MODELS: NEW HYBRID HAMILTONIAN VER-
SUS OLD PURE-TUNNELING HAMILTONIAN

ISABELLE KLEINER, *Laboratoire Interuniversitaire des Systèmes Atmosphériques (LISA), CNRS et Universités Paris Est et Paris Diderot, Créteil, France*; JON T. HOUGEN, *Sensor Science Division, National Institute
of Standards and Technology, Gaithersburg, MD, USA.*

In this talk we report on our progress in trying to make the hybrid Hamiltonian competitive with the pure-tunneling
Hamiltonian for treating large-amplitude motions in methylamine. A treatment using the pure-tunneling model has the advantages of: (i) requiring relatively little computer time, (ii) working with relatively uncorrelated fitting parameters, and (iii)
yielding in the vast majority of cases fits to experimental measurement accuracy. These advantages are all illustrated in the
work published this past year on a gigantic $v_t = 1$ data set for the torsional fundamental band in methyl amine[a]. A treatment
using the hybrid model has the advantages of: (i) being able to carry out a global fit involving both $v_t = 0$ and $v_t = 1$ energy
levels and (ii) working with fitting parameters that have a clearer physical interpretation. Unfortunately, a treatment using the
hybrid model has the great disadvantage of requiring a highly correlated set of fitting parameters to achieve reasonable fitting
accuracy, which complicates the search for a good set of molecular fitting parameters and a fit to experimental accuracy. At
the time of writing this abstract, we have been able to carry out a fit with J up to 15 that includes all available infrared data in
the $v_t = 1$-0 torsional fundamental band, all ground-state microwave data with K up to 10 and J up to 15, and about a hundred
microwave lines within the $v_t = 1$ torsional state, achieving weighted root-mean-square (rms) deviations of about 1.4, 2.8, and
4.2 for these three categories of data. We will give an update of this situation at the meeting.

[a]I. Gulaczyk, M. Kreglewski, V.-M. Horneman, J. Mol. Spectrosc., in Press (2017).

TI09   **4:18 – 4:33**

TUNNELING EFFECTS AND CONFORMATION DETERMINATION OF THE POLAR FORMS OF 1,3,5-TRISILAPENTANE

FRANK E MARSHALL, *Department of Chemistry, Missouri University of Science and Technology, Rolla, MO, USA*; WILLIAM RAYMOND NEAL TONKS, *Chemistry, College of Charleston, Charleston, SC, USA*; DAVID JOSEPH GILLCRIST, *Department of Chemistry, Missouri University of Science and Technology, Rolla, MO, USA*; CHARLES J. WURREY, *Department of Chemistry, University of Missouri - Kansas City, Kansas City, MO, USA*; GAMIL A GUIRGIS, *Chemistry, College of Charleston, Charleston, SC, USA*; G. S. GRUBBS II, *Department of Chemistry, Missouri University of Science and Technology, Rolla, MO, USA.*

1,3,5-trisilapentane has been synthesized and studied in the microwave region for the first time using CP-FTMW spectroscopy. The lowest calculated energy structure, C_2 is essentially non-polar with a calculated dipole of 0.063 D. However, slightly higher in energy at 145 cm^{-1} and 196 cm^{-1} are the calculated energies for the C_1 and C_{2v} conformations, respectively. These structures have much larger dipoles calculated at 1.07 D for C_1 and 4.88 D for C_{2v}. Both of these structures have been confirmed using experiment and the details of such analysis will be discussed.

In addition to the structure determination, 1,3,5-trisilapentane has two terminal $-SiH_3$ groups. The calculated barrier to internal rotation of these groups are calculated to be 327.5 cm^{-1} for C_{2v} and 343.2 cm^{-1} for C_1. This barrier is low enough to exhibit internal rotation splitting in the spectra and treatment of these motions in the analysis will be discussed.

TI10   **4:35 – 4:50**

A COMPARISON OF THE MOLECULAR STRUCTURES OF $C_4H_9OCH_3$, $C_4H_9SCH_3$, $C_5H_{11}OCH_3$, AND $C_5H_{11}SCH_3$ USING MICROWAVE SPECTROSCOPY

BRITTANY E. LONG, *Chemistry Department, Trinity University, San Antonio, TX, USA*; JUAN BETANCUR, *Natural and Social Science, Purchase College SUNY, Purchase, NY, USA*; YOON JEONG CHOI, *Department of Chemistry, Wesleyan University, Middletown, CT, USA*; S. A. COOKE, *Natural and Social Science, Purchase College SUNY, Purchase, NY, USA*; G. S. GRUBBS II, *Department of Chemistry, Missouri University of Science and Technology, Rolla, MO, USA*; JONATHAN OGULNICK, TARA HOLMES, *Natural and Social Science, Purchase College SUNY, Purchase, NY, USA.*

Pure rotational spectra of the title molecules have been recorded using chirped pulse Fourier transform microwave spectroscopy. Under our experimental conditions only one conformer has been observed for each of the four compounds. These conformers have torsional angles of CXCC = 180°, XCCC = 60°, CCCC = 180°, and, for the C_5H_{11}-X-CH_3 species, $CCCC_{Methyl}$ = 180°. These angles correspond to anti-gauche-anti conformations for the butyl methyl ether/thioether species, and anti-gauche-anti-anti conformations for the pentyl methyl ether/thioether species. Splittings due to the internal rotation of the X-CH_3 group are observed in both butyl species but are not observed in the pentyl species. The barrier to the X-CH_3 internal rotation has been investigated through spectral analyses and quantum chemical calculations. The differences in the internal rotation barrier between the ethers and thioethers will be discussed and will further be compared to the barriers obtained for similar molecules.

TJ. Linelists

Tuesday, June 20, 2017 – 1:45 PM

Room: 161 Noyes Laboratory

Chair: Brian Drouin, California Institute of Technology, Pasadena, CA, USA

TJ01   1:45 – 2:00

ACCURACY and COMPLETENESS of MOLECULAR LINE LISTS

<u>OLEG POLYANSKY</u>, JONATHAN TENNYSON, *Department of Physics and Astronomy, University College London, Gower Street, London WC1E 6BT, United Kingdom.*

We review recent progress in the calculation of the global, high-temperature and accurate room-temperature linelists of various molecules relevant for the analysis of both Earth and exoplanet atmospheres and cool stars. These global line lists can be constructed based on progress in calculation of energy levels up to dissociation and the fitting of the molecular PESs to the experimental data close to dissociation. Sub-percent accuracy in the intensity of calculated absorption lines is achieved thanks to progress in *ab initio* electronic structure calculations which is aided by the possibility of characterizing their accuracy by the comparison with experimental intensities measured with sub-percent accuracy for some molecular lines. The advantage of variational calculations over experiment though is the ability to produce billions of lines covering all the isotopologues, which is clearly impossible for the experimental observations. Atmospherically and astrophysically important molecules such as H_2O, CO_2, CO and H_3^+ will be considered together with some examples of the other molecules.

TJ02   2:02 – 2:17

A RIGOROUS COMPARISON OF THEORETICAL AND MEASURED CARBON DIOXIDE LINE INTENSITIES

HONGMING YI, ADAM J. FLEISHER, LYN GAMESON, *Chemical Sciences Division, National Institute of Standards and Technology, Gaithersburg, MD, USA*; EMIL J ZAK, OLEG POLYANSKY, JONATHAN TENNYSON, *Department of Physics and Astronomy, University College London, Gower Street, London WC1E 6BT, United Kingdom*; <u>JOSEPH T. HODGES</u>, *Chemical Sciences Division, National Institute of Standards and Technology, Gaithersburg, MD, USA.*

The ability to calculate molecular line intensities from first principles plays an increasingly important role in populating line-by-line spectroscopic databases because of its generality and extensibility to various isotopologues, spectral ranges and temperature conditions. Such calculations require a spectroscopically determined potential energy surface, and an accurate dipole moment surface that can be either fully *ab initio* [a][b] or an effective quantity based on fits to measurements [c]. Following our recent work where we used high-precision measurements of intensities in the (30013 →00001) band of $^{12}C^{16}O_2$ to bound the uncertainty of calculated line lists [d], here we carry out high-precision, frequency-stabilized cavity ring-down spectroscopy measurements in the R-branch of the $^{12}C^{16}O_2$ (20012 →00001) band from J = 16 to 52. Gas samples consisted of 50 μmol mol^{-1} or 100 μmol mol^{-1} of nitrogen-broadened carbon dioxide with gravimetrically determined SI-traceable molar composition. We demonstrate relative measurement precision (Type A) at the 0.15 % level and estimate systematic (Type B) uncertainty contributions in % of: isotopic abundance 0.01; sample density, 0.016; cavity free spectral rang,e 0.03; line shape, 0.05; line interferences, 0.05; and carbon dioxide molar fraction, 0.06. Combined in quadrature, these components yield a relative standard uncertainty in measured line intensity less than 0.2 % for most observed transitions. These intensities differ by more than 2 % from those measured by Fourier transform spectroscopy and archived in HITRAN 2012 but differ by less than 0.5 % with the calculations of Zak et al.

---

[a]E. Zak et al., *J. Quant. Spectrosc. Radiat. Transf.* **177**, (2016) 31.
[b]Huang et al., *J. Quant. Spectrosc. Radiat. Transf.* **130**, (2013) 134.
[c]Tashkun et al., *J. Quant. Spectrosc. Radiat. Transf.* **152**, (2015) 45.
[d]Polyansky et al., *Phys Rev. Lett.* **114**, (2015) 243001.

TJ03

PRECISION CAVITY-ENHANCED DUAL-COMB SPECTROSCOPY: APPLICATION TO THE GAS METROLOGY OF CO_2, H_2O, and N_2O.

ADAM J. FLEISHER, DAVID A. LONG, JOSEPH T. HODGES, *Chemical Sciences Division, National Institute of Standards and Technology, Gaithersburg, MD, USA.*

With inherent simplicity, mutual phase coherence, and a high degree of user control, electro-optic frequency combs are amenable to both dual-comb spectroscopy[a] and cavity-enhanced comb spectroscopy.[b] This combination of fast, multiplexed spectroscopy, with an effective absorption pathlength >1 km, is used here to perform line-by-line metrology of the gas-phase absorption spectra of CO_2, H_2O, and N_2O in the near-infrared. We report absolute transition frequency with precision better than 1 MHz in 1 s of spectral acquisition per transition using a comb with an instantaneous optical bandwidth of 6 GHz, tunable over the entire 6240-6370 cm^{-1} range. A full model for the electric field transmitted through the enhancement cavity (even in the presence of strong molecular absorption and dispersion) will be discussed.

[a]I. Coddington et al., *Optica* **3**, 414 (2016)
[b]B. Bernhardt et al., *Nat. Photonics* **4**, 55 (2010)

TJ04

EXPERIMENTAL LINE LIST OF WATER VAPOR ABSORPTION LINES IN THE SPECTRAL RANGES 1850 – 2280 CM^{-1} AND 2390 – 4000 CM^{-1}

JOEP LOOS, *Remote Sensing Technology Institute, German Aerospace Center (DLR), Wessling, Germany*; MANFRED BIRK, *Remote Sensing Technology Institute, Experimental Methods, German Aerospace Center DLR, Oberpfaffenhofen, Germany*; GEORG WAGNER, *Remote Sensing Technology Institute, DLR, Wessling, Germany.*

A new experimental line parameter list of water vapor absorption lines in the spectral ranges 1850 – 2280 cm^{-1} and 2390 – 4000 cm^{-1} is presented. The line list is based on the analysis of several transmittance spectra measured using a Bruker IFS 125 HR high resolution Fourier transform spectrometer. A total of 54 measurements of pure water and water/air-mixtures at 296 K as well as water/air-mixtures at high and low temperatures were performed. A multispectrum fitting approach was used applying a quadratic speed-dependent hard collision line shape model in the Hartmann-Tran implementation[ab] extended to account for line mixing in the Rosenkranz approximation in order to retrieve line positions, intensities, self- and air-broadening parameters, their speed-dependence, self- and air-shifts as well as line mixing and in some cases collisional narrowing parameters. Additionally, temperature dependence parameters for widths, shifts and in a few cases line mixing were retrieved. For every parameter an extensive error estimation calculation was performed identifying and specifying systematic error sources. The resulting parameters are compared to the databases HITRAN12[c] and GEISA15[d] as well as experimental values. For intensities, a detailed comparison to results of recent ab initio calculations performed at University College London was done showing an agreement within 2 % for a majority of the data. However, for some bands there are systematic deviations attributed to ab initio calculation errors.

[a]N.H. Ngo *et al.* JQSRT **129**, 89-100 (2013) doi:10.1016/j.jqsrt.2013.05.034; JQSRT **134**, 105 (2014) doi:10.1016/j.jqsrt.2013.10.016.
[b]H. Tran *et al.* JQSRT **129**, 199-203 (2013) doi:10.1016/j.jqsrt.2013.06.015; JQSRT **134**, 104 (2014) doi:10.1016/j.jqsrt.2013.10.015.
[c]L.S. Rothman *et al.* JQSRT **130**, 4–50 (2013) doi:10.1016/j.jqsrt.2013.07.002.
[d]N. Jacquinet-Husson *et al.* JMS **112**, 2395–2445 (2016) doi:10.1016/j.jms.2016.06.007.

TJ05 2:53 – 3:08

PROGRESS IN THE MEASUREMENT ON TEMPERATURE-DEPENDENCE OF H_2-BROADENING OF COLD AND HOT CH_4

KEEYOON SUNG, *Jet Propulsion Laboratory, California Institute of Technology, Pasadena, CA, USA*; V. MALATHY DEVI, D. CHRIS BENNER, *Department of Physics, College of William and Mary, Williamsburg, VA, USA*; TIMOTHY J. CRAWFORD, *Jet Propulsion Laboratory, California Institute of Technology, Pasadena, CA, USA*; ARLAN MANTZ, *Department of Physics, Astronomy and Geophysics, Connecticut College, New London, CT, USA*; MARY ANN H. SMITH, *Science Directorate, NASA Langley Research Center, Hampton, VA, USA*.

We report preliminary measurements on the temperature dependence of H_2-broadening of CH_4 in the near infrared at temperatures between 100 and 370 K. In support of the Jovian and exoplanet atmospheric remote sensing in the near infrared, we have measured the temperature dependence of H_2-broadened half width and pressure shift coefficients of CH_4, both of which are known to be rotational quantum number dependent. We studied both cold and hot CH_4 in the atmospheric K band (2.2 μm) with the focus on a) weaker lines in the $\nu_2+\nu_3$ band at low temperatures for cold giant planets and b) stronger lines in the $\nu_3+\nu_4$ band at elevated temperatures for extra-solar planets (e.g., hot-Jupiters). Three custom-built gas absorption cells (two cold and one hot) were used to obtain the spectra of CH_4 and H_2 mixtures at temperatures between 100 and 370 K. We will discuss our on-going spectrum analysis for a few select J manifolds and provide comparisons with published values, which are available only at room temperature.

TJ06 3:10 – 3:25

HIGH ACCURACY POTENTIAL ENERGY SURFACE, DIPOLE MOMENT SURFACE, ROVIBRATIONAL ENERGIES AND LINE LIST CALCULATIONS FOR $^{14}NH_3$

PHILLIP COLES, *Department of Physics and Astronomy, University College London, Gower Street, London WC1E 6BT, United Kingdom*; SERGEI N. YURCHENKO, *Department of Physics and Astronomy, University College London, Gower Street, London WC1E 6BT, United Kingdom*; OLEG POLYANSKY, *Department of Physics and Astronomy, University College London, Gower Street, London WC1E 6BT, United Kingdom*; ALEKSANDRA KYUBERIS, ROMAN I. OVSYANNIKOV, NIKOLAY FEDOROVICH ZOBOV, *Microwave Spectroscopy, Institute of Applied Physics, Nizhny Novgorod, Russia*; JONATHAN TENNYSON, *Department of Physics and Astronomy, University College London, Gower Street, London WC1E 6BT, United Kingdom*.

We present a new spectroscopic potential energy surface (PES) for $^{14}NH_3$, produced by refining a high accuracy *ab initio* PES[a] to experimental energy levels taken predominantly from MARVEL[b]. The PES reproduces 1722 matched J=0-8 experimental energies with a root-mean-square error of 0.035 cm-1 under 6000 cm^{-1} and 0.059 under 7200 cm^{-1}. In conjunction with a new DMS calculated using multi reference configuration interaction (MRCI) and H=aug-cc-pVQZ, N=aug-cc-pWCVQZ basis sets, an infrared (IR) line list has been computed which is suitable for use up to 2000 K. The line list is used to assign experimental lines in the 7500 - 10,500 cm^{-1} region and previously unassigned lines in HITRAN in the 6000-7000 cm^{-1} region.

[a]Oleg L. Polyansky, Roman I. Ovsyannikov, Aleksandra A. Kyuberis, Lorenzo Lodi, Jonathan Tennyson, Andrey Yachmenev, Sergei N. Yurchenko, Nikolai F. Zobov, *J. Mol. Spec.*, 327 (2016) 21-30

[b]Afaf R. Al Derzia, Tibor Furtenbacher, Jonathan Tennyson, Sergei N. Yurchenko, Attila G. Császár, *J. Quant. Spectrosc. Rad. Trans.*, 161 (2015) 117-130

TJ07 3:27 – 3:42

A NEW LINELIST FOR OH $A^2\Sigma$-$X^2\Pi$ ELECTRONIC TRANSITION

MAHDI YOUSEFI, *Department of Physics, Old Dominion University, Norfolk, VA, USA*; PETER F. BERNATH, *Department of Chemistry and Biochemistry, Old Dominion University, Norfolk, VA, USA*.

The OH radical is observed in cool stars, interstellar medium, comets and is an important oxidizer in the Earth's atmosphere. A new linelist for the ($A^2\Sigma^+$-$X^2\Pi$) transition of OH has been calculated. The line positions have been obtained from the literature and the line intensities were calculated from a new ab initio transition dipole moment function obtained from Molpro quantum chemistry package. This dipole moment function along with the RKR potentials have been used in LeRoy's LEVEL program in order to calculate transition dipole matrix elements. These matrix elements are transformed from Hund's case (b) to Hund's case (a) as required for Western's PGopher program. The linelist was calculated with PGopher.

Intermission

TJ08 4:01 – 4:16

HITRAN2016: Part I. Line lists for H_2O, CO_2, O_3, N_2O, CO, CH_4, and O_2

IOULI E GORDON, LAURENCE S. ROTHMAN, YAN TAN, ROMAN V KOCHANOV, *Atomic and Molecular Physics, Harvard-Smithsonian Center for Astrophysics, Cambridge, MA, USA*; CHRISTIAN HILL, *Department of Physics and Astronomy, University College London, Gower Street, London WC1E 6BT, United Kingdom.*

The HITRAN2016[a] database is now officially released[b]. Plethora of experimental and theoretical molecular spectroscopic data were collected, evaluated and vetted before compiling the new edition of the database. The database is now distributed through the dynamic user interface HITRAN*online* (available at www.hitran.org) which offers many flexible options for browsing and downloading the data[c]. In addition HITRAN Application Programming Interface (HAPI) offers modern ways to download the HITRAN data and use it to carry out sophisticated calculations[d]. The line-by-line lists for almost all of the 47 HITRAN molecules were updated in comparison with the previous compilation (HITRAN2012[e]). Some of the most important updates for major atmospheric absorbers, such as H_2O, CO_2, O_3, N_2O, CO, CH_4, and O_2, will be presented in this talk, while the trace gases will be presented in the next talk by Y. Tan. The HITRAN2016 database now provides alternative line-shape representations for a number of molecules, as well as broadening by gases dominant in planetary atmospheres. In addition, substantial extension and improvement of cross-section data is featured, which will be described in a dedicated talk by R. V. Kochanov. The new edition of the database is a substantial step forward to improve retrievals of the planetary atmospheric constituents in comparison with previous editions, while offering new ways of working with the data. The HITRAN database is supported by the NASA AURA and PDART program grants NNX14AI55G and NNX16AG51G.

[a]I. E. Gordon, L. S. Rothman, C. Hill, R. V. Kochanov, Y. Tan, et al. The HITRAN2016 Molecular Spectroscopic Database. JQSRT 2017;submitted.

[b]Many spectroscopists and atmospheric scientists worldwide have contributed data to the database or provided invaluable validations.

[c]C. Hill, I. E. Gordon, R. V. Kochanov, L. Barrett, J.S. Wilzewski, L.S. Rothman, JQSRT. 177 (2016) 4--14.

[d]R.V. Kochanov, I. E. Gordon, L. S. Rothman, P. Wcislo, C. Hill, J. S. Wilzewski, JQSRT. 177 (2016) 15—30.

[e]L. S. Rothman, I. E. Gordon et al. The HITRAN2012 Molecular Spectroscopic Database. JQSRT, 113 (2013) 4–50.

TJ09 4:18 – 4:33

HITRAN2016 DATABASE PART II: OVERVIEW OF THE SPECTROSCOPIC PARAMETERS OF THE TRACE GASES

YAN TAN, *Atomic and Molecular Physics , Harvard-Smithsonian Center for Astrophysics, Cambridge, MA, USA*; IOULI E GORDON, LAURENCE S. ROTHMAN, ROMAN V KOCHANOV, CHRISTIAN HILL, *Atomic and Molecular Physics, Harvard-Smithsonian Center for Astrophysics, Cambridge, MA, USA.*

The 2016 edition of HITRAN database [a] is available now [b]. This new edition of the database takes advantage of the new structure and can be accessed through HITRANonline (www.hitran.org) [c].

The line-by-line lists for almost all of the trace atmospheric species were updated in comparison with the previous edition HITRAN2012. These extended update covers not only updating few transitions of the certain molecules, but also complete replacements of the whole line lists, and as well as introduction of new spectroscopic parameters for non-Voigt line shape. The new line lists for NH_3, HNO_3, OCS, HCN, CH_3Cl, C_2H_2, C_2H_6, PH_3, C_2H_4, CH_3CN, CF_4, C_4H_2, and SO_3 feature substantial expansion of the spectral and dynamic ranges in addition of the improved accuracy of the parameters for already existing lines. A semi-empirical procedure was developed to update the air-broadening and self-broadening coefficients of N_2O, SO_2, NH_3, CH_3Cl, H_2S, and HO_2. We draw particular attention to flaws in the commonly used expression $n_{air} = 0.79n_{N_2} + 0.21n_{O_2}$ to determine the air-broadening temperature dependence exponent in the power law from those for nitrogen and oxygen broadening. A more meaningful approach will be presented. The semi-empirical line width, pressure shifts and temperature-dependence exponents of CO, NH_3, HF, HCl, OCS, C_2H_2, SO_2 perturbed by H_2, He, and CO_2 have been added to the database based on the algorithm described in Wilzewski et al.[d]. The new spectroscopic parameters for HT profile were implemented into the database for hydrogen molecule[e].

[a]The HITRAN database is supported by the NASA AURA program grant NNX14AI55G and NASA PDART grant NNX16AG51G.

[b]I. E. Gordon, L. S. Rothman, et al., J Quant Spectrosc Radiat Transf 2017; submitted.

[c]Hill C, et al., J Quant Spectrosc Radiat Transf 2013;130:51–61.

[d]Wilzewski JS,et al., J Quant Spectrosc Radiat Transf 2016;168:193–206.

[e]Wcisło P, et al., J Quant Spectrosc Radiat Transf 2016;177:75–91.

TJ10   **4:35 – 4:50**

ABSORPTION CROSS-SECTIONS IN HITRAN2016: MAJOR DATABASE UPDATE FOR ATMOSPHERIC, INDUSTRIAL, AND CLIMATE APPLICATIONS

ROMAN V KOCHANOV[a], IOULI E GORDON, LAURENCE S. ROTHMAN, *Atomic and Molecular Physics, Harvard-Smithsonian Center for Astrophysics, Cambridge, MA, USA*; KEITH SHINE, *Department of Chemistry, University of Reading, Reading, United Kingdom*; STEVEN W. SHARPE, TIMOTHY J. JOHNSON, *Chemical Physics and Analysis, Pacific Northwest National Laboratory, Richland, WA, USA*; JEREMY J. HARRISON[b], *Department of Physics and Astronomy, University of Leicester, Leicester, United Kingdom*; PETER F. BERNATH, *Department of Chemistry and Biochemistry, Old Dominion University, Norfolk, VA, USA*; TIMOTHY WALLINGTON, *Research and Advanced Engineering, Ford Motor Company, Dearborn, MI, USA*; MANFRED BIRK, GEORG WAGNER, *Remote Sensing Technology Institute, DLR, Wessling, Germany*; CHRISTIAN HILL, *Department of Physics and Astronomy, University College London, Gower Street, London WC1E 6BT, United Kingdom.*

In this talk, an overview is given for the recent absorption cross-section update in the new HITRAN2016 spectroscopic database release. The updated cross-sections include data for around 330 molecules for applications in atmospheric remote sensing, industrial pollution tracking, climate change monitoring, remote sensing, spectral calibration, and more. These cross-sections come from high-resolution laboratory observations, predominantly using FT-IR technique. The update largely relies on spectra from the PNNL quantitative spectroscopic database and the Hodnebrog et al. (Rev Geophys 2013) compilation, but also on other recently published data for many applications such as biomass burning detection, remote sensing in the UTLS, environment monitoring, etc. (references will be given in the talk). The described data are available via the HITRANonline website[c] and HITRAN Application Programing Interface (HAPI)[d]. This work is supported by NASA AURA (NNX14AI55G) and NASA PDART (NNX16AG51G) grants.

[a]Laboratory of Quantum Mechanics of Molecules and Radiative Processes, Tomsk State University, Tomsk, Russia

[b]National Centre for Earth Observation and Leicester Institute for Space and Earth Observation, University of Leicester, Leicester, UK;

[c]Hill C. et al. JQSRT 2016;177:4–14.

[d]Kochanov RV et al.JQSRT 2016;177:15–30.

TJ11   **4:52 – 5:07**

A NEW LINE LIST FOR $A^3\Pi$ - $X^3\Sigma^-$ TRANSITION OF NH RADICAL

ANTON MADUSHANKA FERNANDO, *Department of Physics, Old Dominion University, Norfolk, VA, USA*; PETER F. BERNATH, *Department of Chemistry and Biochemistry, Old Dominion University, Norfolk, VA, USA.*

The NH radical is important in astronomy as it is observed in cool stars and the interstellar medium. A new line list for $A^3\Pi$ - $X^3\Sigma^-$ electronic transition has been prepared using line positions from the literature and calculated line intensities. High level ab-initio calculations are performed with Molpro to obtain the A-X transition dipole moment function. Potential energy curves and the line strengths are calculated by Le Roy's RKR and LEVEL programs. Line intensities and Einstein A values were calculated with Western's PGOPHER program after converting Hund's case (b) output of LEVEL to Hund's case (a) input needed for PGOPHER.

LINE LISTS FOR LiF AND LiCl IN THE $X^1\Sigma^+$ STATE

DROR M. BITTNER, PETER F. BERNATH, *Department of Chemistry and Biochemistry, Old Dominion University, Norfolk, VA, USA.*

Alkali-containing molecules are expected to be present in the atmospheres of exoplanets such as rocky super-Earths[a] as well as in cool dwarf stars.[b] Line lists for LiF and LiCl in their $X^1\Sigma^+$ ground states have been calculated using LeRoy's LEVEL program.[c] The potential energy functions, including the effects of the breakdown of the Born-Oppenheimer approximation, are obtained by direct fitting the experimental infrared vibration-rotation and microwave pure rotation data with extended Morse oscillator potentials with LeRoy's dPotFit program.[d] The transition dipole matrix elements and line intensities were obtained with LEVEL using a dipole moment function from a high level *ab initio* calculation.

---

[a] Phil. Trans. R. Soc. A **372**, 20130087 (2014)

[b] Astrophys. J. **519**, 793 (1999)

[c] J. Quant. Spectrosc. Radiat. Transfer **186**, 167 (2017)

[d] J. Quant. Spectrosc. Radiat. Transfer **186**, 179 (2017)

TK. Small molecules

Tuesday, June 20, 2017 – 1:45 PM

Room: 140 Burrill Hall

Chair: Vincent Boudon, CNRS / Université Bourgogne Franche-Comté, Dijon, France

---

TK01   1:45 – 2:00

RELATIVE INTENSITY OF A CROSS-OVER RESONANCE TO LAMB DIPS OBSERVED IN STARK SPECTROSCOPY OF METHANE

SHOKO OKUDA, HIROYUKI SASADA, *Department of Physics, Faculty of Science and Technology, Keio University, Yokohama, Japan.*

Last ISMS, we reported on Stark effects of the ν_3 band of methane observed with a sub-Doppler resolution spectrometer. We determined the rotation-induced permanent dipole moment (PEDM) in the vibrational ground state and the vibration-, rotation-, and Coriolis-type-interaction-induced PEDMs in the $v_3 = 1$ state.

Figure illustrates Stark modulation spectrum of the $Q(6)E$ with the external electric field of 31.0 kV/cm and the selection rule of $\Delta M = \pm 1$, where M is the magnetic quantum number. The $\Delta M = 1$ and -1 components of the Lamb dips labeled by A and B are resolved, and the central component C is identified with the cross-over resonance. The Lamb dips are assigned to the magnetic quantum numbers of the lower and upper states, (M'', M') according to the Clebsch-Gordan coefficients. We found that the relative intensity of the cross-over resonance to the associated Lamb dips depends on the P, Q, and R branches. We ascribe the dependence to the collisional relaxation processes.

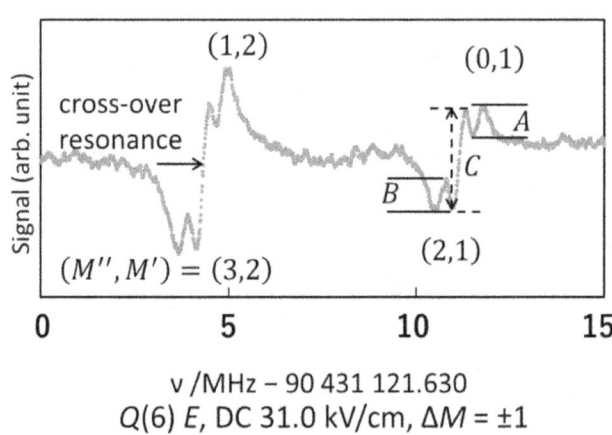

ν /MHz – 90 431 121.630
$Q(6)\ E$, DC 31.0 kV/cm, $\Delta M = \pm 1$

TK02   2:02 – 2:17

CORIOLIS PERTURBATIONS TO THE $3\nu_4$ LEVEL OF THE $\tilde{A}$ STATE OF FORMALDEHYDE

BARRATT PARK, BASTIAN C. KRUEGER, SVEN MEYER, TIM SCHAEFER, *Institute of Physical Chemistry, Georg-August-Universität Göttingen, Göttingen, Germany.*

Formaldehyde is the smallest stable organic molecule containing the carbonyl functional group, and is commonly considered to be a prototype for the study of high-resolution spectroscopy of polyatomic molecules. The a-axis Coriolis interaction between the near-degenerate ν_4 and ν_6 (out-of-plane and in-plane wagging modes, respectively) of the ground electronic state has received extensive attention and is thoroughly understood. In the first excited singlet $\tilde{A}$ $^1A_2$ electronic state, the analogous Coriolis interaction does not occur, because the $\tilde{A}$ state suffers from a pseudo Jahn-Teller distortion, which causes a double-well potential energy structure in the q'_4 out-of-plane coordinate, and which dramatically reduces the effective ν'_4 frequency. The ν'_4 frequency is reduced by so great an extent in the $\tilde{A}$ state that it is the $3\nu'_4$ overtone which is near degenerate with ν'_6. In the current work, we report the precise ν'_6 fundamental frequency in the $\tilde{A}$ state, and we determine the strength of the a-axis Coriolis interaction between $3\nu'_4$ and ν'_6. We also provide a rotational analysis of the $\nu'_4 + \nu'_6$ combination band, which interacts with $3\nu'_4$ via an additional c-axis Coriolis perturbation, and which allows us to provide a complete deperturbed fit to the $3\nu'_4$ rotational structure. Knowledge of the Coriolis interaction strengths among the lowest-lying levels in the $\tilde{A}$ state will aid the interpretation of the spectroscopy and dynamics of many higher-lying band structures, which are perturbed by analogous interactions.

TK03   2:19 – 2:34

A 1 + 1′ RESONANCE-ENHANCED MULTIPHOTON IONIZATION SCHEME FOR ROTATIONALLY STATE-SELECTIVE DETECTION OF FORMALDEHYDE VIA THE $\tilde{A}\,^1A_2 \leftarrow \tilde{X}\,^1A_1$ TRANSITION

BARRATT PARK, BASTIAN C. KRUEGER, SVEN MEYER, *Institute of Physical Chemistry, Georg-August-Universität Göttingen, Göttingen, Germany*; ALEC WODTKE, *Dynamics at Surfaces, Max Planck Institute for Biophysical Chemistry, Göttingen, Germany*; TIM SCHAEFER, *Institute of Physical Chemistry, Georg-August-Universität Göttingen, Göttingen, Germany*.

The formaldehyde molecule is an important model system for understanding dynamical processes in small polyatomic molecules. However, prior to this work, there have been no reports of a resonance-enhanced multiphoton ionization (REMPI) detection scheme for formaldehyde suitable for rovibrationally state-selective detection in molecular beam scattering experiments. Previously reported tunable REMPI schemes are either non rotationally resolved, involve multiple resonant steps, or involve many-photon ionization steps. In the current work, we present a new 1 + 1′ REMPI scheme for formaldehyde. The first photon is tunable and provides rotational resolution via the vibronically allowed $\tilde{A}\,^1A_2 \leftarrow \tilde{X}\,^1A_1$ transition. Molecules are then directly ionized from the $\tilde{A}$ state by one photon of 157 nm. The results indicate that the ionization cross section from the 4^1 vibrational level of the $\tilde{A}$ state is independent of the rotational level used as intermediate, to within experimental uncertainty. The 1 + 1′ REMPI intensities are therefore directly proportional to the $\tilde{A} \leftarrow \tilde{X}$ absorption intensities and can be used for quantitative measurement of $\tilde{X}$-state population distributions.

TK04   2:36 – 2:51

AN EMPIRICAL SPECTROSCOPIC DATABASE FOR ACETYLENE IN THE REGIONS OF 5850-9415 CM^{-1}

ALAIN CAMPARGUE, *UMR5588 LIPhy, Université Grenoble Alpes/CNRS, Saint Martin d'Hères, France*; OLEG LYULIN, *Laboratory of Theoretical Spectroscopy, Institute of Atmospheric Optics, Tomsk, Russia*.

Six studies have been recently devoted to a systematic analysis of the high-resolution near infrared absorption spectrum of acetylene recorded by Cavity Ring Down spectroscopy (CRDS) in Grenoble and by Fourier-transform spectroscopy (FTS) in Brussels and Hefei. On the basis of these works, in the present contribution, we construct an empirical database for acetylene in the 5850 - 9415 cm^{-1} region excluding the 6341-7000 cm^{-1} interval corresponding to the very strong $\nu_1 + \nu_3$ manifold. The database gathers and extends information included in our CRDS and FTS studies. In particular, the intensities of about 1700 lines measured by CRDS in the 7244-7920 cm^{-1} are reported for the first time together with those of several bands of $^{12}C^{13}CH_2$ present in natural isotopic abundance in the acetylene sample. The Herman-Wallis coefficients of most of the bands are derived from a fit of the measured intensity values. A recommended line list is provided with positions calculated using empirical spectroscopic parameters of the lower and upper energy vibrational levels and intensities calculated using the derived Herman-Wallis coefficients. This approach allows completing the experimental list by adding missing lines and improving poorly determined positions and intensities. As a result the constructed line list includes a total of 10973 lines belonging to 146 bands of $^{12}C_2H_2$ and 29 bands of $^{12}C^{13}CH_2$. For comparison the HITRAN2012 database in the same region includes 869 lines of 14 bands, all belonging to $^{12}C_2H_2$. Our weakest lines have an intensity on the order of 10^{-29} cm/molecule, about three orders of magnitude smaller than the HITRAN intensity cut off. Line profile parameters are added to the line list which is provided in HITRAN format. The comparison to the HITRAN2012 line list or to results obtained using the global effective operator approach is discussed in terms of completeness and accuracy.

TK05 2:53 – 3:08

IDENTIFICATION OF PHOTOFRAGMENTS FROM ONE-COLOR RESONANTLY-ENHANCED ($\tilde{A} - \tilde{X}$) MULTI-PHOTON PHOTODISSOCIATION OF ACETYLENE

JUN JIANG, *Department of Chemistry, MIT, Cambridge, MA, USA*; ANGELAR K MUTHIKE, *Department of Chemistry, Spelman College, Atlanta, GA, USA*; ROBERT W FIELD, *Department of Chemistry, MIT, Cambridge, MA, USA*.

One-color (212-220 nm) multi-photon photodissociation of acetylene, resonantly enhanced by the $\tilde{A}(S_1)-\tilde{X}$ transition, gives rise to strong photofragment fluorescence signals in the visible and near UV regions. In this work, fluorescence signals from the photofragments, generated with three intermediate S_1 levels ($trans\ 3^4$, $trans\ 3^5$, and $cis\ 3^1 6^1$), are studied, both in the flow cell and supersonic jet conditions. In the flow cell ($\sim$3 torr), the dispersed fluorescence (DF) spectra of the photofragments are obtained. For all three S_1 levels, we observe C_2 Swan band ($d^3\Pi_g - a^3\Pi_u$) and C_2 Deslandres-d'Azambuja band ($C^1\Pi_g - A^1\Pi_u$) emissions, with the former approximately four times more intense than the latter. In the supersonic jet condition (collision-free), fluorescence time-traces at selected wavelength regions are analyzed. We confirm the presence of the two C_2 emission bands and their relative intensity observed in the DF spectra. In the supersonic jet condition, we also observe long lifetime visible fluorescence signal (>3 μs lifetime), which is likely due to emissions from C_2H fragment, based on previous vacuum UV photolysis studies of acetylene. The photodissociation mechanism is inferred, based on our analysis of the flow cell DF spectra and the fluorescence time-traces obtained in the supersonic jet condition. The C_2H fragment is likely generated from one-photon photodissociation of S_1 acetylene, and an additional photon dissociates the C_2H fragment into the C_2 C and d states.

Intermission

TK06 3:27 – 3:42

CONFORMATIONAL ANALYSIS OF 3,3,3-TRIFLUORO-2-(TRIFLUOROMETHYL)PROPANOIC ACID

JAVIX THOMAS, *Department of Chemistry, University of Alberta, Edmonton, AB, Canada*; MICHAEL J CARRILLO, AGAPITO SERRATO III, *Chemistry, University of Texas Rio Grande Valley, Brownsville, TX, USA*; ELIJAH G SCHNITZLER, WOLFGANG JÄGER, YUNJIE XU, *Department of Chemistry, University of Alberta, Edmonton, AB, Canada*; WEI LIN, *Chemistry, University of Texas Rio Grande Valley, Brownsville, TX, USA*.

Partially fluorinated carboxylic acids exhibit rich conformational landscapes. We report the first high-resolution spectroscopic study of 3,3,3-trifluoro-2-(trifluoromethyl)propanoic acid. Its rotational spectrum was measured using both broadband chirped-pulse and narrow-band cavity-based Fourier transform microwave spectrometers. Two dominant conformers were observed, and their structures confirmed with the aid of quantum chemical calculations. Both conformers take on the Z form of the carboxylic acid group. Similarities and differences between this and other fluorinated carboxylic acids are discussed.

TK07   **3:44 – 3:59**

STRUCTURE AND TUNNELING DYNAMICS OF *gauche*-1,3-BUTADIENE

BRYAN CHANGALA, *JILA, National Institute of Standards and Technology and Univ. of Colorado Department of Physics, University of Colorado, Boulder, CO, USA*; JOSHUA H BARABAN, *Department of Chemistry and Biochemistry, University of Colorado, Boulder, CO, USA*; MARIE-ALINE MARTIN-DRUMEL, *CNRS, Institut des Sciences Moleculaires d'Orsay, Orsay, France*; SANDRA EIBENBERGER, DAVID PATTERSON, *Department of Physics, Harvard University, Cambridge, MA, USA*; JOHN F. STANTON, *Department of Chemistry, The University of Texas, Austin, TX, USA*; BARNEY ELLISON, *Department of Chemistry and Biochemistry, University of Colorado, Boulder, CO, USA*; MICHAEL C McCARTHY, *Atomic and Molecular Physics, Harvard-Smithsonian Center for Astrophysics, Cambridge, MA, USA.*

We have recently shown that *gauche*-1,3-butadiene is unambiguously non planar, with a C–C–C=C dihedral angle of about $34°$, and readily tunnels between two equivalent *gauche* structures.[a] In this talk, subsequent microwave studies of *gauche*-1,3-butadiene and its isotopologues as well as the empirical equilibrium structure will be summarized. The experiments have utilized the complementary techniques of cavity enhanced Fourier transform microwave (FTMW) spectroscopy with a supersonic expansion and chirped-pulse FTMW in a cryogenic buffer gas cell. The structural characterization is complicated by the effects of facile tunneling, and full dimensional *ab initio* rotational-VMP2 calculations have been performed to address this issue. We will show how the tunneling splitting frequency, which ranges between about 0.5 and 2.0 cm^{-1} (depending on the isotopologue), can be extracted from the experimental spectra by careful examination of tunneling-rotation perturbations.

[a]M.-A. Martin-Drumel *et al.*, ISMS 2016, MI11

TK08   **4:01 – 4:16**

FIRST HIGH RESOLUTION IR SPECTRA OF 2,2-D_2-PROPANE. THE ν_{15} (B_1) A-TYPE BAND NEAR 954.709 cm^{-1}. DETERMINATION OF GROUND AND UPPER STATE CONSTANTS.

DANIEL GJURAJ, *Department of Physics, Iona College, New Rochelle, NY, USA*; S.J. DAUNT, ROBERT GRZYWACZ, *Department of Physics & Astronomy, The University of Tennessee-Knoxville, Knoxville, TN, USA*; WALTER LAFFERTY, *Optical Technology Division, National Institute of Standards and Technology, Gaithersburg, MD, USA*; JEAN-MARIE FLAUD, *CNRS, Universités Paris Est Créteil et Paris Diderot, LISA, Créteil, France*; BRANT E. BILLINGHURST, *EFD, Canadian Light Source Inc., Saskatoon, Saskatchewan, Canada.*

As part of our project on the study of isotopologues of propane we have taken the spectra of the 2-D and 2,2-D_2 substituted species. There have been no studies of these species since the early IR studies. [a] [b] [c] [d]

We recorded high resolution ($\Delta\nu = 0.0009$ cm^{-1}) FTS data on the Canadian Light Source Far-IR beamline. The spectra of all bands of both species in the region examined (500 - 1250 cm^{-1}) show torsionally perturbed lines, all but one band appearing globally perturbed. Virtually all bands were not amenable to analysis at present except for the ν_{15} (B_1) A-type band centered at 954.709 cm^{-1}. One can still see a few perturbed lines with torsional components but overall most lines were single and could be readily assigned using traditional methods. The spectrum is modelled well using PGOPHER.[e] No MW determined GS constants were available so we have analyzed about 3500 levels to determine both ground state and upper state rotational constants.

[a]Friedman & Turkevich, J. Chem. Phys. **17**, 1012 ff. (1949); McMurry, Thornton & Condon, J. Chem. Phys. **17**, 918 ff. (1949).
[b]McMurry & Thornton, J. Chem. Phys. **19**, 1014 ff.(1951).
[c]Gayles & King, Spectrochim. Acta **21**, 543 ff.(1965).
[d]Kondo & Saeki, Spectrochim. Acta **29A**, 735 ff. (1973).
[e]Western, J. Quant. Spectrosc. Rad. Transf. **186**, 221 ff. (2017).

132

TK09 4:18 – 4:33

HIGH-RESOLUTION INFRARED SPECTROSCOPY OF IMIDAZOLE CLUSTERS IN HELIUM DROPLETS USING QUANTUM CASCADE LASERS

DEVENDRA MANI, CIHAD CAN, NITISH PAL, GERHARD SCHWAAB, MARTINA HAVENITH, *Physikalische Chemie II, Ruhr University Bochum, Bochum, Germany.*

Imidazole ring is a part of many biologically important molecules and drugs. Imidazole monomer, dimer and its complexes with water have earlier been studied using infrared spectroscopy in helium droplets[1,2] and molecular beams[3]. These studies were focussed on the N-H and O-H stretch regions, covering the spectral region of 3200-3800 cm^{-1}.

We have extended the studies on imidazole clusters into the ring vibration region. The imidazole clusters were isolated in helium droplets and were probed using a combination of infrared spectroscopy and mass spectrometry. The spectra in the region of 1000-1100 cm^{-1} and 1300-1460 cm^{-1} were recorded using quantum cascade lasers. Some of the observed bands could be assigned to imidazole monomer and higher order imidazole clusters, using pickup curve analysis and ab initio calculations. Work is still in progress. The results will be discussed in detail in the talk.

References: 1) M.Y. Choi and R.E. Miller, *J. Phys. Chem. A*, **110**, 9344 (2006). 2) M.Y. Choi and R.E. Miller, *Chem. Phys. Lett.*, **477**, 276 (2009). 3) J. Zischang, J. J. Lee and M. Suhm, *J. Chem. Phys.*, **135**, 061102 (2011).

Note: This work was supported by the Cluster of Excellence RESOLV (Ruhr-Universitat EXC1069) funded by the Deutsche Forschungsgemeinschaft.

TK10 4:35 – 4:50

LUMINESCENCE OF ADENINE MOLECULES IN GAS PHASE UNDER THE LOW ENERGY ELECTRON BEAM

Y.Y. SVYDA, M.I. SHAFRANYOSH, M.O. MARGITYCH, M.I. SUKHOVIYA, I.I. SHAFRANYOSH, *Physics, Uzhhorod National University, Uzhhorod, Ukraine.*

This report presents results of experimental study of the excitation processes of nucleic acid base adenine by electron impact. This research is a continuation of our previous studies, conducted for other nucleic acid bases – thymine[a] and uracil[b].

The adenine samples (Sigma Aldrich Company) were in form of polycrystalline powder. The gas phase of molecules was formed by heating of adenine polycrystalline powder in a separate metal container. The temperature of the container with adenine powder did not exceed 370 K. Formed gas phase of adenine proceeded by steam pipeline into cell of cubic form. A diaphragm (diameter 1.5 mm) was mounted in one of the outer edges of the cell and was used for input of the electron beam and beam source and the receiver of electron beam (a Faraday Cup) was mounted on the opposite side. Electron beam was formed by three-electrode gun with tungsten cathode. Cell was placed into magnetic field so that its field lines were parallel to the electron beam. The magnetic field was $\sim 1.2 \cdot 10^{-2}$ T. To remove the radiation from the cell two quartz windows were mounted on its two edges, which are parallel to the electron beam. Experiments were performed at following conditions: current intensity of the electron beam was within the $7 \cdot 10^{-5}$ A and $\Delta E_{1/2} \sim 0.4$ eV (FWHM) energy spread; the pressure in the chamber of the cell $\sim 1 \cdot 10^{-5}$ Pa.

Emission spectra of adenine in the wavelength range from 200 to 500 nm for different energies of exciting electrons were obtained in the study. Most intensive molecular bands whose maxima are at following wavelengths: 327, 338, 354, 388, 435, 486 nm clearly manifested in spectra. Excitation thresholds of bands are determined and their identification is performed.

[a] I.I. Shafranyosh, M.I. Sukhoviya, Opt. Spectrosc. (2007) 102: 500
[b] I.I. Shafranyosh, M.I. Sukhoviya, J.Chem.Phys. 137, 184303 (2012)

WA. Astronomy
Wednesday, June 21, 2017 – 8:30 AM
Room: 274 Medical Sciences Building

Chair: Brett A. McGuire, National Radio Astronomy Observatory, Charlottesville, VA, USA

WA01   8:30 – 8:45

THE GIGAHERTZ AND TERAHERTZ SPECTRUM of MONO-DEUTERATED OXIRANE (c-C_2H_3DO)

SIEGHARD ALBERT, ZIQIU CHEN, KAREN KEPPLER, *Laboratory of Physical Chemistry, ETH Zurich, Zürich, Switzerland*; PHILIPPE LERCH, *Swiss Light Source, Paul Scherrer Institute, Villigen, Switzerland*; MARTIN QUACK, *Laboratory of Physical Chemistry, ETH Zurich, Zürich, Switzerland*; VOLKER SCHURIG, *Institute of Organic Chemistry, University of Tubingen, Tubingen, Germany*; OLIVER TRAPP, *Department of Chemistry, Ludwig Maximilians University, Munich, Germany*.

The rotational spectrum of the chiral mono-deuterated oxirane c-C_2H_3DO, an isotopomer of oxirane (ethylenoxide), of which the normal isotopomer has already been detected in interstellar clouds, was measured in the ranges 78 to 108 GHz and 25 to 70 cm^{-1}. Thus one can expect that c-C_2H_3DO will be detectable in space in the future given the current accurate laboratory data.c-C_2H_3DO is also of interest as a simple prototypical molecule for isotopic chirality and parity violation.[a,b,c,d] The Zurich GHz spectrometer and a high resolution FTIR interferometer using synchrotron radiation was used for the THz spectrum.[d,e,f] Previous laboratory work on the rotational spectrum of deuterated oxirane extended only to the frequency of 45 GHz. A total of 119 transitions have been newly assigned in the GHz range (extended to 119 GHz) up to J=18 and 900 transitions in the THz region at most to J=70. The analyses of the rotational spectra shall be discussed in detail in relation to their astrophysical importance.

---

[a] M. Quack, *Angew. Chem. Int. Ed.* **28**, 571-586 (1989).

[b] M. Quack, *Fundamental Symmetries and Symmetry Violations from High-resolution Spectroscopy, Handbook of High Resolution Spectroscopy*, M. Quack and F. Merkt eds.,John Wiley & Sons Ltd, Chichester, New York, 2001, vol. 1, ch. 18, pp. 659-722.

[c] R. Berger, G. Laubender, M. Quack, A. Sieben, J. Stohner and M. Willeke, *Angew. Chem. Int. Ed.* **44**, 3623-3626 (2005).

[d] S. Albert, I. Bolotova, Z. Chen, C. Fábri, L. Horný, M. Quack, G. Seyfang and D. Zindel, *Phys.Chem.Chem.Phys.***18**, 21976-21993 (2016).

[e] S. Albert, I. Bolotova, Z. Chen, C. Fábri, L. Horný, M. Quack, G. Seyfang and D. Zindel,Proceedings of the 20th Symposium on Atomic, Cluster and Surface Physics (SASP 2016), Innsbruck University Press, 2016, pp. 127-130, ISBN:978-3-903122-04-8. and to be published.

[f] S. Albert, F. Arn, I. Bolotova, Z. Chen, C. Fábri, G. Grassi, Ph. Lerch, M. Quack, G. Seyfang, A. Wokaun and D. Zindel, *J.Phys.Chem.Lett*,**7**, 3847-3853 (2016).

WA02   8:47 – 9:02

THE MICROWAVE SPECTROSCOPY OF HCOO^{13}CH$_3$ IN THE SECOND TORSIONAL EXCITED STATE

KAORI KOBAYASHI, TAKURO KUWAHARA, YUKI URATA, *Department of Physics, University of Toyama, Toyama, Japan*; NOBUKIMI OHASHI, , *Kanazawa University, Kanazawa, Japan*; MASAHARU FUJITAKE, *Division of Mathematical and Physical Sciences, Graduate School of Natural Science & Technology, Kanazawa University, Kanazawa, Japan*.

Methyl formate (HCOOCH$_3$) is an abundant interstellar molecule, found almost everywhere in the star-forming region. The interstellar abundance of the ^{13}C is about 1/50 of ^{12}C. The ^{13}C substituted methyl formate in the ground and first excited states were already found toward massive star-forming regions including Orion KL. [a] With the aid of the state-of-the-art telescope like ALMA, the pure rotational transitions in the second torsional excited may be identified in the near future and laboratory data are necessary. We recorded the spectra of HCOOCH$_3$ below 340 GHz by using conventional source-modulation microwave spectrometer. The assignment of the pure rotational spectra in the second torsional excited state and the analysis by using pseudo-PAM Hamiltonian, which was effective to analyze the normal species, will be reported.

---

[a] C. Favre, M. Carvajal, D. Field, J. K. Jørgensen, S. E. Bisschop, N. Brouillet, D. Despois, A. Baudry, I. Kleiner, E. A. Bergin, N. R. Crockett, J. L. Neill, L. Marguès, T. R. Huet, and J. Demaison, *Astrophys. J. Suppl. Ser.* **215**, 25 (2014).

ROVIBRATIONAL INTERACTIONS IN THE GROUND AND TWO LOWEST EXCITED VIBRATIONAL STATES OF METHOXY ISOCYANATE

A. PIENKINA, L. MARGULÈS, R. A. MOTIYENKO, *Laboratoire PhLAM, UMR 8523 CNRS - Université Lille 1, Villeneuve d'Ascq, France*; J.-C. GUILLEMIN, *Institut des Sciences Chimiques de Rennes, UMR 6226 CNRS - ENSCR, Rennes, France.*

Recent detection of methyl isocyanate (CH_3NCO) in the Orion[a], towards Sgr B2(N)[b] and on the surface of the comet 67P/Churyumov-Gerasimenko[c] motivated us to study another isocyanate, methoxy isocyanate (CH_3ONCO) as a possible candidate molecule for searches in the interstellar clouds. Neither identification or laboratory rotational spectra of CH_3ONCO has been reported up to now.

Methoxy isocyanate was synthesized by the flash vacuum pyrolysis of N-Methoxycarbonyl-O-methyl-hydroxylamine (MeOC(O)NHOMe) at a temperature of 800 K. Experimental spectrum of CH_3ONCO was recorded in situ in the millimeter-wave range (75-105 GHz and 150-330 GHz) using Lille's fast-scan fully solid-state DDS spectrometer. The recorded spectrum is strongly perturbed due to the interaction between the overall rotation and the skeletal torsion. Perturbations affect even rotational transitions with low K_a levels of the ground vibrational state, appearing in shifting frequency predictions and intensities distortions of the lines. The interactions are significant due to the relatively small vibrational energy difference (≈ 50 cm^{-1}) between the states and different representations of the C_s symmetry point group for the ground (A'), $\nu_{18} = 1$ (A'') and $\nu_{18} = 2$ (A') vibrational states, thus leading to a "ladder" of multiple resonances by means of a-, and b-type Coriolis coupling. The global fit analysis of the rotational spectrum of methoxy isocyanate using Coriolis coupling terms in the ground and two lowest vibrational states ($\nu_{18} = 1$ and $\nu_{18} = 2$) will be presented.

This work was funded by the French ANR under the Contract No. ANR-13-BS05-0008-02 IMOLABS.

[a]J. Cernicharo, N. Marcelino, E. Roueff *et al.* 2012, *ApJ*, **759**, L43
[b]D. T. Halfen, V. V. Ilyushin, & L. M. Ziurys, 2015, *ApJ*, **812**, L5
[c]F. Goesmann, H. Rosenbauer, J. H. Bredehöft *et al.* 2015, *Science*, **349.6247**, aab0689

MILLIMETER AND SUBMILLIMETER WAVE SPECTROSCOPY OF HIGHER ENERGY CONFORMERS OF 1,2-PROPANEDIOL

OLENA ZAKHARENKO, JEAN-BAPTISTE BOSSA, FRANK LEWEN, STEPHAN SCHLEMMER, HOLGER S. P. MÜLLER, *I. Physikalisches Institut, Universität zu Köln, Köln, Germany.*

We have performed a study of the millimeter/submillimeter wave spectrum of four higher energy conformers of 1,2-propanediol (continuation of the previous study on the three lowest energy conformers[a]). The present analysis of rotational transitions carried out in the frequency range 38 – 400 GHz represents a significant extension of previous microwave work. The new data were combined with previously-measured microwave transitions and fitted using a Watson's S-reduced Hamiltonian. The final fits were within experimental accuracy, and included spectroscopic parameters up to sixth order of angular momentum, for the ground states of the four higher energy conformers following previously studied ones: $g'Ga$, $gG'g'$, aGg' and $g'Gg$. The present analysis provides reliable frequency predictions for astrophysical detection of 1,2-propanediol by radio telescope arrays at millimeter wavelengths.

[a]J.-B. Bossa, M.H. Ordu, H.S.P. Müller, F. Lewen, S. Schlemmer, A&A 570 (2014) A12

WA05   9:38 – 9:53

DETERMINATION OF METHANOL PHOTOLYSIS BRANCHING RATIOS VIA ROTATIONAL SPECTROSCOPY

CARSON REED POWERS, MORGAN N McCABE, SUSANNA L. WIDICUS WEAVER, *Department of Chemistry, Emory University, Atlanta, GA, USA.*

Methanol, a ubiquitous molecule in the interstellar medium (ISM), has an important role in the production of more complex organic molecules (COMs) in both grain-surface and gas-phase interstellar chemistry. Some of the direct products of methanol photolysis, including radicals such as methoxy, hydroxymethyl, hydroxyl, and methyl, are believed to directly influence the relative abundances of important COMs that are both detected and theorized to be in the ISM. However, no laboratory study has been performed to date which has determined the individual branching ratios of these photolysis products, because many of the channels cannot be distinguished using traditional techniques. To address this problem, we used a 193 nm excimer laser to photolyze methanol in the throat of a supersonic expansion, and probed the resultant products using a millimeter/submillimeter direct absorption spectrometer. Each product channel has a unique rotational spectrum, allowing quantitative density and temperature information to be determined. This information can in turn be used to calculate the full set of branching ratios for methanol photolysis. In this talk we will present the results of this experiment and discuss the implications for astrochemistry.

WA06   9:55 – 10:10

PHOTOPROCESSING OF METHANOL ICE: FORMATION AND LIBERATION OF CO

HOUSTON H SMITH, AJ MESKO, SAMUEL ZINGA, *Department of Chemistry, Emory University, Atlanta, GA, USA*; STEFANIE N MILAM, *Astrochemistry, NASA Goddard Space Flight Center, Greenbelt, MD, USA*; SUSANNA L. WIDICUS WEAVER, *Department of Chemistry, Emory University, Atlanta, GA, USA.*

The relevance of interstellar ice to the chemical complexity of the interstellar medium has dramatically increased over the past 15 years. Previous astrochemical models including only gas-phase reactions were unable to explain the abundances of many complex organics observed in the interstellar medium. To correct for this, current models have added grain-surface chemistry as a source for some organic molecules that serve as building blocks to biologically-relevant complex organic compounds. We have therefore built a new experiment to investigate the gas-phase chemistry above interstellar ice analogs during thermal and photoprocessing using millimeter/submillimeter spectroscopy. Our first experiments have focused on pure methanol ices to 1) demonstrate this unique technique 2) optimize the experiment and 3) to compare our results with recent work by Cruz-Diaz et al. and Beltran et al. and do further analysis of products they are unable to measure (e.g. isomers CH3O and CH2OH). We have detected CO as a major product of methanol photoprocessing. But it is unclear from our initial results how the formation and photodesorption of CO from methanol ice is related to the ice temperature during the photoprocessing. We have therefore conducted two experiments: simultaneous photoprocessing and thermal desorption, and photoprocessing at a low temperature followed by temperature programmed desorption to liberate the CO. The initial results from both of these experiments will be presented in this talk, as well as the implications of these results for astrochemistry.

WA07

INFRARED SPECTROSCOPY OF DISILICON-CARBIDE, Si$_2$C

DANIEL WITSCH, *Institute of Physics, University of Kassel, Kassel, Germany*; VOLKER LUTTER, *Institute of Physics, University Kassel, Kassel, Germany*; GUIDO W FUCHS, *Physics Department, University of Kassel, Kassel, Germany*; JÜRGEN GAUSS, *Institut für Physikalische Chemie, Universität Mainz, Mainz, Germany*; THOMAS GIESEN, *Institute of Physics, University Kassel, Kassel, Germany.*

Small silicon and carbon containing molecules are thought to be important building blocks of interstellar grains. Some of them have been detected in circumstellar environments of late-type stars by means of rotational spectroscopy e.g., SiC, SiC$_2$, Si$_2$C, c-SiC$_3$, SiC$_4$, while centro-symmetric species, e.g., C$_3$, C$_4$, C$_5$, Si$_2$C$_2$, Si$_2$C$_3$, can only be detected by vibrational transitions, mainly in the infrared. In view of a new generation of high resolution infrared telescope instruments, e.g., EXES (Echelon-Cross-Echelle Spectrograph) onboard SOFIA (Observatory for Infrared Astronomy) and TEXES (Texas Echelon Cross Echelle Spectrograph) at the Gemini-North observatory, accurate laboratory data of small silicon-carbides in the infrared region are of high demand. In this talk we present first laboratory data of the Si$_2$C asymmetric stretching mode at 1200 cm^{-1}. A pulsed Nd:YAG-laser is used to vaporize a solid target of silicon exposed to a dilute sample of methane in helium buffer gas. Si$_2$C is formed in an adiabatic expansion of a supersonic jet and radiation of a quantum cascade laser is used to record rotationally resolved spectra. To date, 160 ro-vibrational lines and have been assigned to the asymmetric stretching vibration of Si$_2$C, and derived molecular parameters are in excellent agreement with ab initio calculations. In our global fit analysis recently published microwave laboratory data (McCarthy *et al.* 2015)[a] and astronomical data (Cernicharo *et al.* 2015)[b] were taken into account. Our new results allow for the identification of Si$_2$C by means of high resolution infrared astronomy towards the warm background of carbon-rich stars.

[a]McCarthy M.C., Baraban J.H., Changala P.B., Stanton J.F., Martin-Drumel M.A, Thorwirth S., et al., *J. Chem. Phys. Lett.* **6**, 2107–2111 (2015).

[b]Cernicharo J., McCarthy M.C., Gottlieb C.A., Agundez M., Velilla Prieto L., Baraban J.H., et al. *Astrophys. J. Lett.* **806**,L3 (2015).

Intermission

WA08

FOURIER TRANSFORM SPECTROSCOPY OF THE A$^3\Pi$–X$^3\Sigma^-$ TRANSITION OF OH$^+$

JAMES NEIL HODGES, PETER F. BERNATH, *Department of Chemistry and Biochemistry, Old Dominion University, Norfolk, VA, USA.*

The OH$^+$ ion is an important species in the Interstellar Medium. It has been used to infer the cosmic ray ionization rate and is an important intermediate for generation of more complex astrochemical species. OH$^+$ observations are typically performed in the sub-millimeter and near-UV ranges, and rely on laboratory spectroscopy to provide transition frequencies. Observations of the A$^3\Pi$–X$^3\Sigma^-$ bands are used to both identify OH$^+$ and determine the column densities along sight lines.[a]

These A-X observations have relied on previous measurements with a grating spectrometer and photographic plates.[b] Here, we present data recorded at Kitt Peak using a Fourier transform spectrometer of the A-X band system. This data and other available data are combined to determine new molecular constants for the A and X electronic states. These new data are between one and two orders of magnitude more precise and should be used in support of observations in lieu of the older transition frequencies. We also intend to calculate improved line intensities in support of astronomical observations.

[a]Zhao, D. *et al.* 2015, ApJ, 805, L12

[b]Merer, A.J. *et al.* 1975, CaJPh, 53, 251

WA09 11:03 – 11:18

ASTROCHEMICAL LABORATORY EXPERIMENTS AS ANALOGS TO PLUTONIAN CHEMISTRY: USING FTIR SPECTROSCOPY TO MONITOR THE SUBLIMATION OF IRRADIATED 1:1:100 CO+H_2O+N_2 AND 1:1:100 CH_4+H_2O+N_2 ICES

KAMIL BARTŁOMIEJ STELMACH, YUKIKO YARNALL, PAUL COOPER, *Department of Chemistry and Biochemistry, George Mason University, Fairfax, VA, USA.*

Pluto is a large icy body composed of N_2, CH_4, and H_2O ices. In many ways, Pluto can be seen as one large matrix isolation experiment where N_2 is the inert matrix that can act to trap and isolate reactive species. The temperature changes on the dwarf planet induce sublimation of N_2 from the surface. Any previously trapped reactive species could then react with the new ice or neighboring molecules. To see if this process might lead to a significant formation of molecules, Fourier-Transform Infrared (FTIR) Spectroscopy (4 cm^{-1} resolution) was used to study and monitor the sublimation of ices created from irradiated gas mixtures of 1:1:100 CO+H_2O+N_2 or 1:1:100 CH_4+H_2O+N_2. The gas mixtures were initially prepared and deposited on a cold finger at a temperature of 6 K and a baseline vacuum of about 1 x 10^{-7} Torr. Gas mixtures were irradiated using an electric discharge or a microwave discharge before deposition to create the unstable chemical species. To sublimate the matrix, the temperature was brought up step-wise in 5-10 K intervals to 45 K. Slow sublimation (10 min per step) resulted in the new species being trapped in a water ice. In addition to (FTIR) spectroscopy, chemical species were also identified or monitored using ultraviolet-visible (UV-Vis) spectroscopy and a residual gas analyzer (RGA). Carbon suboxide (C_3O_2), a common component found in meteorites and a potentially important prebiotic molecule, was formed only after the sublimation step. Other products formed included deprotonated versions of products formed in the original matrix ice. C_3O_2's potential importance in Pluto's surface chemistry and its overall astrobiological significance will be discussed.

WA10 11:20 – 11:35

PHOTOCHEMICAL GENERATION OF H_2NCNX, H_2NNCX, H_2NC(NX) (X = O, S) IN LOW-TEMPERATURE MATRICES

TAMAS VOROS, GYOZO GYORGY LAJGUT, GABOR MAGYARFALVI, GYORGY TARCZAY, *Institute of Chemistry, Eotvos University, Budapest, Hungary.*

The [NH$_2$, C, N, O] and the [NH$_2$, C, N, S] systems were investigated by quantum-chemical computations and matrix-isolation spectroscopic methods. The equilibrium structures of the isomers and their relative energies were determined by CCSD(T) method. This was followed by the computation of the harmonic and anharmonic vibrational wavenumbers, infrared intensities, relative Raman activities and UV excitation energies. These computed data were used to assist the identification of products obtained by UV laser photolysis of 3,4-diaminofurazan, 3,4-diaminothiadiazole and 1,2,4-thiadiazole-3,5-diamine in low-temperature Ar and Kr matrices.[a] Experimentally, first the precursors were studied by matrix-isolation IR and UV spectroscopic methods. Based on these UV spectra, different wavelengths were selected for photolysis. The irradiations, carried out by a tunable UV laser-light source, resulted in the decomposition of the precursors, and in the appearance of new bands in the IR spectra. Some of these bands were assigned to cyanamide (H_2NCN) and its isomer, the carbodiimide molecule (HNCNH), generated from H_2NCN. By the analysis of the relative absorbance vs. photolysis time curves, the other bands were grouped to three different species both for the O- and the S-containing systems. In the case of the O-containing isomers, these bands were assigned to the H_2NNCO:H_2NCN, and H_2NCNO:H_2NCN complexes, and to the ring-structure H_2NC(NO) isomer. In a similar way, the complexes of H_2NNCS and H_2NCNS with the H_2NCN, and H_2NC(NS) were also identified. 1,2,4-thiadiazole-3,5-diamine was also investigated in similar way like the above mentioned precursors. The results of this study also support the identification of the new S-containing isomers. Except for H_2NNCO and H_2NCNS, these molecules were not identified previously. It is expected that at least some of these species, like the methyl isocyanate (CH_3CNO) isomer[b,c], are present and could be identified in astrophysical objects.

[a] T. Voros, Gy. Gy. Lajgut, G. Magyarfalvi, Gy. Tarczay, J. Chem. Phys., 146, 024305, 2017.
[b] D. T. Halfen, V. V. Ilyushin, L. M. Ziurys, Astrophys. J., 812, L5, 2015.
[c] J. Cernicharo et. al., Astron. Astrophys., 587, L4, 2016.

WA11 11:37 – 11:52

INFRARED SPECTRUM OF N-OXIDOHYDROXYLAMINE [•ONH(OH)] PRODUCED IN REACTION H + HONO IN SOLID *PARA*-HYDROGEN

KAROLINA ANNA HAUPA, *Applied Chemistry, National Chiao Tung University, Hsinchu, Taiwan*; YUAN-PERN LEE, *Applied Chemistry, National Chiao Tung University, Hsinchu, Taiwan, Institute of Atomic and Molecular Sciences, Academia Sinica, Taipei, Taiwan.*

Hydrogenation reactions in the N/O chemical network are important for an understanding of the mechanism of formation of organic molecules in dark interstellar clouds, but many reactions remain unknown. We present the results of the reaction H + HONO in solid *para*-hydrogen (p-H_2) at 3.3 K investigated with infrared spectra. Two methods that produced hydrogen atoms were the irradiation of HONO molecules in p-H_2 at 365 nm to produce OH radicals that reacted readily with nearby H_2 to produce mobile H atoms, and irradiation of Cl_2 molecules (co-deposited with HONO) in p-H_2 at 405 nm to produce Cl atoms that reacted readily with nearby H_2 to produce mobile H atoms. In both experiments, we assigned IR lines at 3549.6 (ν_1), 1465.0 (ν_3), 1372.2 (ν_4), 895.6/898.5 (ν_6), and 630.9 (ν_7) cm^{-1} to N-oxidohydroxylamine [•ONH(OH)], the primary product of HONO hydrogenation. The assignments were derived according to the consideration of possible reactions and comparison of observed vibrational wavenumbers and their IR intensities with values predicted with the B3LYP/aug-cc-pVTZ method of quantum-chemical calculations. The agreement between observed and calculated D/H- and $^{15}N/^{14}N$-isotopic ratios further supports these assignments. The role of this reaction in the N/O chemical network in dark interstellar clouds is discussed.

WA12 11:54 – 12:09

INFRARED SPECTRA OF PROTONATED QUINOLINE (1-$C_9H_7NH^+$) IN SOLID *PARA*-HYDROGEN

CHIH-YU TSENG, *Department of Applied Chemistry, National Chiao Tung University, Hsinchu, Taiwan*; YUAN-PERN LEE, *Applied Chemistry, National Chiao Tung University, Hsinchu, Taiwan, Institute of Atomic and Molecular Sciences, Academia Sinica, Taipei, Taiwan.*

Large protonated polycyclic aromatic hydrocarbons (H^+PAH) and polycyclic aromatic nitrogen heterocycles (H^+PANH) have been proposed as possible carriers of unidentified infrared (UIR) emission bands from galactic objects. The nitrogen atom in H^+PANH is expected to induce a blue shift of the C=C stretching band near 6.2 μm so that their emission bands might agree with the UIR band better than those of H^+PAH.

In this work, we report the IR spectrum of protonated quinoline and its neutral species measured upon electron bombardment during deposition of a mixture of quinoline and *para*-hydrogen at 3.2 K. New features were assigned to 1-$C_9H_7NH^+$ and 1-C_9H_7NH, indicating that the protonation and hydrogenation occur at the N-atom site. The intensities of features of 1-$C_9H_7NH^+$ diminished when the matrix was maintained in darkness for 10 h, whereas those of 1-C_9H_7NH increased. Spectral assignments were made according to comparison of experimental results with anharmonic vibrational wavenumbers and IR intensities calculated with the B3LYP/6-311++G(d,p) method. Although agreement between the observed spectrum of 1-$C_9H_7NH^+$ and the UIR emission bands is unsatisfactory, presumably because of the small size of quinoline, we did observe C=C stretching bands at 1641.4, 1598.4, 1562.0 cm^{-1}, blue-shifted from those at 1618.7, 1580.8, 1510.0 cm^{-1} of the corresponding protonated PAH ($C_{10}H_9^+$), pointing to the direction of the UIR bands.

WB. Mini-symposium: Multiple Potential Energy Surfaces

Wednesday, June 21, 2017 – 8:30 AM

Room: 116 Roger Adams Lab

Chair: Jinjun Liu, University of Louisville, Louisville, KY, USA

WB01   *INVITED TALK*   8:30 – 9:00

LIGHT, MOLECULES, ACTION: USING ULTRAFAST UV-VISIBLE AND X-RAY SPECTROSCOPY TO PROBE EXCITED STATE DYNAMICS IN PHOTOACTIVE MOLECULES

R.J. SENSION, *Department of Chemistry, University of Michigan, Ann Arbor, MI, USA.*

Light provides a versatile energy source capable of precise manipulation of material systems on size scales ranging from molecular to macroscopic. Photochemistry provides the means for transforming light energy from photon to process via movement of charge, a change in shape, a change in size, or the cleavage of a bond. Photochemistry produces action. In the work to be presented here ultrafast UV-Visible pump-probe, and pump-repump-probe methods have been used to probe the excited state dynamics of stilbene-based molecular motors, cyclohexadiene-based switches, and polyene-based photoacids. Both ultrafast UV-Visible and X-ray absorption spectroscopies have been applied to the study of cobalamin (vitamin B_{12}) based compounds. Optical measurements provide precise characterization of spectroscopic signatures of the intermediate species on the S_1 surface, while time-resolved XANES spectra at the Co K-edge probe the structural changes that accompany these transformations.

WB02   9:04 – 9:19

BLACK BOX REAL-TIME TRANSIENT ABSORPTION SPECTROSCOPY AND ELECTRON CORRELATION

JOHN PARKHILL, *Chemistry, The University of Notre Dame, Notre Dame, IN , USA.*

We introduce an atomistic, all-electron, black-box electronic structure code to simulate transient absorption (TA) spectra and apply it to simulate pyrazole and a GFP- chromophore derivative1. The method is an application of OSCF2, our dissipative exten- sion of time-dependent density-functional theory. We compare our simulated spectra directly with recent ultra-fast spectroscopic experiments. We identify features in the TA spectra to Pauli-blocking which may be missed without a first-principles model. An important ingredient in this method is the stationary-TDDFT correction scheme recently put forwards by Fischer, Govind, and Cramer which allows us to overcome a limitation of adiabatic TDDFT. We demonstrate that OSCF2 is able to reproduce the energies of bleaches and induced absorptions, as well as the decay of the transient spectrum, with only the molecular structure as input. We show that the treatment of electron correlation is the biggest hurdle for TA simulations, which motivates the second half of the talk a new method for realtime electron correlation.

We continue to derive and propagate self-consistent electronic dynamics. Extending our derivation of OSCF2 to include electron correlation we obtain a non-linear correlated one-body equation of motion which corrects TDHF. Similar equations are known in quantum kinetic theory, but rare in electronic structure. We introduce approximations that stabilize the theory and reduce its computational cost. We compare the resulting dynamics with well-known exact and approximate theories showing improvements over TDHF. When propagated EE2 changes occupation numbers like exact theory, an important feature missing from TDHF or TDDFT. We introduce a rotating wave approximation to reduce the scaling of the model to $O(N^4)$, and enable propagation on realistically large systems. The equation-of-motion does not rely on a pure-state model for the electronic state, and could be used to study the relationship between electron correlation and relaxation/dephasing or as a non-adiabatic kernel for TDDFT. We show that a quasi-thermal Fermi-Dirac population of one-particle states is a stationary state of the method reached as the endpoint of propagation in some limits. We discuss this 'thermalization' of an isolated quantum many-body system in the context of the eigenstate thermalization hypothesis.

WB03 9:21 – 9:36

RESONANCE-ENHANCED EXCITED-STATE RAMAN SPECTROSCOPY OF CONJUGATED THIOPHENE DERIVATIVES: COMBINING EXPERIMENT WITH THEORY

MATTHEW S. BARCLAY, TIMOTHY J QUINCY, MARCO CARICATO, CHRISTOPHER G. ELLES, *Department of Chemistry, University of Kansas, Lawrence, KS, USA.*

Resonance-enhanced Femtosecond Stimulated Raman Spectroscopy (FSRS) is an ultrafast experimental method that allows for the study of excited-state structural behaviors, as well as the characterization of higher electronically excited states accessible through the resonant conditions of the observed vibrations. However, interpretation of the experiment is difficult without an accurate vibrational assignment of the resonance-enhanced spectra. We therefore utilize simulations of off-resonant excited-state Raman spectra, in which we employ a numerical derivative of the analytical excited-state polarizabilities along the normal mode displacements, in order to identify and interpret the resonance-enhanced vibrations observed in experiment. We present results for a benchmark series of conjugated organic thiophene derivatives, wherein we have computed the off-resonant excited-state Raman spectra for each molecule and matched it with its resonance-enhanced experimental spectrum. This comparison allows us to successfully identify the vibrational displacements of the observed FSRS bands, as well as validate the accuracy of the theoretical results through an experimental benchmark. The agreement between the experimental and computed results demonstrates that we are able to predict qualitatively accurate excited-state Raman spectra for these conjugated thiophenes, allowing for a more thorough interpretation of excited-state Raman signals at relatively low computational cost.

WB04 9:38 – 9:53

RESONANT FEMTOSECOND STIMULATED RAMAN BAND INTENSITY AND S_n STATE ELECTRONIC STRUCTURE

TIMOTHY J QUINCY, MATTHEW S. BARCLAY, MARCO CARICATO, CHRISTOPHER G. ELLES, *Department of Chemistry, University of Kansas, Lawrence, KS, USA.*

Femtosecond Stimulated Raman Spectroscopy (FSRS) is a powerful technique capable of providing dynamic vibrational information on molecular excited states. When combined with transient electronic spectroscopies such as Pump-Probe or Pump-Repump-Probe, the excited state dynamics can be viewed with greater clarity. Due to the low intensities of Raman scattering typical for FSRS, experiments are commonly performed with the Raman pump in resonance with the excited state absorption to take advantage of resonance enhancement. However, the inherent information about the resonant state embedded in the Raman scattering is not a well explored component of the technique. 2,5-diphenylthiophene (DPT) in solution is used as a model system to study the wavelength dependence of the excited state Raman resonance enhancement. DPT has strong excited state absorption and stimulated emission bands within the tunable range of the Raman pump, allowing a wide variety of resonance conditions to be probed. Varying the Raman pump wavelength across the excited state absorption band produces different trends in both the absolute and relative magnitudes of the resulting FSRS vibrational modes. Comparing with calculations of the S_1 vibrational modes, we determine the structure of the resonant S_n state potential energy surface based on the motions of the resonantly enhanced vibrations.

Intermission

WB05 10:12 – 10:27

FEMTOSECOND ELEMENT-SPECIFIC XUV SPECTROSCOPY OF COMPLEX MOLECULES AND MATERIALS

JOSH VURA-WEIS, *Department of Chemistry, University of Illinois at Urbana-Champaign, Urbana, IL, USA.*

Systems with multiple heavy atoms, such as the multimetallic clusters favored by Nature for redox catalysis and emerging photovoltaic materials such as $CH_3NH_3PbI_3$, pose challenges for traditional spectroscopic techniques. The growing field of high-harmonic extreme ultraviolet spectroscopy combines the element-, oxidation state-, spin state-, and ligand field specificity of XANES spectroscopy with the femtosecond time resolution of tabletop Ti:Sapphire lasers. We will show that this technique can be used to measure the photophysics of transition metal complexes, organohalide perovskites, and even small metalloproteins, extending the technique to mainstream problems in physical and inorganic chemistry.

WB06 10:29 – 10:44

ULTRAFAST TRANSIENT ABSORPTION SPECTROSCOPY INVESTIGATION OF EXCITED-STATE DYNAMICS OF METHYL AMMONIUM LEAD BROMIDE PEROVSKITE NANOSTRUCTURES

ABDELQADER JAMHAWI, HAMZEH TELFAH, *Department of Chemistry, University of Louisville, Louisville, KY, USA*; MEGHAN B TEUNIS, RAJESH SARDAR, *Department of Chemistry, Indiana University-Purdue University Indianapolis, Indianapolis, IN, USA*; JINJUN LIU, *Department of Chemistry, University of Louisville, Louisville, KY, USA.*

Metal halide perovskites are promising materials for light harvesting. Power conversion efficiency (PCE) of such materials can be improved by tuning the band gap energy and suppressing the trap states. In particular, quantum confinement and shape control of nanostructures are two effective approaches to enhance the photovoltaic properties. Here we report femtosecond transient absorption (TA) spectroscopy studies on the photo-induced dynamics of a series of different nanostructures of methyl ammonium lead bromide (MALB) $CH_3NH_3PbBr_3$: nanoplatelets (2D), nanowires (1D), nonoparticles (0D), and nanocubes (0D). Experimentally obtained TA spectra are simulated using a global model in both the time and wavelength domains. The fit values of center wavelengths and time constants for various processes demonstrate that dimensional and structural confinement affects not only band structure but also exciton dynamics: Sub-picosecond electron and hole relaxation (in the conduction and valence band, respectively) have been observed, while the exciton recombination process is on the timescale of hundreds of picoseconds. Comparison between TA spectra of different nanostructures suggest that the confinement effect plays a significant role in tuning band gaps and minimizing trap states, which can be utilized to improve the PCE of photovoltaic devices.

WB07 10:46 – 11:01

ULTRAFAST TRANSIENT ABSORPTION SPECTROSCOPY INVESTIGATION OF PHOTOINDUCED DYNAMICS IN POLY(3-HEXYLTHIOPHENE)-BLOCK-OLIGO(ANTHRACENE-9,10-DIYL)

JACOB STRAIN, *Department of Chemistry, University of Louisville, Louisville, KY, USA*; HEMALI RATH-NAYAKE, *Chemistry, Western Kentucky University, Bowling Green, KY, USA*; JINJUN LIU, *Department of Chemistry, University of Louisville, Louisville, KY, USA.*

Semiconducting polymer nanostructures featuring bulk heterojunction (BHJ) architecture are promising light harvesters in photovoltaic (PV) devices because they allow control of individual domain sizes, internal structure and ordering, as well as well-defined contact between the electron donor and acceptor. Power conversion efficiency (PCE) of PV devices strongly depends on photoinduced dynamics. Understanding and optimizing photoinduced charge transfer processes in BHJ's hence help improve the performance of PV devices and increase their PCE in particular. We have investigated the photoinduced dynamics of a block polymer containing moieties of poly-3-hexylthiophene (P3HT) and polyanthracene (PANT) in solution and in solid state with femtosecond transient absorption (TA) spectroscopy. The dynamics of the polymer PANT alone are also studied as a control. The TA spectra of PANT includes a strong excited state absorption centered at 610 (nm) along with a stimulated emission signal stretching past the detection limit into the UV region which is absent in the monomer's spectra in the detection window. The block polymer's TA spectra strongly resembles that of P3HT but a noticeable positive pull on P3HT's stimulated emission signal residing at 575-620 (nm) is indicative of the excited state absorption of PANT in the adjacent spectral region. The doubling of the lifetime exciton delocalization on the block polymer versus P3HT alone have alluded that the lifetime of P3HT is extended by the covalent addition of PANT. The current spectroscopic investigation represents an interesting example of photoinduced processes in systems with complex energy level structure. Studies of dependence of change generation and separation on composition, dimension, and morphology of the heterojunctions are in process.

142

PHOTOCHEMICAL DYNAMICS OF INTRAMOLECULAR SINGLET FISSION

<u>ZHOU LIN</u>, HIKARI IWASAKI, TROY VAN VOORHIS, *Department of Chemistry, Massachusetts Institute of Technology, Cambridge, MA, USA.*

Singlet fission (SF) converts a singlet exciton (S_1) into a pair of triplet ones (T_1) via a "multi-exciton" (ME) intermediate: $S_1 \longleftrightarrow {}^1ME \longleftrightarrow {}^1(T_1T_1) \longrightarrow 2T_1$.[a] In exothermic cases, *e.g.*, crystalline pentacene or its derivatives, the quantum yield of SF can reach 200%. With SF doubling the electric current generated by an incident high-energy photon, the solar conversion efficiency in pentacene-based organic photovoltaics (OPVs) can exceed the Shockley–Queisser limit of 33.7%.[b] The ME state is popularly considered to be a dimeric state with significant charge transfer (CT) character that is strongly coupled to both S_1 and ${}^1(T_1T_1)$,[c] while this local model lacks strong support from full quantum dynamics studies. Intramolecular SF (ISF) occurring to covalently-bound dimers in the solution phase is an excellent model for a straightforward dynamics simulation of local excitons. In the present study, we investigate the ISF mechanisms for three covalently-bound dimers of pentacene derivatives, including *ortho*-, *meta*-, and *para*-bis(6,13-bis(triisopropylsilylethynyl)pentacene)benzene, in non-protic solvents. Specifically, we propagate the real-time, non-adiabatic quantum mechanical/molecular mechanical (QM/MM) dynamics on the potential energy surfaces associated with the states of S_1, ${}^1(T_1T_1)$ and CT.[d,e] We explore how the energies of these ISF-relevant states and the non-adiabatic couplings between each other fluctuate with time and the instantaneous molecular configuration (*e.g.*, intermonomer distance and orientation). We also quantitatively compare Condon and non-Condon ISF dynamics with solution-phase spectroscopic data. Our results allow us to understand the roles of CT energy levels in the ISF mechanism and propose a design strategy to maximize ISF efficiency.

[a]M. B. Smith and J. Michl, *Chem. Rev.* **110**, 6891 (2010).

[b]W. Shockley and H. J. Queisser, *J. Appl. Phys.* **32**, 510 (1961).

[c]T. C. Berkelbach, M. S. Hybertsen, and D. R. Reichman, *J. Chem. Phys.* **141**, 074705 (2014).

[d]M. G. Mavros, D. Hait, and T. A. Van Voorhis, *J. Chem. Phys.* **145**, 214105 (2016).

[e]V. Vaissier, and T. A. Van Voorhis, *in preparation*

KEY INTERMEDIATES OF CARBON DIOXIDE REDUCTION ON SILVER FROM VIBRATIONAL NANOSPEC-TROSCOPY

<u>PRASHANT JAIN</u>, *Department of Chemistry, University of Illinois at Urbana-Champaign, Urbana, IL, USA.*

The design of efficacious, selective heterogeneous catalysts relies on the knowledge of the nature of active sites and reactive intermediates involved in the catalytic transformation. This is also true in the case of carbon dioxide reduction, an important scientific and technological problem. With the goal of furthering mechanistic understanding of a complex transformation that yields multiple products, we are employing surface enhanced Raman scattering (SERS) to image carbon dioxide photoreduction on individual Ag nanoparticles within a heterogeneous dispersion. The lack of ensemble-averaging is allowing us to detect fleeting intermediates in the adsorption and catalytic photoreduction processes. In particular, we have detected on some sites physisorbed CO_2 and at others chemisorbed CO_2^- anion radical, a critical intermediate in carbon dioxide reduction. The primary product formed also appears to vary from one catalytic nanoparticle to another: CO, formaldehyde, or formic acid. The origin of such heterogeneities in adsorption and photoreduction behavior are being traced to differences in nanoparticle structure or surface composition, from which structure/activity relationships will be established, with aid from electronic structure theory. This single-nanoparticle approach is providing molecular-level insights into a broad range of industrially and environmentally relevant catalytic transformations.

TWO-PHOTON EXCITATION OF CONJUGATED MOLECULES IN SOLUTION: SPECTROSCOPY AND EXCITED-STATE DYNAMICS

CHRISTOPHER G. ELLES, AMANDA L. HOUK, *Department of Chemistry, University of Kansas, Lawrence, KS, USA*; MARC DE WERGIFOSSE, ANNA KRYLOV, *Department of Chemistry, University of Southern California, Los Angeles, CA, USA.*

We examine the two-photon absorption (2PA) spectroscopy and ultrafast excited-state dynamics of several conjugated molecules in solution. By controlling the relative wavelength and polarization of the two photons, the 2PA measurements provide a more sensitive means of probing the electronic structure of a molecule compared with traditional linear absorption spectra. We compare experimental spectra of trans-stilbene, cis-stilbene, and phenanthrene in solution with the calculated spectra of the isolated molecules using EOM-EE-CCSD. The calculated spectra show good agreement with the low-energy region of the experimental spectra (below 6 eV) after suppressing transitions with strong Rydberg character and accounting for solvent and method-dependent shifts of the valence transitions. We also monitor the excited state dynamics following two-photon excitation to high-lying valence states of trans-stilbene up to 6.5 eV. The initially excited states rapidly relax to the lowest singlet excited state and then follow the same reaction path as observed following direct one-photon excitation to the lowest absorption band at 4.0 eV.

WC. Conformers, isomers, chirality, stereochemistry
Wednesday, June 21, 2017 – 8:30 AM
Room: 1024 Chemistry Annex

Chair: Josh Newby, Hobart and William Smith Colleges, Geneva, NY, USA

WC01   8:30 – 8:45

ROTATIONAL SPECTROSCOPY AND CONFORMATIONAL STUDIES OF 4-PENTYNENITRILE, 4-PENTENENITRILE, AND GLUTARONITRILE

BRIAN M HAYS, DEEPALI MEHTA-HURT, KHADIJA M. JAWAD, ALICIA O. HERNANDEZ-CASTILLO, CHAMARA ABEYSEKERA, DI ZHANG, TIMOTHY S. ZWIER, *Department of Chemistry, Purdue University, West Lafayette, IN, USA.*

The pure rotational spectra of 4-pentynenitrile, 4-pentenenitrile, and glutaronitrile were acquired using chirped pulse Fouirer transform microwave spectroscopy. 4-pentynenitrile and 4-pentenenitrile are the recombination products of two resonance stabilized radicals, propargyl + cyanomethyl or allyl + cyanomethyl, respectively, and are thus anticipated to be significant among the more complex nitriles in Titan's atmosphere. Indeed, these partially unsaturated alkyl cyanides have been found in laboratory analogs of tholins and are also expected to have interesting photochemistry. The optimized structures of all conformers below predicted energies of 500 cm^{-1} were calculated for each molecule. Both of the conformers, trans and gauche, for 4-pentynenitrile have been identified and assigned. Five conformers were assigned in 4-pentenenitrile. The eclipsed conformers, with respect to the vinyl group, dominate the spectrum but some population was found in the syn conformers including the syn-gauche conformer, calculated to be 324 cm^{-1} above the global minimum. The glutaronitrile spectrum contained only the two conformers below 500 cm^{-1}, with reduced amount of the gauche trans conformer. The assigned spectra and structural assignments will be presented.

WC02   8:47 – 9:02

THE CONFORMER SPECIFIC ROTATIONAL SPECTRUM OF 3-PHENYLPROPIONITRILE UTILIZING STRONG FIELD COHERENCE BREAKING

SEAN FRITZ, ALICIA O. HERNANDEZ-CASTILLO, CHAMARA ABEYSEKERA, TIMOTHY S. ZWIER, *Department of Chemistry, Purdue University, West Lafayette, IN, USA.*

The 8-18 GHz conformer specific rotational spectrum of gauche- and anti-3-phenylpropionitrile (C6H5-CH2-CH2-CN) conformers has been recorded using the strong field coherence breaking (SFCB) technique [1] with a modified line picking scheme for multiple selective excitations (MSE). As the recombination product of benzyl and cyanomethyl resonance-stabilized radicals, 3-phenylpropionitrile is a likely component of the complex organics in Titan's atmosphere, motivating its structural characterization. Details of the modified line picking scheme, hyperfine constants and relative population ratios of the two conformers will be presented.

[1] A.O Hernandez-Castillo, Chamara Abeysekera, Brian M. Hays, Timothy S. Zwier, "Broadband Multi-Resonant Strong Field Coherence Breaking as a Tool for Single Isomer Microwave Spectroscopy." J. Chem. Phys. 145, 114203 (2016).

WC03   9:04 – 9:19

EFFECT OF INTRAMOLECULAR DISPERSION INTERACTIONS ON THE CONFORMATIONAL PREFERENCES OF MONOTERPENOIDS

DONATELLA LORU, ANNALISA VIGORITO, ANDREIA SANTOS, JACKSON TANG, M. EUGENIA SANZ, *Department of Chemistry, King's College London, London, United Kingdom.*

The rotational spectra of several monoterpenoids have been reinvestigated with a 2-8 GHz chirped pulse FTMW spectrometer. Axial conformers, in addition to previously reported equatorial conformers[a,b], have been detected for carvone, perillaldehyde, and limonene. Observation of the ^{13}C isotopologues of these monoterpenoids in their natural abundances allowed the determination of r_s and r_0 structures. Axial conformers are stabilised by dispersion interactions between the six-membered ring of the monoterpenoids and the isopropenyl group. Comparison of experimental data with ab initio and density functional calculations shows that an accurate description of dispersion interactions is still a challenge for theoretical methods.

[a] J. R. Avilés Moreno, F. Partal Ureña, J. J. López González and T. R. Huet, Chem. Phys. Lett., 2009, 473, 17–20.
[b] J. R. Avilés Moreno, T. R. Huet and J. J. López González, Struct. Chem., 2013, 24, 1163–1170.

WC04

CONFORMATIONAL STUDY OF DNA SUGARS: FROM THE GAS PHASE TO SOLUTION

ICIAR URIARTE, MONTSERRAT VALLEJO-LÓPEZ, EMILIO J. COCINERO, *Physical Chemistry Department, Universidad del País Vasco (UPV/EHU), Bilbao, Spain*; FRANCISCO CORZANA, *Department of Chemistry, University of La Rioja, Logroño, Spain*; BENJAMIN G. DAVIS, *Department of Chemistry, Oxford University, Oxford, United Kingdom.*

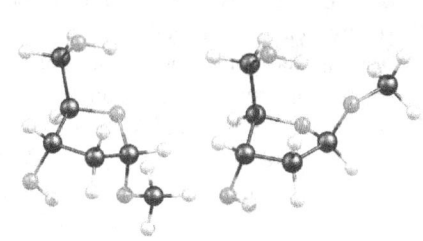

Sugars are versatile molecules that play a variety of roles in the organism. For example, they are important in energy storage processes or as structural scaffolds. Here, we focus on the monosaccharide present in DNA by addressing the conformational and puckering properties in the gas phase of α- and β-methyl-2-deoxy-ribofuranoside and α- and β-methyl-2-deoxy-ribopiranoside. Other sugars have been previously studied in the gas phase[a,b].

The work presented here stems from a combination of chemical synthesis, ultrafast vaporization methods, supersonic expansions, microwave spectroscopy (both chirped-pulsed and Balle-Flygare cavity-based spectrometers) and NMR spectroscopy. Previous studies in the gas phase had been performed on 2-deoxyribose[c], but only piranose forms were detected. However, thanks to the combination of these techniques, we have isolated and characterized for the first time the conformational landscape of the sugar present in DNA in its biologically relevant furanose form. Our gas phase study serves as a probe of the conformational preferences of these biomolecules under isolation conditions. Thanks to the NMR experiments, we can characterize the favored conformations in solution and extract the role of the solvent in the structure and puckering of the monosaccharides.

[a]E. J. Cocinero, A. Lesarri, P. Écija, F. J. Basterretxea, J.-U. Grabow, J. A. Fernández, F. Castaño, *Angew. Chem. Int. Edit.* 2012, **51**, 3119.

[b]P. Écija, I. Uriarte, L. Spada, B. G. Davis, W. Caminati, F. J. Basterretxea, A. Lesarri, E. J. Cocinero, *Chem. Commun.* 2016, **52**, 6241.

[c]I. Peña, E. J. Cocinero, C. Cabezas, A. Lesarri, S. Mata, P. Écija, A. M. Daly, Á. Cimas, C. Bermúdez, F. J. Basterretxea, S. Blanco, J. A. Fernández, J. C. López, F. Castaño, J. L. Alonso, *Angew. Chem. Int. Edit.* 2013, **52**, 11840.

WC05

FOUR STRUCTURES OF TARTARIC ACID REVEALED IN THE GAS PHASE

VANESSA CORTIJO, VERÓNICA DÍEZ, ELENA R. ALONSO, SANTIAGO MATA, JOSÉ L. ALONSO, *Grupo de Espectroscopia Molecular, Lab. de Espectroscopia y Bioespectroscopia, Unidad Asociada CSIC, Universidad de Valladolid, Valladolid, Spain.*

The tartaric acid, one of the most important organic compounds, has been transferred into the gas phase by laser ablation of its natural crystalline form (m.p.174°C) and probed in a supersonic expansion by chirped-pulse Fourier transform microwave spectroscopy (CP-FTMW). Four stable structures, two with an extended (trans) disposition of the carbon chain and two with a bent (gauche) disposition, have been unequivocally identified on the basis of the experimental rotational constants in conjunction with ab initio predictions. The intramolecular interactions that govern the conformational preferences are dominated by cooperative O-H···O=C type and O-H...O hydrogen bonds extended along the entire molecule. The observation of only μc- type spectra for one "trans" and one "gauche" conformers, support the existence of a C2 symmetry for each structure.

Intermission

WC06 10:12 – 10:27

THE CONFORMATIONAL LANDSCAPE OF L-THREONINE: MATRIX ISOLATION INFRARED AND *AB-INITIO* STUDIES

PANKAJ DUBEY[a], ANAMIKA MUKHOPADHYAY[b], K S VISWANATHAN, *Chemical Science, Indian Institute of Science Education and Research, MOHALI, PUNJAB, India.*

Amino acids, containing hydroxy side chains such as L-threonine and tyrosine play an important role in molecular recognition, such as in the docking of propofol, which is a commonly used anaesthetic. A rich conformational landscape of these amino acids makes them interesting candidates in the study of intra and intermolecular interactions. In this work, the conformational landscape of L-threonine was studied, as it can be expected to serve as a basis for understanding structure and functions of polypeptides and other biomolecules. The matrix isolation technique (MI) coupled with a high temperature effusive molecular beam (EMB) nozzle was used to trap conformers of amino acid, which were then characterized using FTIR spectroscopy. The usefulness of MI-EMB-FTIR spectroscopy is that it can trap structures corresponding to the local minima along with the global minimum and hence allows for a better exploration of the potential energy surface. A major challenge in conformational analysis of amino acids using matrix isolation FTIR arises from its non-volatile nature. A home built heating system which was mounted close to the cryotip, was used to evaporate the non-volatile amino acids. Our infrared spectra show that three conformations were trapped in the matrix. Experimental results were supported by *ab-initio* calculations performed using the CCSD(T), MP2 and M06-2X methods together with 6-311++G(d,p) and aug/cc-pVDZ basis sets. The side chains of the amino acids appeared to have an influence on the preferential stabilisation of a particular backbone structure of amino acids. Factors such as entropy, anomeric effect and intramolecular H-bonding were also found to play an important role in determining conformal preferences, which will be discussed.

[a]PD acknowledges fellowship from MHRD, India. Authors thank IISER, Mohali for facilities.

[b]AM thanks CSIR India for the Senior Research Associateship

WC07 10:29 – 10:44

THE ROLE OF THE LOCAL CONFORMATION OF A CYCLICALLY CONSTRAINED β-AMINO ACID IN THE SECONDARY STRUCTURES OF A MIXED α/β DIASTEREOMER PAIR

KARL N. BLODGETT, TIMOTHY S. ZWIER, *Department of Chemistry, Purdue University, West Lafayette, IN, USA.*

Synthetic foldamers are non-natural polymers designed to fold into unique secondary structures that either mimic nature's preferred secondary structures, or expand their possibilities. Among the most studied synthetic foldamers are β-peptides, which lengthen the distance between amide groups from the single substituted carbon spacer in α-peptides by one (β) additional carbon. Cyclically constrained β-amino acids can impart rigidity to the secondary structure of oligomers by locking in a particular conformation. The β-residue *cis*-2-aminocyclohexanecarboxylic acid (*cis*-ACHC) is one such amino acid which has been shown to drive vastly different secondary structures as a function of the local conformation of the cyclohexane ring. We present data on two diastereomers of the mixed α/β tri-peptide Ac-Ala-β_{ACHC}-Ala-NHBn which differ from one another by the chirality along the ACHC residue (SRSS vs. SSRS). The first oligomer is known to crystallize to a 9/11 mixed helix while the second forms no intramolecular hydrogen bonds in the crystal state. This talk will describe the conformation-specific IR and UV spectroscopy of the above two diastereomers under jet cooled conditions in the gas phase. Assignments based on comparison with calculations show the presence of incipient 9/11 mixed helices and competing structures containing more tightly folded hydrogen-bonded networks. The calculated global minimum structures are observed in each case, and in each case these folded structures are reminiscent of a β-turn.

WC08   **10:46 – 11:01**

CONFORMATIONAL EXPLOSION: UNDERSTANDING THE COMPLEXITY OF THE PARA-DIALKYLBENZENE POTENTIAL ENERGY SURFACES

PIYUSH MISHRA, DANIEL M. HEWETT, TIMOTHY S. ZWIER, *Department of Chemistry, Purdue University, West Lafayette, IN, USA.*

This talk focuses on the single-conformation spectroscopy of small-chain para-dialkylbenzenes. This work builds on previous studies from our group on long-chain n-alkylbenzenes that identified the first folded structure in octylbenzene. The dialkylbenzenes are representative of a class of molecules that are common components of coal and aviation fuel and are known to be present in vehicle exhaust. We bring the molecules para-diethylbenzene, para-dipropylbenzene and para-dibutylbenzene into the gas phase and cool the molecules in a supersonic expansion. The jet-cooled molecules are then interrogated using laser-induced fluorescence excitation, fluorescence dip IR spectroscopy (FDIRS) and dispersed fluorescence. The LIF spectra in the S_0-S_1 origin region show dramatic increases in the number of resolved transitions with increasing length of alkyl chains, reflecting an explosion in the number of unique low-energy conformations formed when two independent alkyl chains are present. Since the barriers to isomerization of the alkyl chain are similar in size, this results in an 'egg carton' shape to the potential energy surface. We use a combination of electronic frequency shift and alkyl CH stretch infrared spectra to generate a consistent set of conformational assignments.

WC09   **11:03 – 11:18**

BEYOND THE BEND: EXPLORING THE CONFORMATIONAL LANDSCAPE OF DECYL, UNDECYL, AND DODE-CYLBENZENE

DANIEL M. HEWETT, TIMOTHY S. ZWIER, *Department of Chemistry, Purdue University, West Lafayette, IN, USA.*

Alkylbenzenes are important components in the combustion process: they make up 20-30% of petroleum fuels and are intermediates on the pathway to soot formation. Understanding their conformational preferences is a vital step in understanding the processes by which fuels begin their journey from small, simple hydrocarbons into the large, graphitic masses of soot. Previous work done in our group, in collaboration with the Sibert group, found that the smallest alkylbenzene which folds its chain back over the ring is octylbenzene. The population of the lone folded structure in octylbenzene is low; however, theory predicts a rapid stabilization of the folded conformations relative to more extended structures as the chain length is increased, suggesting a likely shift in population towards folded structures. This talk will focus on our exploration of this possibility by discussing the UV excitation and single conformation IR spectra of decyl, undecyl, and dodecylbenzene, where increasing chain length allows for multiple stable folded configurations.

WC10   **11:20 – 11:35**

CONFORMER-SPECIFIC IR SPECTROSCOPY OF LASER-DESORBED SULFONAMIDE DRUGS: TAUTOMERIC AND CONFORMATIONAL PREFERENCES OF SULFANILAMIDE AND ITS DERIVATIVES

THOMAS UHLEMANN, SEBASTIAN SEIDEL, CHRISTIAN W. MÜLLER, *Physikalische Chemie II, Ruhr University Bochum, Bochum, Germany.*

Molecules containing the sulfonamide group R^1-SO_2-NHR^2 have a longstanding history as antimicrobial agents. Even though nowadays they are not commonly used in treating humans anymore, they continue to be studied as effective inhibitors of metalloenzyme carbonic anhydrases. These enzymes are important targets for a variety of diseases, such as, for instance, breast cancer, glaucoma, and obesity. Here we present the results of our laser desorption single-conformation UV and IR study of sulfanilamide (NH_2Ph-SO_2-NHR, R=H), a variety of singly substituted derivatives, and their monohydrated complexes. Depending on the substituent, the sulfonamide group can either adopt an amino or an imino tautomeric form. The form prevalent in the crystal is not necessarily also the tautomeric form we identified in the molecular beam after laser desorbing the sample. Furthermore, we explored the effect of complexation with a single water molecule on the tautomeric and conformational preferences of the sulfonamides. Our conformer-specific IR spectra in the NH and OH stretch region (3200–$3750\,cm^{-1}$) suggest that the intra- and intermolecular interactions governing the structures of the monomers and water complexes are surprisingly diverse. We have undertaken both Quantum Theory of Atoms in Molecules (QTAIM) and Interacting Quantum Atoms (IQA) analyses of calculated electron densities to quantitatively characterize the nature and strengths of the intra- and intermolecular interactions prevalent in the monomer and water complex structures.

WC11

SODIATED SUGAR STRUCTURES: CRYOGENIC ION VIBRATIONAL SPECTROSCOPY OF Na$^+$(GLUCOSE) ADDUCTS

JONATHAN VOSS, STEVEN J. KREGEL, KAITLYN C FISCHER, ETIENNE GARAND, *Department of Chemistry, University of Wisconsin–Madison, Madison, WI, USA.*

The recent discovery that ionic liquids help facilitate the dissolution of cellulose has renewed interest in understanding how ionic species interact with carbohydrates. Here we present infrared spectra in the $2800 - 3800$ cm^{-1} range of gas-phase mass-selected Na$^+$(Glucose) adducts. These adducts are further probed with IR-dip spectroscopy to yield conformer specific spectra of at least seven unique species. The relative abundances of conformers show that gas-phase interconversion barriers are sufficiently high to preserve the solution-phase populations. Additionally, our results demonstrate that mM concentrations of NaCl do not strongly perturb the anomeric ratio of glucose in solution.

WD. Clusters/Complexes

Wednesday, June 21, 2017 – 8:30 AM

Room: B102 Chemical and Life Sciences

Chair: G. S. Grubbs II, Missouri University of Science and Technology, Rolla, MO, USA

WD01　　　　　　　　　　　　　　　　　　　　　　　　　　　　　　　　　　　　　8:30–8:45

DETECTION OF WATER BINDING TO THE OXYGEN EVOLVING COMPLEX USING LOW FREQUENCY SERS

ANDREW J. WILSON, PRASHANT JAIN, *Department of Chemistry, University of Illinois at Urbana-Champaign, Urbana, IL, USA.*

The oxygen evolving complex (OEC) in Photosystem II (PSII) is a hallmark catalyst for efficiently splitting water to generate molecular oxygen. Much of what is known about the structure of the OEC has been provided by X-ray analysis of PSII at low temperatures, from which the mechanism of water splitting has been inferred. Surface-enhanced Raman scattering (SERS) offers an opportunity to build on our current understanding of this catalytic system as it can provide time-resolved, molecular vibrational information in a physiological environment. With low frequency SERS, we are able to separate the manganese oxide vibrational modes of the OEC from those in a complex, biological environment. With isotopically labelled water, we use SERS to identify water binding to the OEC. Raman spectra calculated by density functional theory support the assignment of water binding to a manganese atom outside of the cuboidal OEC. Detection of water binding sites on the OEC with SERS can not only compliment previous structural studies, but can also provide a powerful platform for in operando mechanistic studies.

WD02　　　　　　　　　　　　　　　　　　　　　　　　　　　　　　　　　　　　　8:47–9:02

MICROSOLVATION AND THE EFFECTS OF NON-COVALENT INTERACTIONS ON INTRAMOLECULAR DYNAMICS

LIDOR FOGUEL, ZACHARY VEALEY, PATRICK VACCARO, *Department of Chemistry, Yale University, New Haven, CT, USA.*

Physicochemical processes brought about by non-covalent interactions between neighboring molecules are undeniably of crucial importance in the world around us, being responsible for effects ranging from the subtle (yet precise) control of biomolecular recognition events to the very existence of condensed phases. Of particular interest is the differential ability of distinct non-covalent forces, such as those mediated by dispersion-dominated aryl (π-π) coupling and electrostatically-driven hydrogen bonding, to affect unimolecular transformations by altering potential surface topographies and the nature of reaction coordinates. A concerted experimental and computational investigation of "microsolvation" (solvation at the molecular level) has been undertaken to elucidate the site-specific coupling between solute and solvent degrees of freedom, as well as attendant consequences for the efficiency and pathway of intrinsic proton-transfer dynamics. Targeted species have been synthesized in situ under "cold" supersonic free-jet expansion conditions ($T_{rot} \approx$ 1-2K) by complexing an active (proton-transfer) substrate with various ligands (e.g., water isotopologs and benzene derivatives) for which competing interaction mechanisms can lead to unique binding motifs. A series of fluorescence-based spectroscopic measurements have been performed on binary adducts formed with the prototypical 6-hydroxy-2-formylfulvene (HFF) system, where a quasi-linear intramolecular O–H···O bond and a zero-point energy that straddles the proton-transfer barrier crest synergistically yield the largest tunneling-induced splitting ever reported for the ground electronic state of an isolated neutral molecule. Such characteristics afford a localized metric for unraveling incipient changes in unimolecular reactivity, with comparison of experimentally observed and quantum-chemical predicted rovibronic landscapes serving to discriminate complexes built upon electrostatic (hydrogen-bonding) and dispersive (aryl-coupling) forces.

WD03

THE JET-COOLED HIGH-RESOLUTION IR SPECTRUM OF FORMIC ACID CYCLIC DIMER

MANUEL GOUBET, SABATH BTEICH, THERESE R. HUET, *Laboratoire PhLAM, UMR 8523 CNRS - Université Lille 1, Villeneuve d'Ascq, France*; OLIVIER PIRALI, *Institut des Sciences Moléculaires d'Orsay, Université Paris-Sud, Orsay, France*; PIERRE ASSELIN, PASCALE SOULARD, ATEF JABRI, *Department of Chemistry, MONARIS, CNRS, UMR 8233, Sorbonne Universités, UPMC Univ Paris 06, Paris, France*; P. ROY, *AILES beamline, Synchrotron SOLEIL, Saint Aubin, France*; ROBERT GEORGES, *IPR UMR6251, CNRS - Université Rennes 1, Rennes, France*.

As the simplest carboxylic acid, formic acid (FA) is an excellent model molecule to investigate the general properties of carboxylic acids. FA is also an atmospherically and astrophysically relevant molecule. It is well known that its dimeric form is predominant in the gas phase at temperatures below 423 K.[a] The cyclic conformation of the dimer (FACD) is an elementary system to be understood for the concerted hydrogen transfer through equivalent hydrogen bonds, an essential process within biomolecules. The IR range is a crucial spectral region, particularly the far-IR, as it gives a direct access to the intermolecular vibrational modes involved in this process. Moreover, due to its centrosymmetric conformation, the FACD exhibits no pure rotation spectrum and, due to spectral line congestion and Doppler broadening, IR bands cannot be rotationally resolved at room temperature.[b] So far, only parts of the ν_5-GS band (C-O stretch) have been observed under jet-cooled conditions using laser techniques.[c]

We present here six rotationally resolved IR bands of FACD recorded under jet-cooled conditions using the Jet-AILES apparatus and the QCL spectrometer at MONARIS, including the far-IR ν_{24}-GS band (intermolecular in-plane bending). Splitting due to vibration-rotation-tunneling motions are clearly observed. A full spectral analysis is in progress starting from the GS constants obtained by Goroya et al. and with the support of electronic structure calculations.[d]

[a]T. Miyazawa and K. S. Pitzer, J. Am. Chem. Soc. 81, 74, 1959

[b]R. Georges, M. Freytes, D. Hurtmans, I. Kleiner, J. Vander Auwera, M. Herman, Chem. Phys. 305, 187, 2004

[c]M. Ortlieb and M. Havenith, J. Phys. Chem. A 111, 7355, 2007; K. G. Goroya, Y. Zhu, P. Sun and C. Duan, J. Chem. Phys. 140, 164311, 2014

[d]This work is supported by the CaPPA project (Chemical and Physical Properties of the Atmosphere) ANR-11-LABX-0005-01

WD04

ROTATIONAL SPECTRA OF 4,4,4-TRIFLUOROBUTYRIC ACID AND THE 4,4,4-TRIFLUOROBUTYRIC ACID-FORMIC ACID COMPLEX

YOON JEONG CHOI, *Department of Chemistry, Wesleyan University, Middletown, CT, USA*; ALEX TREVIÑO, *Department of Chemistry, University of Texas Rio Grande Valley, Brownsville, TX, USA*; SUSANNA L. STEPHENS, *Department of Chemistry, Wesleyan University, Middletown, CT, USA*; S. A. COOKE, *Natural and Social Science, Purchase College SUNY, Purchase, NY, USA*; STEWART E. NOVICK, *Department of Chemistry, Wesleyan University, Middletown, CT, USA*; WEI LIN, *Department of Chemistry, University of Texas Rio Grande Valley, Brownsville, TX, USA*.

The pure rotational spectra of 4,4,4-trifluorobutyric acid, $CF_3CH_2CH_2COOH$, and its complex with formic acid, were studied by a pulsed nozzle, chirped-pulse Fourier transform microwave spectrometer in the frequency range of 8-12 GHz. The rotational constants and centrifugal distortion constants were determined for the first time. Quantum chemical calculations were carried out exploring possible conformations of 4,4,4-trifluorobutyric and the structure of the 4,4,4-trifluorobutyric acid-formic acid complex using B3LYP/aug-cc-pVTZ and MP2/aug-cc-pVTZ calculations. The experimental spectroscopic constants are compared to those obtained from *ab initio* calculations.

WD05   **9:38 – 9:53**

THE THz/FIR SPECTRUM OF SMALL WATER CLUSTERS IN HELIUM NANODROPLETS

GERHARD SCHWAAB, RAFFAEL SCHWAN, DEVENDRA MANI, NITISH PAL, ARGHYA DEY, *Physikalische Chemie II, Ruhr University Bochum, Bochum, Germany*; BRITTA REDLICH, *FELIX Laboratory, Radboud University, Nijmegen, The Netherlands*; LEX VAN DER MEER, *Institute for Molecules and Materials (IMM), Radboud University Nijmegen, Nijmegen, Netherlands*; MARTINA HAVENITH[a], *Physikalische Chemie II, Ruhr University Bochum, Bochum, Germany.*

The microscopic properties of water that are relevant for bulk solvation processes are still not fully understood. Here, we combine mass selective Helium nanodroplet spectroscopy with the powerful Terahertz (THz) and far-infrared (FIR) capabilities of the free electron laser facility FELIX to study the fingerprint of small neutral water clusters in the wavelength range from 90-900cm^{-1}. Helium nanodroplets are a gentle, superfluid matrix and allow aggregation of pre-cooled moieties at ultra-cold temperatures (0.37 K). The fast cooling rate allows in some cases to stabilize not only the global minimum structure but also local minimum structures. The FELIX facility in Nijmegen provides narrowband ($\Delta\nu/\nu = 0.5\%$) pulsed radation covering the frequency range from 80–3300 cm^{-1}. We used a repetition rate of 10 Hz and typical pulse energies from 10 mJ at the 90cm^{-1}and 40 mJ at 900cm^{-1}. This corresponds to average powers of 100–400 mW far beyond those available using other radiation sources in this frequency range. The observed spectrum is exceptionally rich and includes lines that are close to or below our resolution limit. By mass selective detection and by varying the pickup pressure, we were able to identify contributions from dimer, trimer, tetramer and pentamer. The number of resonances indicates stabilization of at least two trimer structures in He nanodroplets. A comparison with theoretical predictions is on the way. We are confident that our experiments will contribute to understand the very special behavior of water in a bottom up approach.

[a]This work was supported by the Cluster of Excellence RESOLV (Ruhr-Universität, EXC1069) funded by the Deutsche Forschungsgemeinschaft

WD06   **9:55 – 10:10**

BROADBAND MICROWAVE SPECTROSCOPY AS A TOOL TO STUDY INTERMOLECULAR INTERACTIONS IN THE DIPHENYL ETHER - WATER SYSTEM

MARIYAM FATIMA, CRISTOBAL PEREZ, MELANIE SCHNELL, *CoCoMol, Max-Planck-Institut für Struktur und Dynamik der Materie, Hamburg, Germany.*

Many biological processes, such as chemical recognition and protein folding, are mainly controlled by the interplay of hydrogen bonds and dispersive forces. This interplay also occurs between organic molecules and solvent water molecules. Broadband rotational spectroscopy studies of weakly bound complexes are able to accurately reveal the structures and internal dynamics of molecular clusters isolated in the gas phase. Amongst them, water clusters with organic molecules are of particular interest. In this work, we investigate the interplay between different types of weak intermolecular interactions and how it controls the preferred interaction sites of aromatic ethers, where dispersive interactions may play a significant role. We present our results on diphenyl ether ($C_{12}H_{10}O$, 1,1'-Oxydibenzene) complexed with up to three molecules of water. Diphenyl ether is a flexible molecule, and it offers two competing binding sites for water: the ether oxygen and the aromatic π system. In order to determine the structure of the diphenyl ether-water complexes, we targeted transitions in the 2-8 GHz range using broadband rotational spectroscopy. We identify two isomers with one water, one with two water, and one with three water molecules. Further analysis from isotopic substitution measurements provided accurate structural information. The preferred interactions, as well as the observed structural changes induced upon complexation, will be presented and discussed.

Intermission

WD07   **10:29 – 10:44**

INVESTIGATION OF THE HYDANTOIN MONOMER AND ITS INTERACTION WITH WATER MOLECULES

SÉBASTIEN GRUET, *CUI, The Hamburg Centre for Ultrafast Imaging, Hamburg, Germany*; CRISTOBAL PEREZ, MELANIE SCHNELL, *CoCoMol, Max-Planck-Institut für Struktur und Dynamik der Materie, Hamburg, Germany.*

Hydantoin (Imidazolidine-2,4-dione, $C_3H_4N_2O_2$) is a five-membered heterocyclic compound of astrobiological interest. This molecule has been detected in carbonaceous chondrites [1], and its formation can rise from the presence of glycolic acid and urea, two prebiotic molecules [2]. The hydrolysis of hydantoin under acidic conditions can also produce glycine [3], an amino acid actively searched for in the interstellar medium.

Spectroscopic data of hydantoin is very limited and mostly dedicated to the solid phase. The high resolution study in gas phase is restricted to the work recently published by Ozeki et al. reporting the pure rotational spectra of the ground state and two vibrational states of the molecule in the millimeter-wave region (90-370 GHz)[4].

Using chirped-pulse Fourier-transform microwave (CP-FTMW) spectroscopy, we recorded the jet-cooled rotational spectra of hydantoin with water between 2 to 8 GHz. We observed the ground state of hydantoin monomer and several water complexes with one or two water molecules. All the observed species exhibit a hyperfine structure due to the two nitrogen atoms present in the molecule, which were fully resolved and analyzed. Additional experiments with a ^{18}O enriched water sample were realized to determine the oxygen-atom positions of the water monomers. These experiments yielded accurate structural information on the preferred water binding sites. The observed complexes and the interactions that hold them together, mainly strong directional hydrogen bonds, will be presented and discussed.

[1] Shimoyama, A. and Ogasawara, R., *Orig. Life Evol. Biosph.*, 32, 165-179, 2002. DOI:10.1023/A:1016015319112.

[2] Menor-Salván, C. and Marín-Yaseli, M.R., *Chem. Soc. Rev.*, 41(16), 5404-5415, 2012. DOI:10.1039/c2cs35060b.

[3] De Marcellus P., Bertrand M., Nuevo M., Westall F. and Le Sergeant d'Hendecourt L., *Astrobiology*. 11(9), 847-854, 2011. DOI:10.1089/ast.2011.0677.

[4] Ozeki, H., Miyahara R., Ihara H., Todaka S., Kobayashi K., and Ohishi M., *Astron. Astrophys.*, Forthcoming article (Accepted: 12 January 2017), DOI:10.1051/0004-6361/201629880.

WD08   **10:46 – 11:01**

HYDRATION OF AN ACID ANHYDRIDE: THE WATER COMPLEX OF ACETIC SULFURIC ANHYDRIDE

CJ SMITH, ANNA HUFF, BECCA MACKENZIE, KEN LEOPOLD, *Chemistry Department, University of Minnesota, Minneapolis, MN, USA.*

The water complex of acetic sulfuric anhydride (ASA, CH_3COOSO_2OH) has been observed by pulsed nozzle Fourier transform microwave spectroscopy. ASA is formed in situ in the supersonic jet via the reaction of SO_3 and acetic acid and subsequently forms a complex with water during the expansion. Spectra of the parent and fully deuterated form, as well as those of the species derived from $CH_3{}^{13}COOH$, have been observed. The fitted internal rotation barrier of the methyl group is 219.599(21), cm^{-1} indicating the complexation with water lowers the internal rotation barrier of the methyl group by 9% relative to that of free ASA. The observed species is one of several isomers identified theoretically in which the water inserts into the intramolecular hydrogen bond of the ASA. Aspects of the intermolecular potential energy surface are discussed.

WD09   **11:03 – 11:18**

VIBRATIONAL COUPLING IN SOLVATED FORM OF EIGEN PROTON

JER-LAI KUO, *Institute of Atomic and Molecular Sciences, Academia Sinica, Taipei, Taiwan.*

Recent studies have shown that features in the vibrational spectra of solvated H_3O^+ can be modulated not only by the type messengers[a], but also by the number of messengers[b]. Vibrational spectra can be simulated with accurate theoretical simulations and obtain the peak position and absorption intensity by solving the quantum vibrational Schrodinger equation using the potential and dipole moment obtained ab initio methods. In this work, we studied vibrational coupling between intra- and inter-molecular modes of this ionic cluster to glean into the details of the vibrational couplings manifested in the spectra region of 600-7000 cm^{-1}.

[a] J. A. Tan, J-W Li, C-c Chiu, H. T. Huynh, H-Y Liao and J-L Kuo, Phys. Chem. Chem. Phys., 18, 30721 (2016)

[b] J-W Li, M. Morita, T. Takahashi and J-L Kuo, J. Phys. Chem. A, 119, 10887 (2015)

WD10   **11:20 – 11:35**

INFRARED PHOTODISSOCIATION CLUSTER STUDIES ON CO_2 INTERACTION WITH TITANIUM OXIDE CATALYST MODELS

LEAH G DODSON, *JILA and NIST, University of Colorado, Boulder, CO, USA*; MICHAEL C THOMPSON, J. MATHIAS WEBER, *JILA and the Department of Chemistry and Biochemistry, University of Colorado-Boulder, Boulder, CO, USA.*

Titanium oxide catalysts are some of the most promising photocatalyst candidates for renewable energy storage applications via production of solar fuels. To contribute to a molecular-level understanding of the interaction of CO_2 with titanium oxide, we turn to cluster models in order to circumvent the challenges posed by speciation in the condensed phase. In this work, we use infrared photodissociation spectroscopy ($950 - 2400$ cm^{-1}) in concert with density functional theory calculations to identify and characterize $[TiO_x(CO_2)_y]^-$ ($x = 1 - 3$, $y = 3 - 7$) clusters. We use these model systems to study the interaction of CO_2 with TiO, TiO_2, and TiO_3, and we find that each species exhibits unique infrared signatures and binding motifs. We will discuss the structures of these cluster ions, and how the coordination of the titanium atom plays a role in reduction of CO_2.

WD11   **11:37 – 11:52**

OXALATE FORMATION IN TITANIUM–CARBON DIOXIDE ANIONIC CLUSTERS STUDIED BY INFRARED PHOTODISSOCIATION SPECTROSCOPY

LEAH G DODSON, *JILA and NIST, University of Colorado, Boulder, CO, USA*; MICHAEL C THOMPSON, J. MATHIAS WEBER, *JILA and the Department of Chemistry and Biochemistry, University of Colorado-Boulder, Boulder, CO, USA.*

Carbon-carbon bond formation during carbon dioxide fixation would enable bulk synthesis of hydrocarbon chains, generally through formation of an oxalate intermediate. In this talk, we demonstrate the formation of $[Ti(CO_2)_y]^-$ ($y = 4 - 6$) gas phase clusters with an oxalate ligand bearing significant (> 1 e$^-$) negative charge. Gas phase anionic clusters were generated using laser ablation of a titanium metal target in the presence of a CO_2 expansion, and the infrared photodissociation spectra were measured from $950 - 2400$ cm^{-1}, revealing vibrations characteristic of the oxalate anion. The molecular structure of these clusters was identified by comparing the experimental vibrational spectra with density functional theory calculations.

WE. Spectroscopy as an analytical tool

Wednesday, June 21, 2017 – 8:30 AM

Room: 161 Noyes Laboratory

Chair: Brooks Pate, The University of Virginia, Charlottesville, VA, USA

WE01   8:30 – 8:45

PYROLYSIS AND MATRIX-ISOLATION FTIR OF ACETOIN

SARAH COLE, MARTHA ELLIS, JOHN SOWARDS, <u>LAURA R. McCUNN</u>, *Department of Chemistry, Marshall University, Huntington, WV, USA.*

Acetoin, $CH_3C(O)CH(OH)CH_3$, is an additive used in foods and cigarettes as well as a common component of biomass pyrolysate during the production of biofuels, yet little is known about its thermal decomposition mechanism. In order to identify thermal decomposition products of acetoin, a gas-phase mixture of approximately 0.3% acetoin in argon was subject to pyrolysis in a resistively heated SiC microtubular reactor at 1100-1500 K. Matrix-isolation FTIR spectroscopy was used to identify pyrolysis products. Many products were observed in analysis of the spectra, including acetylene, propyne, ethylene, and vinyl alcohol. These results provide clues to the overall mechanism of thermal decomposition and are important for predicting emissions from many industrial and residential processes.

WE02   8:47 – 9:02

EMISSION SPECTROSCOPY OF ATMOSPHERIC-PRESSURE BALL PLASMOIDS: HIGHER ENERGY REVEALS A RICH CHEMISTRY

<u>SCOTT E. DUBOWSKY</u>, AMBER NICOLE ROSE, *Department of Chemistry, University of Illinois at Urbana-Champaign, Urbana, IL, USA*; NICK GLUMAC, *Mechanical Science and Engineering, University of Illinois at Urbana-Champaign, Urbana, IL, USA*; BENJAMIN J. McCALL, *Departments of Chemistry and Astronomy, University of Illinois at Urbana-Champaign, Urbana, IL, USA.*

Ball plasmoids (self-sustaining spherical plasmas) are a particularly unique example of a non-equilibrium air plasma. These plasmoids have lifetimes on the order of hundreds of milliseconds without an external power source, however, current models dictate that a ball plasmoid should recombine in a millisecond or less. Ball plasmoids are considered to be a laboratory analogue of natural ball lightning, a phenomenon that has eluded scientific explanation for centuries. We are searching for the underlying physicochemical mechanism(s) by which ball plasmoids and (by extension) ball lightning are stabilized using a variety of diagnostic techniques.

This presentation will focus on optical emission spectroscopy (OES) of ball plasmoid discharges between 190-850 nm. The previous generation of OES measurements[a,b] of this system showed emission from only a few atomic and molecular species, however, the energy available for the discharges in these experiments was limited by the size of the capacitor banks and voltages to which the capacitor banks were charged. We are capable of generating plasmoids at much higher energies, and as a result we are the first to report a very rich chemistry previously not observed in ball plasmoids. We have identified signals from species including NO $A^2\Sigma^+{\rightarrow}X^2\Pi$, OH $A^2\Sigma^+{\rightarrow}X^2\Pi$, NH $A^3\Pi{\rightarrow}X^3\Sigma^-$, AlO $A^2\Pi{\rightarrow}X^2\Sigma^+$, NH^+ $B^2\Delta{\rightarrow}X^2\Pi$, W I, Al I, Cu I, and H_α, all of which have not yet been reported for this system. Analysis of the emission spectra and fitting procedures will be discussed, rotational temperatures of constituent species will be reported, and theories of ball plasmoid stabilization based upon these new results will be presented.

[a] Versteegh, A.; Behringer, K.; Fantz, U.; Fussman, G.; Jüttner, B.; Noack, S. *Plas. Sour. Sci. Technol.* **2008**, 17(2), 024014
[b] Stephan, K. D.; Dumas, S.; Komala-Noor, L.; McMinn, J. *Plas. Sour. Sci. Technol.* **2013**, 22(2), 025018

WE03

S-NITROSOTHIOLS OBSERVED USING CAVITY RING-DOWN SPECTROSCOPY

MARY LYNN RAD, *Department of Chemistry, The University of Virginia, Charlottesville, VA, USA*; BEN-JAMIN M GASTON, *Department of Pediatrics, Case Western Reserve University, Cleveland, OH, USA*; KEVIN LEHMANN, *Department of Chemistry and Physics, The University of Virginia, Charlottesville, VA, USA.*

The biological importance of nitric oxide has been known for nearly forty years due to its role in cardiovascular and nervous signaling. The main carrier molecules, s-nitrosothiols (RSNOs), are of additional interest due to their role in signaling reactions. Additionally, these compounds are related to several diseases including muscular dystrophy, stroke, myocardial infarction, Alzheimer's disease, Parkinson's disease, cystic fibrosis, asthma, and pulmonary arterial hypertension. One of the main barriers to elucidating the role of these RSNOs is the low (nanomolar) concentration present in samples of low volume (typically $\sim$100 μL). To this end we have set up a cavity ring-down spectrometer tuned to observe ^{14}NO and ^{15}NO released from cell growth samples. To decrease the limit of detection we have implemented a laser locking scheme employing Zeeman modulation of NO in a reference cell and have tuned the polarization of the laser using a half wave plate to optimize the polarization for the inherent birefringence of the CRDS mirrors. Progress toward measuring RSNO concentration in biological samples will be presented.

WE04

SI-TRACEABLE SCALE FOR MEASUREMENTS OF RADIOCARBON CONCENTRATION

JOSEPH T. HODGES, ADAM J. FLEISHER, QINGNAN LIU, DAVID A. LONG, *Chemical Sciences Division, National Institute of Standards and Technology, Gaithersburg, MD, USA.*

Radiocarbon (^{14}C) dating of organic materials is based on measuring the ^{14}C/^{12}C atomic fraction relative to the nascent value that existed when the material was formed by photosynthetic conversion of carbon dioxide present in the atmosphere. This field of measurement has numerous applications including source apportionment of anthropogenic and biogenic fuels and combustion emissions, carbon cycle dynamics, archaeology, and forensics.

Accelerator mass spectrometry (AMS) is the most widely used method for radiocarbon detection because it can measure extremely small amounts of radiocarbon (background of nominally 1.2 parts-per-trillion) with high relative precision (0.4 %). AMS measurements of radiocarbon are typically calibrated by reference to standard oxalic-acid ($C_2H_2O_4$) samples of known radioactivity that are derived from plant matter. Specifically, the internationally accepted absolute dating reference for so-called "modern-equivalent" radiocarbon is 95 % of the specific radioactivity in AD 1950 of the National Bureau of Standards (NBS) oxalic acid standard reference material and normalized to $\delta^{13}C_{VPDB} = 19$ per mil [a]. With this definition, a "modern-equivalent" corresponds to 1.176(70) parts-per-trillion of ^{14}C relative to total carbon content.

As an alternative radiocarbon scale, we propose an SI-traceable method to determine ^{14}C absolute concentration which is based on linear Beer-Lambert-law absorption measurements of selected ^{14}C^{16}O$_2$ ν_3-band line areas. This approach is attractive because line intensities of chosen radiocarbon dioxide transitions can be determined by *ab initio* calculations with relative uncertainties below 0.5 %. This assumption is justified by the excellent agreement between theoretical values of line intensities and measurements for stable isotopologues of CO_2 [b]. In the case of cavity ring-down spectroscopy (CRDS) measurements of ^{14}C^{16}O$_2$ peak areas, we show that absolute, SI-traceable concentrations of radiocarbon can be determined through measurements of time, frequency, pressure and temperature. Notably, this approach will not require knowledge of the radiocarbon half-life and is expected to provide a stable scale that does not require an artifact standard.

[a] M. Stuiver and H. A. Polach, *Radiocarbon* **19**, (1977) 355
[b] O. L. Polyansky et al., *Phys. Rev. Lett.* **114**, (2015) 243001

WE05   9:38 – 9:53

LINEAR AND NON-LINEAR THERMAL LENS SIGNAL OF THE FIFTH C-H VIBRATIONAL OVERTONE OF NAPH-
THALENE IN LIQUID SOLUTIONS OF HEXANE

CARLOS MANZANARES, MARLON DIAZ, ANN BARTON, PARASHU R NYAUPANE, *Department of
Chemistry and Biochemistry, Baylor University, Waco, TX, USA.*

The thermal lens technique is applied to vibrational overtone spectroscopy of solutions of naphthalene in n-hexane. The
pump and probe thermal lens technique is found to be very sensitive for detecting samples of low composition (ppm) in
transparent solvents. In this experiment two different probe lasers: one at 488 nm and another 568 nm were used. The
C-H fifth vibrational overtone spectrum of benzene is detected at room temperature for different concentrations. A plot of
normalized integrated intensity as a function of concentration of naphthalene in solution reveals a non-linear behavior at
low concentrations when using the 488 nm probe and a linear behavior over the entire range of concentrations when using
the 568 nm probe. The non-linearity cannot be explained assuming solvent enhancement at low concentrations. A two
color absorption model that includes the simultaneous absorption of the pump and probe lasers could explain the enhanced
magnitude and the non-linear behavior of the thermal lens signal. Other possible mechanisms will also be discussed.

WE06   9:55 – 10:10

STUDY OF THE IMIDAZOLIUM-BASED IONIC LIQUID – Ag ELECTRIFIED INTERFACE ON THE CO_2 ELEC-
TROREDUCTION BY SUM FREQUENCY SPECTROSCOPY.

NATALIA GARCIA REY, DANA DLOTT, *Department of Chemistry, University of Illinois at Urbana-
Champaign, Urbana, IL, USA.*

Imidazolium based ionic liquids (ILs) have been used as a promising system to improve the CO_2 electroreduction at lower
overpotential than other organic or aqueous electrolytes[1]. Although the detailed mechanism of the CO_2 electroreduction
on Ag has not been elucidated yet, we have developed a methodology to study the electrified interface during the CO_2
electroreduction using sum frequency generation (SFG) spectroscopy in combination with cyclic voltammetry[2]. In this work,
we tuned the composition of imidazolium-based ILs by exchanging the anion or the functional groups of the imidazolium. We
use the nonresonant SFG (NR-SFG) to study the IL-Ag interface and resonant SFG (RES-SFG) to identify the CO adsorbed
on the electrode and monitor the Stark shift as a function of cell potential. In previous studies on CO_2 electroreduction in the
IL: 1-ethyl-3-methylimidazolium tetrafluorborate (EMIM-BF$_4$) on Ag, we showed three events occurred at the same potential
(-1.33 V vs. Ag/AgCl): the current associated with CO_2 electroreduction increased, the Stark shift of the adsorbed atop CO
doubled in magnitude and the EMIM-BF$_4$ underwent a structural transition[3]. In addition, we also observed how the structural
transition of the EMIM-BF$_4$ electrolyte shift to lower potentials when the IL is mixed with water. It is known that water
enhances the CO_2 electroreduction producing more CO[4]. Moreover, the CO is adsorbed in multi-bonded and in atop sites
when more water is present in the electrolyte.

[1]Lau, G. P. S.; Schreier, M.; Vasilyev, D.; Scopelliti, R.; Grätzel, M.; Dyson, P. J., New Insights into the Role of
Imidazolium-Based Promoters for the Electroreduction of CO_2 on a Silver Electrode. J. Am. Chem. Soc. 2016, 138, 7820-
7823. [2]García Rey, N.; Dlott, D. D., Studies of Electrochemical Interfaces by Broadband Sum Frequency Generation. J.
Electroanal. Chem. 2016. DOI:10.1016/j.jelechem.2016.12.023. [3]García Rey, N.; Dlott, D. D., Structural Transition in an
Ionic Liquid Controls CO_2 Electrochemical Reduction. J. Phys. Chem. C 2015, 119, 20892–20899. [4]Rosen, B. A.; Zhu, W.;
Kaul, G.; Salehi-Khojin, A.; Masel, R. I., Water Enhancement of CO_2 Conversion on Silver in 1-Ethyl-3-Methylimidazolium
Tetrafluoroborate. J. Electrochem. Soc. 2013, 160, H138-H141.

Intermission

WE07

SPECDATA: AUTOMATED ANALYSIS SOFTWARE FOR BROADBAND SPECTRA

JASMINE N OLIVEIRA, *Atomic and Molecular Physics , Harvard-Smithsonian Center for Astrophysics, Cambridge, MA, USA*; MARIE-ALINE MARTIN-DRUMEL, *CNRS, Institut des Sciences Moleculaires d'Orsay, Orsay, France*; MICHAEL C McCARTHY, *Atomic and Molecular Physics, Harvard-Smithsonian Center for Astrophysics, Cambridge, MA, USA*.

With the advancement of chirped-pulse techniques, broadband rotational spectra with a few tens to several hundred GHz of spectral coverage are now routinely recorded. When studying multi-component mixtures that might result, for example, with the use of an electrical discharge, lines of new chemical species are often obscured by those of known compounds, and analysis can be laborious.

To address this issue, we have developed SPECdata, an open source, interactive tool which is designed to simplify and greatly accelerate the spectral analysis and discovery. Our software tool combines both automated and manual components that free the user from computation, while giving him/her considerable flexibility to assign, manipulate, interpret and export their analysis.

The automated – and key – component of the new software is a database query system that rapidly assigns transitions of known species in an experimental spectrum. For each experiment, the software identifies spectral features, and subsequently assigns them to known molecules within an in-house database (Pickett .cat files, list of frequencies...), or those catalogued in Splatalogue (using automatic on-line queries).

With suggested assignments, the control is then handed over to the user who can choose to accept, decline or add additional species. Data visualization, statistical information, and interactive widgets assist the user in making decisions about their data. SPECdata has several other useful features intended to improve the user experience. Exporting a full report of the analysis, or a peak file in which assigned lines are removed are among several options. A user may also save their progress to continue at another time. Additional features of SPECdata help the user to maintain and expand their database for future use. A user-friendly interface allows one to search, upload, edit or update catalog or experiment entries.

WE08

IDENTIFYING BROADBAND ROTATIONAL SPECTRA WITH NEURAL NETWORKS

DANIEL P. ZALESKI, KIRILL PROZUMENT, *Chemical Sciences and Engineering Division, Argonne National Laboratory, Argonne, IL, USA*.

A typical broadband rotational spectrum may contain several thousand observable transitions, spanning many species[a]. Identifying the individual spectra, particularly when the dynamic range reaches 1,000:1 or even 10,000:1, can be challenging. One approach is to apply automated fitting routines[b]. In this approach, combinations of 3 transitions can be created to form a "triple", which allows fitting of the A, B, and C rotational constants in a Watson-type Hamiltonian. On a standard desktop computer, with a target molecule of interest, a typical AUTOFIT routine takes 2–12 hours depending on the spectral density. A new approach is to utilize machine learning[c] to train a computer to recognize the patterns (frequency spacing and relative intensities) inherit in rotational spectra and to identify the individual spectra in a raw broadband rotational spectrum. Here, recurrent neural networks have been trained to identify different types of rotational spectra and classify them accordingly. Furthermore, early results in applying convolutional neural networks for spectral object recognition in broadband rotational spectra appear promising.

[a] Perez et al. "Broadband Fourier transform rotational spectroscopy for structure determination: The water heptamer." Chem. Phys. Lett., 2013, 571, 1–15.
[b] Seifert et al. "AUTOFIT, an Automated Fitting Tool for Broadband Rotational Spectra, and Applications to 1-Hexanal." J. Mol. Spectrosc., 2015, 312, 13–21.
[c] Bishop. "Neural networks for pattern recognition." Oxford university press, 1995.

WE09 11:03 – 11:18

ADVANCES IN MOLECULAR ROTATIONAL SPECTROSCOPY FOR APPLIED SCIENCE

BRENT HARRIS, SHELBY S. FIELDS, ROBIN PULLIAM, MATT MUCKLE, JUSTIN L. NEILL, *BrightSpec Labs, BrightSpec, Inc., Charlottesville, VA, USA.*

Advances in chemical sensitivity and robust, solid-state designs for microwave/millimeter-wave instrumentation compel the expansion of molecular rotational spectroscopy as research tool into applied science. It is familiar to consider molecular rotational spectroscopy for air analysis. Those techniques for molecular rotational spectroscopy are included in our presentation of a more broad application space for materials analysis using Fourier Transform Molecular Rotational Resonance (FT-MRR) spectrometers. There are potentially transformative advantages for direct gas analysis of complex mixtures, determination of unknown evolved gases with parts per trillion detection limits in solid materials, and unambiguous chiral determination. The introduction of FT-MRR as an alternative detection principle for analytical chemistry has created a ripe research space for the development of new analytical methods and sampling equipment to fully enable FT-MRR. We present the current state of purpose-built FT-MRR instrumentation and the latest application measurements that make use of new sampling methods.

WE10 11:20 – 11:35

FOURIER TRANSFORM MICROWAVE SPECTROSCOPIC STUDIES OF DIMETHYL ETHER AND ETHYLENE FLAMES

DANIEL A. OBENCHAIN, *Institut für Physikalische Chemie und Elektrochemie, Gottfried-Wilhelm-Leibniz-Universität, Hannover, Germany*; JULIA WULLENKORD, KATHARINA KOHSE-HÖINGHAUS, *Physikalische Chemie I, University of Bielefeld, Bielefeld, Germany*; JENS-UWE GRABOW, *Institut für Physikalische Chemie und Elektrochemie, Gottfried-Wilhelm-Leibniz-Universität, Hannover, Germany*; NILS HANSEN, *Combustion Research Facility, Sandia National Laboratories, Livermore, CA, USA.*

Microwave spectroscopy has been a proven technique for the detection of short-lived molecules produced from a variety of molecular sources. With the goal of observing more reactive intermediates produced in combustion reactions, the products of a home-built flat flame burner were measured on a coaxially oriented beam resonator arrangement (COBRA) Fourier transform microwave spectrometer.[a] The products are coupled into a molecular beam using a fast-mixing nozzle styled after the work of Gutowsky and co-workers.[b]

Probing the flame at various positions, the relative abundance of products can be observed as a function of flame depth. One dimensional intensity profiles are available for formaldehyde, ketene, acetaldehyde, and dimethyl ether, where either a dimethyl ether fuel or an ethylene fuel was burned in the presence of oxygen. The current arrangement allows only for stable species produced in the flame to be observed in the molecular beam. This combination of species source and detection shows promise for future work in observing new, short-lived, combustion intermediates.

[a] J.-U. Grabow, W. Stahl, H. Dreizler, Rev. Sci. Instrum. 67, 4072, 1996

[b] T. Emilsson, T. D. Klots, R. S. Ruoff, H.S. Gutowsky, J. Chem. Phys. 93, 6971, 1990

WE11 11:37 – 11:52

STRATEGIES FOR INTERPRETING TWO DIMENSIONAL MICROWAVE SPECTRA

MARIE-ALINE MARTIN-DRUMEL, *CNRS, Institut des Sciences Moleculaires d'Orsay, Orsay, France*; KYLE N. CRABTREE, ZACHARY BUCHANAN, *Department of Chemistry, The University of California, Davis, CA, USA.*

Microwave spectroscopy can uniquely identify molecules because their rotational energy levels are sensitive to the three principal moments of inertia. However, a priori predictions of a molecule's structure have traditionally been required to enable efficient assignment of the rotational spectrum. Recently, automated microwave double resonance spectroscopy (AMDOR) has been employed to rapidly generate two dimensional spectra based on transitions that share a common rotational level, which may enable automated extraction of rotational constants without any prior estimates of molecular structure. Algorithms used to date for AMDOR have relied on making several initial assumptions about the nature of a subset of the linked transitions, followed by testing possible assignments by "brute force." In this talk, we will discuss new strategies for interpreting AMDOR spectra, using eugenol as a test case, as well as prospects for library-free, automated identification of the molecules in a volatile mixture.

WF. Mini-symposium: ALMA's Molecular View
Wednesday, June 21, 2017 – 1:45 PM
Room: 274 Medical Sciences Building

Chair: Amanda Steber, Universität Hamburg, Hamburg, Germany

WF01   *INVITED TALK*   1:45 – 2:15

PROBING CO FREEZE-OUT AND DESORPTION IN PROTOPLANETARY DISKS

CHUNHUA QI, *Radio and Geoastronomy Division, Harvard-Smithsonian Center for Astrophysics, Cambridge, MA, USA.*

Snow lines, the boundaries where the most abundant volatiles such as H_2O, CO_2 and CO freeze out from the gas phase onto dust grains in the midplane of protoplanetary disks, are believed to play an important role in planet formation and composition. Locating the CO snow line is challenging in disks. This has prompted an exploration of chemical signatures of CO freeze-out and desorption. We present ALMA observations of the CO, N_2H^+ and DCO^+ emission to probe the CO freeze-out and desorption in protoplanetary disks, and evaluate their utility as tracers of the CO snow line location.

WF02   2:19 – 2:34

AN UPDATED GAS/GRAIN SULFUR NETWORK FOR ASTROCHEMICAL MODELS

JACOB LAAS, PAOLA CASELLI, *The Center for Astrochemical Studies, Max-Planck-Institut für extraterrestrische Physik, Garching, Germany.*

Sulfur is a chemical element that enjoys one of the highest cosmic abundances. However, it has traditionally played a relatively minor role in the field of astrochemistry, being drowned out by other chemistries after it depletes from the gas phase during the transition from a diffuse cloud to a dense one. A wealth of laboratory studies have provided clues to its rich chemistry in the condensed phase, and most recently, a report by a team behind the Rosetta spacecraft has significantly helped to unveil its rich cometary chemistry. We have set forth to use this information to greatly update/extend the sulfur reactions within the OSU gas/grain astrochemical network in a systematic way, to provide more realistic chemical models of sulfur for a variety of interstellar environments. We present here some results and implications of these models.

WF03   2:36 – 2:51

A NEW MODEL OF THE CHEMISTRY OF IONIZING RADIATION IN SOLIDS

CHRISTOPHER N SHINGLEDECKER, ERIC HERBST, *Department of Chemistry, The University of Virginia, Charlottesville, VA, USA.*

Cosmic rays are a form of high energy radiation found throughout the galaxy that can cause significant physio-chemical changes in solids, such as interstellar dust grain ice-mantles. These particles consist mostly of protons and can initiate a solid-state irradiation chemistry of significant astrochemical interest. In order to better understand the chemical effects of long-term exposure to ionizing radiation, we have written a new Monte Carlo model, CIRIS: the Chemistry of Ionizing Radiation in Solids, which is, to the best of our knowledge, the first successful program of its kind to follow the damage and subsequent chemistry of an irradiated material over time. In our code, two distinct regimes are considered. One is dominated by the atomic physics of track calculations in which both the irradiating proton and the subsequently generated secondary electrons are followed on a collision by collision basis. The other regime occurs after the ion-target collision, in which mobile species are free to randomly hop throughout the bulk of the ice and react via a diffusive mechanism. Here, we will present an initial test of our code in which we have successfully modeled previous experimental work. In these simulations, we are able to reproduce the measured abundances and predict the approximate ice thickness used in that study.

WF04 2:53 – 3:08

THE KEY ROLE OF NUCLEAR-SPIN ASTROCHEMISTRY

ROMANE LE GAL, ERIC HERBST, *Department of Chemistry, The University of Virginia, Charlottesville, VA, USA*; CHANGJIAN XIE, HUA GUO, *Department of Chemistry and Chemical Biology, University of New Mexico, Albuquerque, NM, USA*; DAHBIA TALBI, *Laboratoire Univers et Particules Montpellier, CNRS-Universite de Montpellier 2, Montpellier, France*; SEBASTIEN MULLER, CARINA PERSSON, *Onsala Space Observatory, Chalmers University of Technology, Onsala, Sweden.*

Thanks to the new spectroscopic windows opened by the recent generation of telescopes, a large number of molecular lines have been detected. In particular, nuclear-spin astrochemistry has gained interest owing to numerous ortho-to-para ratio (OPR) measurements for species including H_3^+, CH_2, C_3H_2, H_2O, NH_3, NH_2, H_2S, H_2CS, H_2O^+ and H_2Cl^+. Any multi-hydrogenated species can indeed present different spin configurations, if some of their hydrogen nuclei are identical, and the species thus exist in distinguishable forms, such as ortho and para. In thermal equilibrium, OPRs are only functions of the temperature and since spontaneous conversion between ortho and para states is extremely slow in comparison with typical molecular cloud lifetimes, OPRs were commonly believed to reflect a "formation temperature". However, observed OPRs are not always consistent with their thermal equilibrium values, as for the NH_3 and NH_2 cases. It is thus crucial to understand how interstellar OPRs are formed to constrain the information such new probes can provide. This involves a comprehensive analysis of the processes governing the interstellar nuclear-spin chemistry, including the formation and possible conversions of the different spin symmetries both in the gas and solid phases. If well understood, OPRs might afford new powerful astrophysical diagnostics on the chemical and physical conditions of their environments, and in particular could trace their thermal history. In this context, observations of non-thermal values for the OPR of the radical NH_2 toward four high-mass star-forming regions[a], and a 3:1 value measured for the H_2Cl^+ OPR toward diffuse[b] and denser gas, led us to develop detailed studies of the mechanisms involved in obtaining such OPRs with the aid of quasi-classical trajectory calculations[c]. We will present these new promising results, improving our understanding of the interstellar medium.

---

[a]Persson et al. 2016, A&A, 586, A128

[b]Neufeld et al. 2016, ApJ, 807, 54

[c]Le Gal et al. 2016, A&A, 596, A35 and Le Gal et al., in prep

WF05 3:10 – 3:25

ROTATIONAL SPECTROSCOPY OF REACTIVE SPECIES AT THE CENTER FOR ASTROCHEMICAL STUDIES.

VALERIO LATTANZI, SILVIA SPEZZANO, PAOLA CASELLI, *The Center for Astrochemical Studies, Max-Planck-Institut für extraterrestrische Physik, Garching, Germany.*

The Center for Astrochemical Studies at the Max Planck Institute for Extraterrestrial Physics in Garching, is a recently established group which collects scientists with very diverse backgrounds. In the same group observers, theoreticians, chemists and molecular astrophysicists join their efforts with the ultimate goal of properly interpreting observations with the new generation telescopes and unveiling our astrochemical/physical heritage. Among these tasks, the gas-phase spectroscopic characterisation of molecular species of astrophysical relevance is one of the main goals of the laboratory sub-group. This talk will mainly focus on the first experiment built in our center, the CASAC (CAS Absorption Cell) spectrometer: this experiment has been optimised on the production and probe of small molecular ions and radicals. The main laboratory techniques along with the more prominent outcomes of recent studies will be presented. Finally, a brief update on the status of the other instruments available in our center will be given, including their planned upgrades.

WF06 3:27 – 3:42

A PRESTELLAR CORE 3MM LINE SURVEY: MOLECULAR COMPLEXITY IN L183

VALERIO LATTANZI, LUCA BIZZOCCHI, PAOLA CASELLI, *The Center for Astrochemical Studies, Max-Planck-Institut für extraterrestrische Physik, Garching, Germany.*

Cold dark clouds represent a very unique environment to test our knowledge of the chemical and physical evolution of the structures that ultimately led to life. Starless cores, such as L183, are indeed the first phase of the star formation process and the nursery of chemical complexity. In this work we present the detection of several large astronomical molecules in the prestellar core L183, as a result of a 3mm single-pointing survey performed with the IRAM 30m antenna. The abundances of the observed species will be then compared to those found in similar environments, highlighting correspondences and uniquenesses of the different sources.

Intermission

WF07 4:01 – 4:16

MILLIMETER WAVE SPECTRUM OF THE TWO MONOSULFUR DERIVATIVES OF METHYL FORMATE: S- AND O-METHYL THIOFORMATE, IN THE GROUND AND THE FIRST EXCITED TORSIONAL STATES

<u>ATEF JABRI</u>, *Department of Chemistry, MONARIS, CNRS, UMR 8233, Sorbonne Universités, UPMC Univ Paris 06, Paris, France*; R. A. MOTIYENKO, L. MARGULÈS, *Laboratoire PhLAM, UMR 8523 CNRS - Université Lille 1, Villeneuve d'Ascq, France*; J.-C. GUILLEMIN, *Institut des Sciences Chimiques de Rennes, UMR 6226 CNRS - ENSCR, Rennes, France*; E. A. ALEKSEEV, *Radiospectrometry Department, Institute of Radio Astronomy of NASU, Kharkov, Ukraine*; ISABELLE KLEINER, *CNRS et Universités Paris Est et Paris Diderot, Laboratoire Interuniversitaire des Systèmes Atmosphériques (LISA), Créteil, France*; BELÉN TERCERO, JOSE CERNICHARO, *Molecular Astrophysics, ICMM, Madrid, Spain.*

Methyl formate $CH_3OC(O)H$ is a relatively abundant component of the interstellar medium (ISM) [a]. Thus, we decided to study its sulfur derivatives as they can be reasonably proposed for detection in the ISM. In fact there is two relatively stable isomers for methyl thioformate, S-Methyl thioformate $CH_3SC(O)H$ and O-Methyl thiofomate $CH_3OC(S)H$. Theoretical investigations on these molecules have been done recently by Senent et al.[b]. Previous experimental investigations were performed only for the S-Methyl thioformate in the 10-41 GHz spectral range by Jones et al.[c] and Caminati et al.[d]. For the present study both isomers were synthesized and the millimeter wave spectrum was then recorded for the first time from 150 to 660 GHz with the Lille's spectrometer based on solid-state sources. The internal rotation effect on the millimeter wave spectra is not the same for these two molecules because the barrier height to internal rotation is relatively low for the S- isomer ($V_3 \approx 140$ cm^{-1}) and rather high for the O- isomer ($V_3 \approx 700$ cm^{-1}). Analysis of the ground and excited torsional states performed with the *BELGI-C$_s$* code[e] will be presented and discussed. We will provide the search for methyl thioformate in different sources.

[a] E. Chruchwell, G. Winnewisser, A&A, 45, 229 (1975)

[b] M. L. Senent, C. Puzzarini, M. Hochlaf, R. Dominguez-Gomez, and M. Carvajal, J. Chem. Phys., 141, 104303 (2014)

[c] G. I. L. Jones, D. G. Lister, N. L. Owen, J. Mol. Spectrosc., 60, 348 (1976)

[d] W. Caminati, B. P. V. Eijck, D. G. Lister, J. Mol. Spectrosc., 90, 15 (1981)

[e] J. T. Hougen, I. Kleiner, and M. Godefroid, J. Mol. Spectrosc. 163, 559 (1994)

VIBRATIONALLY EXCITED c-C_3H_2 RE-VISITED: NEW LABORATORY MEASUREMENTS AND THEORETICAL CALCULATIONS

HARSHAL GUPTA, *Division of Astronomical Sciences, National Science Foundation, Arlington, VA, USA*; J. H. WESTERFIELD, *Department of Chemistry, New College of Florida, Sarasota, FL, USA*; JOSHUA H BARABAN, *Department of Chemistry and Biochemistry, University of Colorado, Boulder, CO, USA*; BRYAN CHANGALA, *JILA, National Institute of Standards and Technology and Univ. of Colorado Department of Physics, University of Colorado, Boulder, CO, USA*; SVEN THORWIRTH, *I. Physikalisches Institut, Universität zu Köln, Köln, Germany*; JOHN F. STANTON, *Department of Chemistry, The University of Texas, Austin, TX, USA*; MARIE-ALINE MARTIN-DRUMEL, *CNRS, Institut des Sciences Moleculaires d'Orsay, Orsay, France*; OLIVIER PIRALI, *Institut des Sciences Moléculaires d'Orsay, Université Paris-Sud, Orsay, France*; CARL A GOTTLIEB, *Radio and Geoastronomy Division, Harvard-Smithsonian Center for Astrophysics, Cambridge, MA, USA*; MICHAEL C McCARTHY, *Atomic and Molecular Physics, Harvard-Smithsonian Center for Astrophysics, Cambridge, MA, USA.*

Cyclopropenylidene, c-C_3H_2, is one of the more abundant organic molecules in the interstellar medium, as evidenced from astronomical detection of its single ^{13}C and both its singly- and doubly-deuterated isotopic species. For this reason, vibrational satellites are of considerable astronomical interest, and were the primary motivation for the earlier laboratory work by Mollaaghababa and co-workers [1].

The recent detection of intense unidentified lines near 18 GHz in a hydrocarbon discharge by FT microwave spectroscopy has spurred a renewed search for the vibrational satellite transitions of c-C_3H_2. Several strong lines have been definitively assigned to the v_6 progression on the basis of follow-up measurements at 3 mm, double resonance and millimeter-wave absorption spectroscopy, and new theoretical calculations using a rovibrational VMP2 method [2] and a high-quality ab initio potential energy surface. The treatment was applied to several excited states as well as the ground state, and included deperturbation of Coriolis interactions.

[1] R. Mollaaghababa, C.A. Gottlieb, J. M. Vrtilek, and P. Thaddeus, *J. Chem. Phys.*, **99**, 890-896 (1992).

[2] P. B. Changala and J. H. Baraban. *J. Chem. Phys.*, **145**, 174106 (2016).

MILLIMETER WAVE SPECTRUM OF METHYL KETENE AND ITS SEARCH IN ORION

CELINA BERMÚDEZ, L. MARGULÈS, R. A. MOTIYENKO, *Laboratoire PhLAM, UMR 8523 CNRS - Université Lille 1, Villeneuve d'Ascq, France*; BELÉN TERCERO, JOSE CERNICHARO, *Molecular Astrophysics, ICMM, Madrid, Spain*; J.-C. GUILLEMIN, *UMR 6226 CNRS - ENSCR, Institut des Sciences Chimiques de Rennes, Rennes, France*; Y. ELLINGER, *Laboratoire de Chimie Théorique (UMR 7616), Université Paris 6, Paris, FRANCE.*

The knowledge of synthetic routes of complex organic molecules is still far to be fully understood. The creation of reliable models is particularly challenging. Hollis et al.[a] pointed out that the observations of molecular isomers provides an excellent tool to evaluate the hypothesis of the synthetic pathways. In the group of isomers C_3H_4O that contains two unsaturations, the three most stable are cyclopropanone, propenal (also known as acrolein) and methyl ketene. Among these isomers, only propenal was tentatively detected in Sgr B2(N)[b]. Spectroscopic measurements of methyl ketene CH_3CHCO are limited to the microwave domain[c]. We extended the measurements into millimeter waves in order to provide accurate frequency predictions suitable for astrophysical purposes. Methyl ketene has one more carbon atom than acetaldehyde (CH_3CHO) and in terms of rotational spectroscopy is quite similar to acetaldehyde. The analysis of the rotational spectrum of methyl ketene is complicated due to internal rotation of the methyl group, that is characterized by the barrier of intermediate height $V_3 = 416$ cm^{-1}, and by quite large value of the coupling parameter $\rho = 0.194$. The spectroscopic results and the searches of methyl ketene in Orion will be presented.

This work was supported by the CNES and the Action sur Projets de l'INSU, PCMI. This work was also done under ANR-13-BS05-0008-02 IMOLABS

[a]Hollis, J. M.; et al., 2006, ApJ **642**, 933

[b]Hollis, J. M.; et al., 2006, ApJ **643**, L25

[c]Bak, B.; et al., 1966, J. Chez. Phys. **45**, 883

WF10

ON THE RELATIVE STABILITY OF CUMULENONE AND ALDEHYDE ISOMERS: WHEN WE HEAT345(Q) THINGS UP

KELVIN LEE, *Radio and Geoastronomy Division, Harvard-Smithsonian Center for Astrophysics, Cambridge, MA, USA*; MICHAEL C McCARTHY, *Atomic and Molecular Physics, Harvard-Smithsonian Center for Astrophysics, Cambridge, MA, USA*; JOHN F. STANTON, *Department of Chemistry, The University of Texas, Austin, TX, USA*.

Isomers of $H_2C_{2n+1}O$ are examples of complex organic molecules that are either known or proposed to exist in the interstellar medium. For the smallest of these chains (H_2C_3O) only two of three isomers are observed in space: propynal ($HC(O)CCH$) and cyclopropenone ($c - C_3H_2O$), while evidence for the remaining isomer propadienone (H_2C_3O) is currently lacking. Potentially, this behaviour may be rationalised by a thermodynamic argument: several studies have provided quantum chemical calculations in an effort to determine the relative thermodynamic stability between these three isomers. An early study by Radom, at the SCF/6-31G** level ranked $HC(O)CCH$ as the thermodynamic minimum, followed by H_2C_3O, and $c - C_3H_2O$. The most recent determination by Karton and Talbi, using W2-F12 theory, places H_2C_3O as the lowest energy isomer; 2.5 kJ mol^{-1} lower than the $HC(O)CCH$ form. In an attempt to resolve this long-standing ambiguity, we were motivated to provide high level calculations based on the HEAT protocol. In this talk, we will discuss the relative stability of H_2C_3O and H_2C_5O isomers, along with their sulfur analogues, as revealed by HEAT345(Q) theory.

WG. Mini-symposium: Chirality-Sensitive Spectroscopy
Wednesday, June 21, 2017 – 1:45 PM
Room: 116 Roger Adams Lab

Chair: Laurent Nahon, Synchrotron SOLEIL, Gif sur Yvette Cedex, France

WG01   *INVITED TALK*   1:45 – 2:15

OPTICAL ROTATORY DISPERSION: NEW TWISTS ON AN OLD TOPIC

PATRICK VACCARO, *Department of Chemistry, Yale University, New Haven, CT, USA.*

Among the many physicochemical properties used to distinguish chiral molecules, perhaps none has had as profound and sustained an impact in the realm of chemistry as the characteristic interactions that take place with polarized light. Of special note is the dispersive (non-resonant) phenomenon of circular birefringence (CB), the manifestation of which first was reported over two centuries ago and which still is employed routinely – in the more familiar guise of specific optical rotation – to gauge the enantiomeric purity of the products emerging from asymmetric syntheses. Concerted experimental and theoretical efforts designed to probe such electronic optical activity in *isolated* chiral molecules will be presented, with special emphasis directed towards the marked influence that intramolecular (vibrational and conformational) dynamics and intermolecular (environmental) perturbations can exert upon the *intrinsic* chiroptical response. Requisite isolated-molecule measurements have been made possible by our continuing development of cavity ring-down polarimetry (CRDP), an ultrasensitive polarimetric scheme that has permitted the first quantitative analyses of optical rotatory dispersion (ORD or wavelength-resolved CB) to be performed in rarefied (gaseous) media. Various technical aspects of CRDP will be discussed to illustrate the unique capabilities and practical limitations afforded by this novel methodology. Comparison of specific rotation values acquired for a broad spectrum of rigid and flexible chiral species under complementary isolated and solvated conditions will highlight the intimate coupling that exists among electronic and nuclear degrees of freedom as well as the pronounced, yet oftentimes counterintuitive, effects incurred by subtle solute-solvent interactions. The disparate nature of optical activity extracted from different surroundings will be demonstrated, with quantum-chemical calculations serving to elucidate the structural, electronic, and environmental provenance of observed behavior. In addition to unraveling basic processes that mediate chiroptical response in condensed media, the vapor-phase ORD benchmarks resulting from these studies afford a critical assessment for computational predictions of dispersive optical activity and for their burgeoning ability to assist in the assignment of absolute stereochemical configuration.

WG02   2:19 – 2:34

A CHIRAL TAG STUDY OF THE ABSOLUTE CONFIGURATION OF CAMPHOR

DAVID PRATT, *Chemistry, University of Vermont, Burlington, VT, USA*; LUCA EVANGELISTI, *Dipartimento di Chimica G. Ciamician, Università di Bologna, Bologna, Italy*; TAYLOR SMART, MARTIN S. HOLDREN, KEVIN J MAYER, CHANNING WEST, BROOKS PATE, *Department of Chemistry, The University of Virginia, Charlottesville, VA, USA.*

The chiral tagging method for rotational spectroscopy uses an established approach in chiral analysis of creating a complex with an enantiopure tag so that enantiomers of the molecule of interest are converted to diastereomer complexes. Since the diastereomers have distinct structure, they give distinguishable rotational spectra. Camphor was chosen as an example for the chiral tag method because it has spectral properties that could pose challenges to the use of three wave mixing rotational spectroscopy to establish absolute configuration. Specifically, one of the dipole moment components of camphor is small making three wave mixing measurements challenging and placing high accuracy requirements on computational chemistry for calculating the dipole moment direction in the principal axis system. The chiral tag measurements of camphor used the hydrogen bond donor 3-butyn-2-ol. Quantum chemistry calculations using the B3LYP-D3BJ method and the def2TZVP basis set identified 7 low energy isomers of the chiral complex. The two lowest energy complexes of the homochiral and heterochiral complexes are observed in a measurement using racemic tag. Absolute configuration is confirmed by the use of an enantiopure tag sample. Spectra with ^{13}C-sensitivity were acquired so that the carbon substitution structure of the complex could be obtained to provide a structure of camphor with correct stereochemistry. The chiral tag complex spectra can also be used to estimate the enantiomeric excess of the sample and analysis of the broadband spectrum indicates that the sample enantiopurity is higher than 99.5%. The structure of the complex is analyzed to determine the extent of geometry modification that occurs upon formation of the complex. These results show that initial isomer searches with fixed geometries will be accurate. The reduction in computation time from fixed geometry assumptions will be discussed.

WG03

ROTATIONAL SPECTROSCOPY OF THE METHYL GLYCIDATE-WATER COMPLEX

JASON GALL, JAVIX THOMAS, ZHIBO WANG, WOLFGANG JÄGER, <u>YUNJIE XU</u>, *Department of Chemistry, University of Alberta, Edmonton, AB, Canada.*

Many biologically important molecules are chiral and perform their biological functions in an aqueous medium. In this study, we investigate the intermolecular interactions of methyl glycidate, a chiral epoxy ester, with water using rotational spectroscopy. We examine the competition among the three hydrogen-bond acceptor sites at methyl glycidate: the epoxy oxygen, the carbonyl oxygen, and the ester oxygen when interacting with water. We also probe how interaction with water modifies the methyl internal rotation barriers and conformational distribution of methyl glycidate. The possible large amplitude and tunnelling motions associated with water are investigated and analyzed.

WG04

VIBRATIONAL CIRCULAR DICHROISM SPECTRA OF METHYL GLYCIDATE IN CHLOROFORM AND WATER: APPLICATION OF THE CLUSTERS-IN-A-LIQUID MODEL

ANGELO SHEHAN PERERA, JAVIX THOMAS, *Department of Chemistry, University of Alberta, Edmonton, AB, Canada*; CHRISTIAN MERTEN, *Physikalische Chemie II, Ruhr University Bochum, Bochum, Germany*; <u>YUNJIE XU</u>, *Department of Chemistry, University of Alberta, Edmonton, AB, Canada.*

Infrared and vibrational circular dichroism (VCD) spectra of methyl glycidate, a chiral epoxy ester, were measured in CCl_4 and water in the 1000 cm^{-1} – 1800 cm^{-1} region. The experimental VCD spectra of methyl glycidate in water and in CCl_4 show noticeable differences. In particular, there are strong VCD signatures at the water bending mode region, which can be attributed to chirality transfer from chiral methyl glycidate to water through hydrogen-bonding interactions. We applied the clusters-in-a-liquid model[1] where both implicit and explicit solute-solvent interactions are considered to simulate the experimental infrared and VCD features of methyl glycidate in CCl_4 and water. All final geometry optimizations, frequency calculations, infrared and VCD intensity calculations were performed at the B3LYP-D3BJ/6-311++G(2d,p) level of theory where D3BJ is Grimme's empirical dispersion correction with damping factor.[2] We emphasize the link between the small methyl glycidate hydrates and the main long-lived species which exist in aqueous solution.

1 A. S. Perera, J. Thomas, M. R. Poopari, Y. Xu, Front. Chem. 2016, 4, 1-17. 2 S. Grimme, S. Ehrlich, L. Goerigk, J. Comp. Chem. 2011, 32, 1456-1465.

WG05

SOLVENT, TEMPERATURE And CONCENTRATION EFFECTS On THE OPTICAL ACTIVITY Of CHIRAL FIVE-And-SIX MEMBERED RING KETONES CONFORMERS

<u>WATHEQ AL-BASHEER</u>, *Department of Physics, King Fahd University of Petroleum & Minerals, Dhahran, Saudi Arabia.*

Chiral five-and-six membered ring ketones are important molecules that are found in many biological systems and can exist in many possible conformers. In this talk, experimental and computational investigation of solvent, temperature and concentration effects on the circular dichroism (CD) and optical rotation (OR) of (R)-3 -methylcyclohexanone (R3MCH), (R)-3-methylcyclopentanone (R3MCP) and carvone conformers will be discussed. CD and OR measurements of these ketones gaseous samples and in ten common solvents of wide polarity range for different concentrations and sample temperatures were recorded and related to molecular conformation. Density functional theoretical calculations were performed using Gaussian09 at B3LYP functions with aug-cc-pVDZ level of theory. Also, CD and OR spectra for the optimized geometries of the ketones dominant conformers were computed over the ultraviolet and visible region in the gas phase as well as in ten solvents of varying polarity range, and under the umbrella of the polarizable continuum model (PCM). By comparing theoretical and experimental results, few thermodynamic parameters were deduced for the individual equatorial and axial conformers of each molecule in gas phase and in solvation.

Intermission

RAPID-ADIABATIC-PASSAGE CONTROL OF RO-VIBRATIONAL POPULATIONS IN POLYATOMIC MOLECULES

EMIL J ZAK, *Department of Physics and Astronomy, University College London, Gower Street, London WC1E 6BT, United Kingdom*; ANDREY YACHMENEV, *Center for Free-Electron Laser Science (CFEL), Deutsches Elektronen-Synchrotron (DESY), Hamburg, Germany.*

We present a simple method for control of ro-vibrational populations in polyatomic molecules in the presence of inhomogeneous electric fields [1]. Cooling and trapping of heavy polar polyatomic molecules has become one of the frontier goals in high-resolution molecular spectroscopy, especially in the context of parity violation measurement in chiral compounds [2]. A key step toward reaching this goal would be development of a robust and efficient protocol for control of populations of ro-vibrational states in polyatomic, often floppy molecules. Here we demonstrate a modification of the stark-chirped rapid-adiabatic-passage technique (SCRAP) [3], designed for achieving high levels of control of ro-vibrational populations over a selected region in space. The new method employs inhomogeneous electric fields to generate space- and time- controlled Stark-shifts of energy levels in molecules. Adiabatic passage between ro-vibrational states is enabled by the pump pulse, which raises the value of the Rabi frequency. This Stark-chirped population transfer can be used in manipulation of population differences between high-field-seeking and low-field-seeking states of molecules in the Stark decelerator [4]. Appropriate timing of voltages on electric rods located along the decelerator combined with a single pump laser renders our method as potentially more efficient than traditional Stark decelerator techniques. Simulations for NH_3 show significant improvement in effectiveness of cooling, with respect to the standard 'moving-potential' method [5]. At the same time a high phase-space acceptance of the molecular packet is maintained.

[1] E. J. Zak, A. Yachmenev (submitted).

[2] C. Medcraft, R. Wolf, M. Schnell, Angew. Chem. Int. Ed., 53, 43, 11656–11659 (2014)

[3] M. Oberst, H. Munch, T. Halfman, PRL 99, 173001 (2007).

[4] K. Wohlfart, F. Grätz, F. Filsinger, H. Haak, G. Meijer, J. Küpper, Phys. Rev. A 77, 031404(R) (2008).

[5] H. L. Bethlem, F. M. H. Crompvoets, R. T. Jongma, S. Y. T. van de Meerakker, G. Meijer, Phys. Rev. A, 65, 053416 (2002).

CHIRAL TAGGING OF VERBENONE WITH 3-BUTYN-2-OL FOR ESTABLISHING ABSOLUTE CONFIGURATION AND DETERMINING ENANTIOMERIC EXCESS

LUCA EVANGELISTI, *Dipartimento di Chimica G. Ciamician, Università di Bologna, Bologna, Italy*; KEVIN J MAYER, MARTIN S. HOLDREN, TAYLOR SMART, CHANNING WEST, BROOKS PATE, *Department of Chemistry, The University of Virginia, Charlottesville, VA, USA*; GALEN SEDO, *Department of Natural Sciences, University of Virginia's College at Wise, Wise, VA, USA*; FRANK E MARSHALL, G. S. GRUBBS II, *Department of Chemistry, Missouri University of Science and Technology, Rolla, MO, USA.*

Chiral analysis of a commercial sample of (1S)-(-)-verbenone has been performed using the chiral tag approach. The chirped-pulse Fourier transform microwave spectrum of the verbenone-butynol complex is measured in the 2-8 GHz frequency range. Verbenone is placed in a nozzle reservoir heated to 333K (about 1 Torr vapor pressure). The complex is formed by using a carrier gas of neon with approximately 0.1% butynol. The expansion pressure is about 2 atm. A measurement using racemic butynol is performed to identify isomers of both diastereomer complexes. Quantum chemistry calculations using the B3LYP-D3BJ method with the def2TZVP basis set provided estimated spectroscopic constants for the homochiral and heterochiral complexes. This analysis included 8 isomers for each diastereomer. Four rotational spectra are identified for isomers of the homochiral complex and correspond to the four lowest energy isomers from the theoretical study. Three heterochiral complexes are identified and also correspond to the lowest energy isomers from theory. Subsequent measurements were made with enantiopure tag (both (R)-(+)-3-buty-2-nol and (S)-(-)-3-butyn-2-ol) to establish the absolute configuration of verbenone. The sensitivity of the measurement was sufficient to perform ^{13}C-isotopologue analysis of three of the homochiral complexes and two of the heterochiral complexes. These results provide definitive structures of verbenone with correct stereochemistry. The commercial sample has relatively low enantiomeric excess with the certificate of analysis reporting an EE of 53.6%. Using the intensities of assigned transitions of the chiral tag complexes, the enantiomeric excess was determined from the broadband rotational spectrum through the ratio of the intensities of pairs of transitions. A total of 2617 pairs of transitions were analyzed. The average EE was found to be 53.6% with a standard deviation of 2%.

WG08 4:18 – 4:33

COHERENT POPULATION TRANSFER IN CHIRAL MOLECULES USING TAILORED MICROWAVE PULSES

CRISTOBAL PEREZ, *CoCoMol, Max-Planck-Institut für Struktur und Dynamik der Materie, Hamburg, Germany*; AMANDA STEBER, *CUI, The Hamburg Centre for Ultrafast Imaging, Hamburg, Germany*; SERGIO R DOMINGOS, ANNA KRIN, DAVID SCHMITZ, MELANIE SCHNELL, *CoCoMol, Max-Planck-Institut für Struktur und Dynamik der Materie, Hamburg, Germany.*

Over the last years, microwave three-wave mixing (M3WM) experiments have been shown to provide a sensitive way to generate and measure enantiomer-specific molecular responses. These experiments opened the door for enantiomeric excess determination in complex samples without previous separation or purification. We present here a new type of experiment, based on M3WM[a], to achieve enantiomeric enrichment of a chiral sample by using microwave pulses. We will show that control over the relative phases and polarizations of pulses provides a way to selectively populate a specific quantum rotational state with an enantiomer of choice. The experimental implementation as well as the characterization of the observed enantiomer-selective responses will be presented and discussed. As a proof of concept and to showcase the applicability of our approach we will present the enantiomer enrichment of several terpenes.

[a]Sandra Eibenberger, John Doyle, and David Patterson, *arXiv:1608.04691* **(2016)**

WG09 4:35 – 4:50

COMPLEXES OF SMALL CHIRAL MOLECULES: PROPYLENE OXIDE AND 3-BUTYN-2OL

LUCA EVANGELISTI, *Dipartimento di Chimica G. Ciamician, Università di Bologna, Bologna, Italy*; CHANNING WEST, ELLIE COLES, BROOKS PATE, *Department of Chemistry, The University of Virginia, Charlottesville, VA, USA.*

Complexes of propylene oxide with 3-butyn-2-ol were observed in the molecular rotational spectra, and isotopologue analysis allowed for structural determination of the complexes. Using a gas mixture of 0.1% propylene oxide and 0.1% 3-butyn-2-ol in neon, the broadband rotational spectrum was measured in the 2-8 GHz frequency range using a chirped-pulse Fourier transform microwave spectrometer. Four isomers of each diastereomer pair, formed by a hydrogen bond between the two monomers, are identified in quantum chemistry study of the complex using B3LYP-D3BJ with the def2TZVP basis set. The initial measurement used racemic samples of both molecules in order to obtain all possible isomers of the complex in the pulsed jet expansion. A total of six distinct spectra were assigned in the racemic measurement - three for both the homochiral and heterochiral complex. Substitution structures for the most intense homochiral and heterochiral complexes were obtained. These complexes use the two lowest energy conformations of butynol despite conformational cooling of the monomer, resulting in a single identified isomer. This result shows that a wide range monomer conformational geometries need to be examined when performing searches for the lowest energy geometry. Analysis of the diastereomer spectra was used to develop a method for determining the enantiomeric excess of 3-butyn-2-ol and propylene oxide for use as a chiral tag, which could be used in subsequent measurements to determine enantiomeric excess. The sensitivity limits for enantiomeric excess determination and the linearity of the rotational spectroscopy signals as a function of sample enantiomeric excess will be presented.

WG10 4:52 – 5:07

CHIRAL PROCESS MONITORING USING FOURIER TRANSFORM MICROWAVE SPECTROSCOPY

JUSTIN L. NEILL, MATT MUCKLE, *BrightSpec Labs, BrightSpec, Inc., Charlottesville, VA, USA*; BROOKS PATE, *Department of Chemistry, The University of Virginia, Charlottesville, VA, USA.*

We present the application of Fourier transform microwave (FTMW) spectroscopy in monitoring the chiral purity of components in a reaction mixture. This is of particular interest due to the increasing use of continuous pharmaceutical manufacturing processes, in which a number of attributes (including the chiral purity of the product) can change on short time scales. Therefore, new techniques that can accomplish this measurement rapidly are desired. The excellent specificity of FTMW spectroscopy, coupled with newly developed techniques for measuring enantiomeric excess in a mixture, have motivated this work.

In collaboration with B. Frank Gupton (Virginia Commonwealth University), we are testing this application first with the synthesis of artemisinin. Artemisinin, a common drug for malaria treatment, is of high global health interest and subject to supply shortages, and therefore a strong candidate for continuous manufacturing. It also has moderately high molecular weight (282 amu) and seven chiral centers, making it a good candidate to test the capabilities of FTMW spectroscopy. Using a miniature cavity-enhanced FTMW spectrometer design,[a] we aim to demonstrate selective component quantification in the reaction mixture. Future work that will be needed to fully realize this application will be discussed.

[a]R.D. Suenram, J.U. Grabow, A.Zuban, and I.Leonov, Rev. Sci. Instrum. 70, 2127 (1999).

WG11 5:09 – 5:24

HIGH RESOLUTION FTIR SPECTROSCOPY OF TRISULFANE HSSSH: A CANDIDATE FOR DETECTING PARITY VIOLATION IN CHIRAL MOLECULES

SIEGHARD ALBERT, IRINA BOLOTOVA, ZIQIU CHEN, CSABA FÁBRI, MARTIN QUACK, GEORG SEY-FANG, DANIEL ZINDEL, *Laboratory of Physical Chemistry, ETH Zurich, Zürich, Switzerland.*

The measurement of the parity violating energy difference $\Delta_{pv}E$ between the enantiomers of chiral molecules is among the major current challenges in high resolution spectroscopy and physical-chemical stereochemistry.[a,b] Theoretical predictions have recently identified dithiine[b] and trisulfane[c] as suitable candidates for such experiments. We report the first successful high-resolution analyses of the Fourier transform infrared (FTIR) spectra of trisulfane. A band centered at 861.0292 cm^{-1} can be assigned unambiguously to the chiral *trans* conformer by means of ground state combination differences in comparison with known pure rotational spectra. [d] A second band near 864.698 cm^{-1} is tentatively assigned to the *cis* conformer by comparison with theory.

[a]M. Quack , *Fundamental Symmetries and Symmetry Violations from High-resolution Spectroscopy, Handbook of High Resolution Spectroscopy, M. Quack and F. Merkt eds.,*John Wiley & Sons Ltd, Chichester, New York, 2001, vol. 1, ch. 18, pp. 659-722.

[b]S. Albert, I. Bolotova, Z. Chen, C. Fábri, L. Horný, M. Quack, G. Seyfang and D. Zindel, *Phys.Chem.Chem.Phys.***18**, 21976-21993 (2016).

[c]C. Fábri, L. Horný and M. Quack, *ChemPhysChem***16**, 3584-3589 (2015).

[d]M. Liedtke, K. M. T. Yamada, G. Winnewisser and J. Hahn, *J.Mol.Struct.***413**, 265-270 (1997).

WH. Dynamics and kinetics

Wednesday, June 21, 2017 – 1:45 PM

Room: 1024 Chemistry Annex

Chair: J. Gary Eden, University of Illinois, Urbana, IL, USA

WH01   1:45 – 2:00

DIRECT MEASUREMENT OF OD+CO→ *cis*-DOCO, *trans*-DOCO, AND D+CO$_2$ BRANCHING KINETICS USING TIME-RESOLVED FREQUENCY COMB SPECTROSCOPY

BRYCE J BJORK, *JILA, National Institute of Standards and Technology and Univ. of Colorado Department of Physics, University of Colorado, Boulder, CO, USA*; THINH QUOC BUI, *JILA, National Institute of Standards and Technology and Univ. of Colorado Department of Physics, University of Colorado, Boulder, Boulder, CO, USA*; BRYAN CHANGALA, BEN SPAUN, *JILA, National Institute of Standards and Technology and Univ. of Colorado Department of Physics, University of Colorado, Boulder, CO, USA*; KANA IWAKUNI, *JILA, National Institute of Standards and Technology and Univ. of Colorado Department of Physics, University of Colorado, Boulder, Colorado, Boulder, CO, USA*; JUN YE, *JILA, National Institute of Standards and Technology and Univ. of Colorado Department of Physics, University of Colorado, Boulder, Boulder, CO, USA*.

The kinetics of the reaction OH+CO→H+CO$_2$ has attracted experimental and theoretical studies for more than 40 years due to its importance in atmospheric and combustion environments. This reaction proceeds on a rich potential energy landscape, first by forming vibrationally excited HOCO*; subsequently, HOCO* either back reacts to OH+CO, dissociates to H+CO$_2$, or is stabilized to ground state HOCO by collisions with a third body. Due to the formation of the HOCO intermediate, the rate coefficient displays anomalous temperature and strong pressure dependences. Time-resolved Frequency Comb Spectroscopy (TRFCS) combines a mid-IR mode-locked femtosecond laser, a broadband optical enhancement cavity, and spatially dispersive detection system to simultaneously provide broad spectral bandwidth, high spectral resolution, high absorption sensitivity, and microsecond time resolution. We have applied this powerful technique to identify the deuterated analogues of HOCO isomers, *trans*-DOCO and *cis*-DOCO, for the first time in the reaction OD+CO under ambient conditions. By directly monitoring the concentrations of OD (reactant), *trans*-DOCO, *cis*-DOCO (intermediates), and CO$_2$(product), we unambiguously measure all pressure-dependent branching rates of the OD+CO reaction.

WH02   2:02 – 2:17

DYNAMIC TIME-RESOLVED CHIRPED-PULSE ROTATIONAL SPECTROSCOPY OF VINYL CYANIDE PHOTO-PRODUCTS IN A ROOM TEMPERATURE FLOW REACTOR

DANIEL P. ZALESKI, KIRILL PROZUMENT, *Chemical Sciences and Engineering Division, Argonne National Laboratory, Argonne, IL, USA*.

Chirped-pulsed (CP) Fourier transform rotational spectroscopy invented by Brooks Pate and coworkers a decade ago is an attractive tool for gas phase chemical dynamics and kinetics studies. A good reactor for such a purpose would have well-defined (and variable) temperature and pressure conditions to be amenable to accurate kinetic modeling. Furthermore, in low pressure samples with large enough number of molecular emitters, reaction dynamics can be observable directly, rather than mediated by supersonic expansion. In the present work, we are evaluating feasibility of *in situ* time-resolved CP spectroscopy in a room temperature flow tube reactor. Vinyl cyanide (CH$_2$CHCN), neat or mixed with inert gasses, flows through the reactor at pressures $1-50$ μbar ($0.76-38$ mTorr) where it is photodissociated by a 193 nm laser. Millimeter-wave beam of the CP spectrometer co-propagates with the laser beam along the reactor tube and interacts with nascent photoproducts. Rotational transitions of HCN, HNC, and HCCCN are detected, with ≥ 10 μs time-steps for 500 ms following photolysis of CH$_2$CHCN. The post-photolysis evolution of the photoproducts' rotational line intensities is investigated for the effects of rotational and vibrational thermalization of energized photoproducts. Possible contributions from bimolecular and wall-mediated chemistry are evaluated as well.

TOWARDS A QUANTUM DYNAMICAL STUDY OF THE H_2O+H_2O INELASTIC COLLISION: REPRESENTATION OF THE POTENTIAL AND PRELIMINARY RESULTS

STEVE ALEXANDRE NDENGUE, RICHARD DAWES, *Department of Chemistry, Missouri University of Science and Technology, Rolla, MO, USA.*

Water, an essential ingredient of life, is prevalent in space and various media. H_2O in the gas phase is the major polyatomic species in the interstellar medium (ISM) and a primary target of current studies of collisional dynamics. In recent years a number of theoretical and experimental studies have been devoted to H_2O-X (with X=He, H_2, D_2, Ar, ...) elastic and inelastic collisions in an effort to understand rotational distributions of H_2O in molecular clouds. Although those studies treated several abundant species, no quantum mechanical calculation has been reported to date for a nonlinear polyatomic collider. We present in this talk the preliminary steps toward this goal, using the H_2O molecule itself as our collider, the very accurate MB-Pol surface to describe the intermolecular interaction and the MultiConfiguration Time Dependent (MCTDH) algorithm to study the dynamics. One main challenge in this effort is the need to express the Potential Energy Surface (PES) in a sum-of-products form optimal for MCTDH calculations. We will describe how this was done and present preliminary results of state-to-state probabilities.

NORMAL MODE ANALYSIS ON THE RELAXATION OF AN EXCITED NITROMETHANE MOLECULE IN ARGON BATH

LUIS A. RIVERA-RIVERA[a], *Department of Chemistry, Texas A & M University, College Station, TX, USA;* ALBERT F. WAGNER, *Chemical Sciences and Engineering Division, Argonne National Laboratory, Argonne, IL, USA.*

In our previous work [Rivera-Rivera *et al. J. Chem. Phys.* 142, 014303 (2015).] classical molecular dynamics simulations followed, in an Ar bath, the relaxation of nitromethane (CH_3NO_2) instantaneously excited by statistically distributing 50 kcal/mol among all its internal degrees of freedom. The 300 K Ar bath was at pressures of 10 to 400 atm. Both rotational and vibrational energies exhibited multi-exponential decay. This study explores mode-specific mechanisms at work in the decay process. With the separation of rotation and vibration developed by Rhee and Kim [*J. Chem. Phys.* 107, 1394 (1997).], one can show that the vibrational kinetic energy decomposes only into vibrational normal modes while the rotational and Coriolis energies decompose into both vibrational and rotational normal modes. Then the saved CH_3NO_2 positions and momenta can be converted into mode-specific energies whose decay over 1000 ps can be monitored. The results identify vibrational and rotational modes that promote/resist energy lost and drive multi-exponential behavior. In addition to mode-specificity, the results show disruption of IVR with increasing pressure.

[a]Present Address: Department of Physical Sciences, Ferris State University, Big Rapids, MI 49307-2225

WH05 2:53 – 3:08

PROTON TRANSFER AND LOW-BARRIER HYDROGEN BONDING: A SHIFTING VIBRATIONAL LANDSCAPE DICTATED BY LARGE AMPLITUDE TUNNELING

ZACHARY VEALEY, LIDOR FOGUEL, PATRICK VACCARO, *Department of Chemistry, Yale University, New Haven, CT, USA.*

Our fundamental understanding of synergistic hydrogen-bonding and proton-transfer phenomena has been advanced immensely by studies of model systems in which the coherent transduction of hydrons is mediated by two degenerate equilibrium configurations that are isolated from one another by a potential barrier of substantial height. This topography advantageously affords unambiguous signatures for the underlying state-resolved dynamics in the form of tunneling-induced spectral bifurcations, the magnitudes of which encode both the overall efficacy and the detailed mechanism of the unimolecular transformation. As a prototypical member of this class of compounds, 6-hydroxy-2-formylfulvene (HFF) supports an unusual quasi-linear O–H$\cdots$O $\leftrightarrow$ O$\cdots$H–O reaction coordinate that presents a minimal impediment to proton migration – a situation commensurate with the concepts of low-barrier hydrogen bonding (which are characterized by great strength, short distance, and a vanishingly small barrier for hydron migration). A variety of fluorescence-based, laser-spectroscopic probes have been deployed in a cold supersonic free-jet expansion to explore the vibrational landscape and anomalously large tunneling-induced shifts that dominate the $\tilde{X}^1A_1$ potential-energy surface of HFF, thus revealing the most rapid proton tunneling ever reported for a molecular ground state ($\tau_{pt} \leq 120$fs). The surprising efficiency of such tunneling-mediated processes stems from proximity of the zero-point level to the barrier crest and produces a dramatic alteration in the canonical pattern of vibrational features that reflects, in part, the subtle transition from quantum-mechanical barrier penetration to classical over-the-barrier dynamics. The ultrafast proton-transfer regime that characterizes the $\tilde{X}^1A_1$ manifold will be juxtaposed against analogous findings for the lowest-lying singlet excited state $\tilde{A}^1B_2$ ($\pi^* \leftarrow \pi$), where a marked change in the nature of the reaction coordinate leads to the near-complete quenching of proton transfer. Experimental results, as well as complementary quantum-chemical analyses, will be discussed and contrasted with those obtained for related hydron-migration systems in an effort to highlight the unique bonding motifs and reaction propensities evinced by HFF.

Intermission

WH06 3:27 – 3:42

MOLECULAR BEAM SURFACE SCATTERING OF FORMALDEHYDE FROM Au(111): CHARACTERIZATION OF THE DIRECT SCATTER AND TRAPPING-DESORPTION CHANNELS

BASTIAN C. KRUEGER, BARRATT PARK, SVEN MEYER, ROMAN J. V. WAGNER, *Institute of Physical Chemistry, Georg-August-Universität Göttingen, Göttingen, Germany*; ALEC WODTKE, *Dynamics at Surfaces, Max Planck Institute for Biophysical Chemistry, Göttingen, Germany*; TIM SCHAEFER, *Institute of Physical Chemistry, Georg-August-Universität Göttingen, Göttingen, Germany.*

Quantum state resolved molecular beam scattering studies of small polyatomic molecules from metal surfaces present new challenges for experimentalists, but provide unprecedented new opportunities for detailed study of polyatomic molecular dynamics at surfaces. In the current work, we report preliminary characterization of the scattering of formaldehyde from the Au(111) surface. We report the measured desorption energy (0.31 eV), and characterize the distinct trapping-desorption and direct scattering channels, via the dependence of the scattered velocity and rotational distributions on surface temperature and incident molecular beam energy. Finally, we estimate the trapping probability as a function of incidence energy, which indicates the importance of molecular degrees of freedom in the mechanism for trapping.

172

WH07 3:44–3:59

ROTATIONALLY-RESOLVED SCATTERING OF FORMALDEHYDE FROM THE Au(111) SURFACE: AN AXIS SPE-CIFIC ROTATIONAL RAINBOW AND ITS ROLE IN TRAPPING PROBABILITY

BARRATT PARK, BASTIAN C. KRUEGER, SVEN MEYER, *Institute of Physical Chemistry, Georg-August-Universität Göttingen, Göttingen, Germany*; ALEXANDER KANDRATSENKA, ALEC WODTKE, *Dynamics at Surfaces, Max Planck Institute for Biophysical Chemistry, Göttingen, Germany*; TIM SCHAEFER, *Institute of Physical Chemistry, Georg-August-Universität Göttingen, Göttingen, Germany.*

The conversion of translational to rotational motion often plays a major role in the trapping of small molecules at surfaces, a crucial first step for a wide variety of chemical processes that occur at gas-surface interfaces. However, to date most quantum-state resolved surface scattering experiments have been performed on diatomic molecules, and very little detailed information is available about how the structure of non-linear polyatomic molecules influences the mechanisms for energy exchange with surfaces. In the current work, we employ a new rotationally-resolved $1 + 1'$ resonance-enhanced multiphoton ionization (REMPI) scheme to measure rotational distribution in formaldehyde molecules directly scattered from the Au(111) surface at incident kinetic energies in the range 0.3–1.2 eV. The results indicate a pronounced propensity to excite a-axis rotation (twirling) rather than b- or c-axis rotation (tumbling or cartwheeling), and are consistent with a rotational rainbow scattering model. Classical trajectory calculations suggest that the effect arises—to zeroth order—from the three-dimensional shape of the molecule (steric effects). The results have broad implications for the enhanced trapping probability of prolate and near-prolate molecules at surfaces.

WH08 *Post-Deadline Abstract - Original Abstract Withdrawn* 4:01–4:16

CHARACTERIZATION OF EXTENDED TIME SCALE 2D IR PROBES OF PROTEINS

SASHARY RAMOS, AMANDA L LE SUEUR, KEITH J SCOTT, MEGAN THIELGES, *Department of Chemistry, Indiana University, Bloomington, IN, USA.*

The role of dynamics in the function of proteins is well appreciated, but not precisely understood due to the difficulty in their measurement. Two-dimensional infrared (2D IR) spectroscopy is a powerful approach for the study of protein dynamics with high spatial and temporal resolution. This approach has led to the development of spectrally resolved IR probes that can be applied towards the measurement of dynamics at specific sites in a protein. However, the experimental time scale is limited by the vibrational lifetime of the probe, as such their remains a need for extended time scale probes. Towards the development of better 2D IR probes for the study of protein dynamics the spectroscopic characterization of p-cyano-seleno-phenylalanine (CNSePhe), isotopically labeled p-($^{13}C^{15}N$-cyano)phenylalanine ($^{13}C^{15}NPhe$) and the site-specific incorporation of $^{13}C^{15}NPhe$ in the protein plastocyanin is discussed. The incorporation of the heavy Se atom and the isotopic labeling are shown to increase the vibrational lifetime of the probe which results in collection of 2D IR spectra for analysis of dynamics on longer timescales.

WH09 4:18–4:33

NONLINEAR PHOTOCHROMIC SWITCHING IN THE PLASMONIC FIELD OF A NANOPARTICLE ARRAY

CHRISTOPHER J OTOLSKI, *Department of Chemistry, University of Kansas, Lawrence, KS, USA*; CHRISTOS ARGYROPOULOS, *Engineering, University of Nebraska–Lincoln, Lincoln, USA*; CHRISTOPHER G. ELLES, *Department of Chemistry, University of Kansas, Lawrence, KS, USA.*

Plasmonic nanostructures provide unique environments for non-resonant excitation and switching of photochromic compounds. In this study, photochromic diarylethene molecules were deposited on top of a periodically ordered array of gold nanorods (170 x 80 nm) and then irradiated with <100 fs laser pulses. Irradiation at 800 nm drives the plasmon resonance of the nanoparticle array and induces the photochromic conversion of molecules via non-resonant two-photon excitation. Transmission measurements using broadband continuum laser pulses probe the progress of the photochemical cycloreversion reaction as molecules switch from a visible-absorbing closed-ring structure to a transparent open-ring structure. The spatial dependence of the two-photon conversion of molecules in the plasmonic near field of the array is modeled using calculated field enhancements, and compared with similar measurements for a film of molecules on a glass substrate. Wavelength-dependent polarization effects in the near field of the array lead to interesting anisotropy results in the transmission signal. The results emphasize the importance of both the spatial dependence and anisotropy of the enhanced electric fields in driving non-resonant photochromic reactions.

WH10 4:35 – 4:50

ENERGY POOLING, ION RECOMBINATION, AND REACTIONS OF RUBIDIUM AND CESIUM IN HYDROCARBON GASSES.

SEAN MICHAEL BRESLER, J. PARK, MICHAEL HEAVEN, *Department of Chemistry, Emory University, Atlanta, GA, USA.*

Diode Pumped Alkali Lasers (DPAL) are continuous wave lasers, potentially capable of megawatt average powers. These lasers exploit the D1 and D2 lines of alkali metals resulting in a 3-level laser with the lasing transition in the near infrared region of the electromagnetic spectrum. Energy pooling processes involving collisions between excited alkali metals cause a fraction of the gain media to be highly excited and eventually ionized. These high energy cesium atoms and ions chemically react with small hydrocarbons utilized as buffer gasses for the system, depleting the gain media. A kinetic model supported by experimental data is introduced to explain the cumulative effects of optical trapping, energy pooling, and chemical reactivity in heavy alkali metal (Rb, Cs) systems. Spectroscopic studies demonstrating metal hydride formation will also be presented.

WH11 4:52 – 5:07

ANALYSIS OF THREE-BODY FORMATION RATES COEFFICIENTS OF Hg^*, Hg_2^*, AND Hg_3^* VIA PHOTOEXCITATION OF Hg VAPOR

WENTING WENDY CHEN, J. GARY EDEN, *Department of Electrical and Computer Engineering, University of Illinois at Urbana-Champaign, Urbana, IL, USA.*

Decay rates of 335 nm emission from Hg_2^* and 485 nm emission from Hg_3^* were recorded under 266 nm photoexciation. A previously unobserved turning point in the decay rates with respect to Hg number density curve was recorded. A new rate equation model was built to reveal the three-body formation rates coefficients of Hg^*, Hg_2^*, and Hg_3^* by matching the simulated decay rates with Hg number density curves with experimental recorded ones:

$Hg^* + Hg + Hg \rightarrow Hg_2^* + Hg$

$Hg_2^* + Hg + Hg \rightarrow Hg_3^* + Hg$

$Hg_3^* + Hg + Hg \rightarrow Hg_4^* + Hg$

Pump and probe experiments with 266 nm and tunable blue laser were also conducted and suppression of both 335 nm and 485 nm emission at different probe laser wavelength were recorded. The delay between occurring time of 335 nm and 485 nm was observed. The suppression intensity of the two cases were also analyzed and compared.

WI. Theory and Computation
Wednesday, June 21, 2017 – 1:45 PM
Room: B102 Chemical and Life Sciences

Chair: Tucker Carrington, Queen's University, Kingston, ON, Canada

---

WI01 $\qquad$ 1:45 – 2:00

A CANONICAL APPROACH TO GENERATE MULTIDIMENSIONAL POTENTIAL ENERGY SURFACES

JAY R. WALTON, *Department of Mathematics, Texas A & M University, College Station, TX, USA*; <u>LUIS A. RIVERA-RIVERA</u>, *Department of Physical Sciences , Ferris State University , Big Rapids, MI, USA*; ROBERT R. LUCCHESE, *Department of Chemistry, Texas A & M University, College Station, TX, USA*.

Previously adaptions of canonical approaches were applied to algebraic forms of the classic Morse, Lennard-Jones, and Kratzer potentials. Using the classic Morse, Lennard-Jones, or Kratzer potential as reference, inverse canonical transformations allow the accurate generation of Born-Oppenheimer potentials for H_2^+ ion, neutral covalently bound H_2, van der Waals bound Ar_2, and the hydrogen bonded 1-dimensional dissociative coordinate in water dimer. This methodology is now extending to multidimensional potential energy surfaces, and as a proof-of-concept, it is applied to the 3-dimensional water molecule potential surface. Canonical transformations previously developed for diatomic molecules are used to construct accurate approximations to the 3-dimensional potential surface of the water molecule from judiciously chosen 1-dimensional planar slices that are shown to have the same canonical shape as the classical Lennard-Jones potential curve. Spline interpolation is then used to piece together the 1-dimensional canonical potential curves, to obtain the full 3-dimensional potential surface of water molecule with a relative error less than 0.008.

WI02 $\qquad$ 2:02 – 2:17

AB INITIO CALCULATIONS OF THE GROUND AND EXCITED STATES OF THE ZNTE MOLECULE AND ITS IONS ZNTE$^+$ AND ZNTE$^-$

NOUR EL HOUDA BENSIRADJ, <u>OURIDA OUAMERALI</u>, AZEDDINE DEKHIRA, *Laboratory lctcp, University USTHB, Algiers, Algeria*; TIMÓN VICENTE, *Molecular Physics, Instituto de Estructura de la Materia (IEM-CSIC), Madrid, Spain*.

The ZnTe system exhibits very interesting optoelectronic properties. It is a promising candidate for the development of detectors of Terahertz (THz) radiation, as well as a growing number of applications, particularly in the area of radiology.

In this work, we report a theoretical study of the ground state and various excited states of ZnTe and its ions ZnTe$^+$ and ZnTe$^-$. The potential energy curves are calculated using CASSCF method, as implemented in Molpro. These curves serve to determine the different spectroscopic constants such as the internuclear distance (R_e), the harmonic vibration frequency (ω_e), the rotation constant (B_e) and the dissociation energy (D_e). The results obtained are in good agreement with the available experimental data.

WI03 $\qquad$ 2:19 – 2:34

THEORETICAL CALCULATION OF THE UV-VIS SPECTRAL BAND LOCATIONS OF PAHS WITH UNKNOWN SYNTHESES PROCEDURES AND PROSPECTIVE CARCINOGENIC ACTIVITY

<u>JORGE OSWALDO ONA-RUALES</u>, *Department of Chemical Engineering, Nazarbayev University, Astana, Kazakhstan*; YOSADARA RUIZ-MORALES, , *Instituto Mexicano del Petroleo, Mexico City, Mexico*.

Annellation Theory and ZINDO/S semiempirical calculations have been used for the calculation of the locations of maximum absorbance (LMA) of the Ultraviolet-Visible (UV-Vis) of 31 $C_{34}H_{16}$ PAHs (molecular mass 424 Da) with unknown protocols of synthesis. The presence of benzo[a]pyrene bay-like regions and dibenzo[a,l]pyrene fjord-like regions in several of the structures that could be linked to an enhancement of the biological behavior and carcinogenic activity stresses the importance of $C_{34}H_{16}$ PAHs in fields like molecular biology and cancer research. In addition, the occurrence of large PAHs in oil asphaltenes exemplifies the importance of these calculations for the characterization of complex systems. The $C_{34}H_{16}$ PAH group is the largest molecular mass group of organic compounds analyzed so far following the Annellation Theory and ZINDO/S methodology. Future analysis using the same approach will provide evidence regarding the LMA of other high molecular mass PAHs.

WI04   2:36 – 2:51

THEORETICAL INVESTIGATION OF PHOTOASSOCIATIVE EXCITATION SPECTROSCOPY OF XENON MONOIO-
DIDE

WENTING WENDY CHEN, *Department of Electrical and Computer Engineering, University of Illinois at Urbana-Champaign, Urbana, IL, USA*; FANG SHEN, *Department of Physics, University of Illinois at Urbana-Champaign, Urbana, IL, USA*; J. GARY EDEN, *Department of Electrical and Computer Engineering, University of Illinois at Urbana-Champaign, Urbana, IL, USA.*

The experimental photoassociation spectrum of B ← X transition of XeI over a broad internuclear distance range have been simulated quantum mechanically using a new spectral simulation technique including an improved potential model of the X state. The photoassociation spectrum generated from the simulated upper and lower potentials reproduces all spectral details of the experimental spectrum. Spectroscopic constants obtained are consistent with but unique compared to previously reported results. The V-R coupled energy structures of XeI molecules are also verified by the simulation results.

WI05   2:53 – 3:08

INSIGHT INTO THE CHARGE TRANSFER MECHANISMS OF HEAVY ATOM SUBSTITUTED MALDI MATRICES

CHELSEA N BRIDGMOHAN, LICHANG WANG, KRISTOPHER M KIRMESS, *Department of Chemistry and Biochemistry, Southern Illinois University Carbondale, Carbondale, IL, USA.*

The underlying mechanism of how MALDI matrices work is poorly understood. Experimental literature suggests that the triplet excited state (T_1) of the matrix plays a significant role in its ability to transfer charge to the analyte effectively. The heavy atom substitution effect predicts that the addition of a heavy atom to an otherwise "dead" matrix, such as 2,4-dihydroxybenzoic acid, would increase the rate of Intersystem Crossing (ISC) to the T_1 state via spin-orbit coupling. This effect was observed experimentally as there was a visible decay in singlet lifetime and an increase in triplet lifetime, as well as a better matrix performance when compared to its original, unsubstituted partner. To provide insight into the photophysical properties of 2,4-dihydroxybenzoic acid and its halogenated isomers, calculations were performed using *Gaussian09*. Geometry optimizations, frequencies, and IR spectra of all isomers were calculated using Density Functional Theory (DFT) with B3LYP functional and the 6-31G+(d,p) basis set. UV-Vis and fluorescence spectra were generated using Time-Dependent DFT (TDDFT). The following values for the singlet ground state (S_0), triplet excited state (T_1), and singlet excited state (S_1) were tabulated and compared: optimization energies, HOMO-LUMO energies and orbital contours, and bond distances. In addition, the energy values for Proton Affinity (PA) and Gas Phase Acidity (GPA) were determined.

WI06   3:10 – 3:25

SCALAR RELATIVISTIC EQUATION-OF-MOTION COUPLED CLUSTER CALCULATIONS OF CORE-
IONIZED/EXCITED STATES

LAN CHENG, *Department of Chemistry, Johns Hopkins University, Baltimore, MD, USA.*

Scalar relativistic equation-of-motion coupled cluster (EOMCC) calculations of core ionization/excitation energies for a set of benchmark molecules are reported. The Arnoldi algorithm as well as the core-valence-separation (CVS) scheme have been used to expedite the convergence of the wave function for the core-ionized/excited states. Scalar relativistic effects have been accounted for using the spin-free exact two-component theory in its one-electron variant (SFX2C-1e) and their importance are assessed. Preliminary calculations of ligand core excitation spectra of transition-metal containing compounds are also presented.

Intermission

WI07 3:44 – 3:59

MULTI-STATE EXTRAPOLATION OF UV/VIS ABSORPTION SPECTRA WITH QM/QM HYBRID METHODS

SIJIN REN, MARCO CARICATO, *Department of Chemistry, University of Kansas, Lawrence, KS, USA.*

In this work, we present a simple approach to obtain absorption spectra from hybrid QM/QM calculations. The goal is to obtain reliable spectra for compounds that are too large to be treated entirely at a high level of theory. The approach is based on the extrapolation of the entire absorption spectrum obtained by individual subcalculations. Our program locates the main spectral features in each subcalculation, e.g. band peaks and shoulders, and fits them to Gaussian functions. Each Gaussian is then extrapolated with a formula similar to that of ONIOM (Our own N-layered Integrated molecular Orbital molecular Mechanics). However, information about individual excitations is not necessary so that difficult state-matching across subcalculations is avoided. This multi-state extrapolation thus requires relatively low implementation effort while affording maximum flexibility in the choice of methods to be combined in the hybrid approach. The test calculations show the efficacy and robustness of this methodology in reproducing the spectrum computed for the entire molecule at a high level of theory.

WI08 4:01 – 4:16

A PROTOCOL FOR HIGH-ACCURACY THEORETICAL THERMOCHEMISTRY

BRADLEY WELCH, RICHARD DAWES, *Department of Chemistry, Missouri University of Science and Technology, Rolla, MO, USA.*

Theoretical studies of spectroscopy and reaction dynamics including the necessary development of potential energy surfaces rely on accurate thermochemical information. The Active Thermochemical Tables (ATcT) approach by Ruscic[1] incorporates data for a large number of chemical species from a variety of sources (both experimental and theoretical) and derives a self-consistent network capable of making extremely accurate estimates of quantities such as temperature dependent enthalpies of formation. The network provides rigorous uncertainties, and since the values don't rely on a single measurement or calculation, the provenance of each quantity is also obtained. To expand and improve the network it is desirable to have a reliable protocol such as the HEAT approach[2] for calculating accurate theoretical data.

Here we present and benchmark an approach based on explicitly-correlated coupled-cluster theory and vibrational perturbation theory (VPT2). Methyldioxy and Methyl Hydroperoxide are important and well-characterized species in combustion processes and begin the family of (ethyl-, propyl-based, etc) similar compounds (much less is known about the larger members). Accurate anharmonic frequencies are essential to accurately describe even the 0 K enthalpies of formation, but are especially important for finite temperature studies. Here we benchmark the spectroscopic and thermochemical accuracy of the approach, comparing with available data for the smallest systems, and comment on the outlook for larger systems that are less well-known and characterized.

[1]B. Ruscic, Active Thermochemical Tables (ATcT) values based on ver. 1.118 of the Thermochemical Network (2015); available at ATcT.anl.gov

[2]A. Tajti, P. G. Szalay, A. G. Császár, M. Kállay, J. Gauss, E. F. Valeev, B. A. Flowers, J. Vázquez, and J. F. Stanton. JCP 121, (2004): 11599.

WI09   4:18 – 4:33

INVESTIGATION OF SOLVATION EFFECTS ON OPTICAL ROTATORY DISPERSION USING THE POLARIZABLE CONTINUUM MODEL

TAL AHARON, *Department of Chemistry, University of Kansas, Lawrence, KS, USA*; PAUL M LEMLER, PATRICK VACCARO, *Department of Chemistry, Yale University, New Haven, CT, USA*; MARCO CARICATO, *Department of Chemistry, University of Kansas, Lawrence, KS, USA.*

The Optical Rotatory Dispersion (ORD) of a chiral solute is heavily affected by solvation, but this effect does not follow the usual correlation with the solvent polarity, i.e., larger solvent polarity does not imply a larger change in the solute's property. Therefore, a great deal of experimental and theoretical effort has been directed towards correlating the solvation effect on the ORD and the solvent properties. This discovery followed from the development of cavity ring down polarimetry (CRPD), which allows measurements of gas-phase ORD. In order to investigate this phenomenon, we chose a set of five rigid molecules to limit the effect of molecular vibrations and isolate the role of solvation. The latter was investigated with the Polarizable Continuum Model (PCM), and compared to experimental results. We used Bondi radii to build the PCM cavity, and performed extensive calculations at multiple frequencies using density functional theory (DFT) with two functionals: B3LYP and CAM-B3LYP, together with the aug-cc-pVDZ basis set. We also performed coupled cluster singles and doubles (CCSD/aug-cc-pVDZ) calculations at the wavelengths where gas-phase data are available, all of which are augmented with zero point vibrational corrections. These results are compared to experimental data and seem to indicate that PCM does not entirely account for the environmental effects on the ORD.

WI10   4:35 – 4:50

A CODE FOR AUTOMATED CONSTRUCTION OF POTENTIAL ENERGY SURFACES FOR VAN DER WAALS SYSTEMS

ERNESTO QUINTAS SÁNCHEZ, RICHARD DAWES, *Department of Chemistry, Missouri University of Science and Technology, Rolla, MO, USA.*

The potential energy surface (PES) constitutes a cornerstone for theoretical studies of spectroscopy and dynamics. We fit PESs using a local interpolating moving least squares (L-IMLS) approach.[a] The L-IMLS method is interpolative and has the flexibility to fit energies or energies and gradients, where inclusion of gradient information significantly reduces the number of points required for an accurate fit.

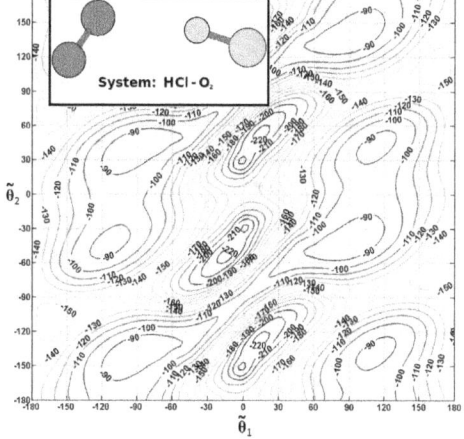

The method permits fully automated PES generation: beginning with an initial set of seed points, an automatic point selection scheme determines where new data are required and, in a series of iterations, computes new ab initio data and updates the fit until a specified accuracy is reached. We have interfaced this fitting approach to popular electronic structure codes such as Molpro and CFOUR to automatically generate ab initio 4D PESs for vdWs systems composed of two (rigid) linear fragments.

We present here our freely distributed code designed to run in parallel on a computing cluster, allowing the user to specify the system (masses, interatomic equilibrium distances, symmetry, energy range of interest, etc.) through an input file. For a selection of benchmark systems, we show that PESs with fitting errors below 1 cm^{-1} can be constructed using only a few hundred ab initio points.

---

[a] M. Majumder, S. Ndengue and R. Dawes, Molecular Physics 114, 1 (2016).

TRIPLET TUNING – A NEW "BLACK-BOX" COMPUTATIONAL SCHEME FOR PHOTOCHEMICALLY ACTIVE MOLECULES

ZHOU LIN, TROY VAN VOORHIS, *Department of Chemistry, Massachusetts Institute of Technology, Cambridge, MA, USA.*

Density functional theory (DFT) is an efficient computational tool that plays an indispensable role in the design and screening of π-conjugated organic molecules with photochemical significance. However, due to intrinsic problems in DFT such as self-interaction error, the accurate prediction of energy levels is still a challenging task.[a] Functionals can be parameterized to correct these problems, but the parameters that make a well-behaved functional are system-dependent rather than universal in most cases. To alleviate both problems, optimally tuned range-separated hybrid functionals were introduced, in which the range-separation parameter, ω, can be adjusted to impose Koopman's theorem, $\varepsilon_{HOMO} = -I$. These functionals turned out to be good estimators for asymptotic properties like ε_{HOMO} and ε_{LUMO}.[b,c] In the present study, we propose a "black-box" procedure that allows an automatic construction of molecule-specific range-separated hybrid functionals following the idea of such optimal tuning. However, instead of focusing on ε_{HOMO} and ε_{LUMO}, we target more local, photochemistry-relevant energy levels such as the lowest triplet state, T_1. In practice, we minimize the difference between two E_{T_1}'s that are obtained from two DFT-based approaches, Δ-SCF and linear-response TDDFT. We achieve this minimization using a non-empirical adjustment of two parameters in the range-separated hybrid functional – ω, and the percentage of Hartree–Fock contribution in the short-range exchange, c_{HF}. We apply this triplet tuning scheme to a variety of organic molecules with important photochemical applications, including laser dyes, photovoltaics, and light-emitting diodes, and achieved good agreements with the spectroscopic measurements for E_{T_1}'s and related local properties.[d]

---

[a] A. Dreuw and M. Head-Gordon, *Chem. Rev.* **105**, 4009 (2015).

[b] O. A. Vydrov and G. E. Scuseria, *J. Chem. Phys.* **125**, 234109 (2006).

[c] L. Kronik, T. Stein, S. Refaely-Abramson, and R. Baer, *J. Chem. Theory Comput.* **8**, 1515 (2012).

[d] Z. Lin and T. A. Van Voorhis, *in preparation for submission to J. Chem. Theory Comput.*

WJ. Lineshapes, collisional effects

Wednesday, June 21, 2017 – 1:45 PM

Room: 161 Noyes Laboratory

Chair: Iouli E Gordon, Harvard-Smithsonian Center for Astrophysics, Cambridge, MA, USA

WJ01 1:45 – 2:00

NUMERICAL EVALUATION OF PARAMETER CORRELATION IN THE HARTMANN-TRAN LINE PROFILE

ERIN M. ADKINS, ZACHARY REED, JOSEPH T. HODGES, *Chemical Sciences Division, National Institute of Standards and Technology, Gaithersburg, MD, USA.*

The partially correlated quadratic, speed-dependent hard-collision profile (pCqSDHCP), for simplicity referred to as the Hartmann-Tran profile (HTP), [a] has been recommended as a generalized lineshape for high resolution spectroscopy.[b] The HTP parameterizes complex collisional effects such as Dicke narrowing, speed dependent narrowing, and correlations between velocity-changing and dephasing collisions, while also simplifying to simpler profiles that are widely used, such as the Voigt profile. As advanced lineshape profiles are adopted by more researchers, it is important to understand the limitations that data quality has on the ability to retrieve physically meaningful parameters using sophisticated lineshapes that are fit to spectra of finite signal-to-noise ratio. In this work, spectra were simulated using the HITRAN Application Programming Interface (HAPI)[c] across a full range of line parameters.[d] Simulated spectra were evaluated to quantify the precision with which fitted lineshape parameters can be determined at a given signal-to-noise ratio, focusing on the numerical correlation between the retrieved Dicke narrowing frequency and the velocity-changing and dephasing collisions correlation parameter.

[a]Tran, H., N. Ngo, and J.-M. Hartmann, *Journal of Quantitative Spectroscopy and Radiative Transfer* 2013. 129: p. 89-100.

[b]Tennyson, et al., *Pure Appl. Chem.* 2014, 86: p. 1931-1943.

[c]Kochanov, R.V., et al., *Journal of Quantitative Spectroscopy and Radiative Transfer* 2016. 177: p. 15-30.

[d]Tran, H., N. Ngo, and J.-M. Hartmann, *Journal of Quantitative Spectroscopy and Radiative Transfer* 2013. 129: p. 199-203.

WJ02 2:02 – 2:17

TEMPERATURE DEPENDENCE OF NEAR-INFRARED CO_2 LINE SHAPES MEASURED BY CAVITY RING-DOWN SPECTROSCOPY

MÉLANIE GHYSELS, ADAM J. FLEISHER, QINGNAN LIU, JOSEPH T. HODGES, *Chemical Sciences Division, National Institute of Standards and Technology, Gaithersburg, MD, USA.*

We present high signal-to-noise ratio, mode-by-mode cavity ring-down spectroscopy (CRDS) line shape measurements of air-broadened transitions in the $30013 \rightarrow 0001$ band of $^{12}C^{16}O_2$ located near $\lambda = 1.6$ μm. Absorption spectra were acquired from (230-290) K with a variable-temperature spectrometer developed in the framework of the NASA Orbiting Carbon Observatory-2 Mission to improve our understanding of carbon dioxide and oxygen line shape parameters. This system comprises a monolithic, thermally stabilized two-mirror, optical resonator exhibiting a mode stability of 200 kHz and a minimum detectable absorption coefficient of 10^{-11} cm^{-1}. Observed spectra were modeled the using the recently recommended Hartmann-Tran line profile (HTP) [a] (and several of its limiting cases) which includes the effects of Dicke narrowing, speed dependent broadening, correlation between velocity- and phase-changing collisions and first-order line mixing effects. At fixed temperature, line shape parameters were determined by constrained multispectrum fitting of spectra acquired over the pressure range (30 - 300) Torr. For each transition considered, analysis of the temperature dependence of the fitted line shape parameters yielded the pressure-broadening temperature exponent and speed dependence parameter, where the latter quantity was found to be in good agreement with theoretical values consistent with the HTP model.

[a]Tennyson, et al., *Pure Appl. Chem.* **86**, (2014) 1931

EXPERIMENTAL STUDY OF TEMPERATURE-DEPENDENCE LAWS OF NON-VOIGT ABSORPTION LINE SHAPE PARAMETERS

JONAS WILZEWSKI[a], MANFRED BIRK, JOEP LOOS, GEORG WAGNER, *Remote Sensing Technology Institute, Experimental Methods, German Aerospace Center DLR, Oberpfaffenhofen, Germany.*

To improve the understanding of temperature-dependence laws of spectral line shape parameters, spectra of the ν_3 rovibrational band of CO_2 perturbed by 10, 30, 100, 300 and 1000 mbar of N_2 were measured at nine temperatures between 190 K and 330 K using a 22 cm long single-pass absorption cell in a Bruker IFS125 HR Fourier Transform spectrometer. The spectra were fitted employing a quadratic speed-dependent hard collision model in the Hartmann-Tran implementation[bc] extended to account for line mixing in the Rosenkranz approximation by means of a multispectrum fitting approach developed at DLR[d]. This enables high accuracy parameter retrievals to reproduce the spectra down to noise level and we will present the behavior of line widths, shifts, speed-dependence-, collisional narrowing- and line mixing-parameters over this 140 K temperature range.

[a]also at Ludwig-Maximilians-Universität, Physics Department, Munich, Germany
[b]Ngo *et al.* JQSRT **29**, 89-100 (2013); JQSRT **134**, 105 (2014).
[c]Tran *et al.* JQSRT **129**, 199-203 (2013); JQSRT **134**, 104 (2014).
[d]Loos *et al.*, 2014; http://doi.org/10.5281/zenodo.11156.

LINE SHAPES AND INTENSITIES OF CARBON MONOXIDE TRANSITIONS IN THE (3→0) AND (4→1) BANDS

ZACHARY REED, *Chemical Sciences Division, National Institute of Standards and Technology, Gaithersburg, MD, USA*; OLEG POLYANSKY, *Department of Physics and Astronomy, University College London, London, IX, United Kingdom*; JOSEPH T. HODGES, *Material Measurement Laboratory, National Institute of Standards and Technology, Gaithersburg, MD, USA.*

We have measured several carbon monoxide transitions in the (3→0) and (4→1) band using frequency stabilized cavity ringdown spectroscopy (FS-CRDS). The measured transitions are compared to the line strength values in HITRAN 2012 [1], those determined by Wojtewitz et al [2], and to theoretical calculations. The cavity length is actively locked to an iodine stabilized HeNe laser, providing long term frequency stability of 10 kHz and is linked to a self-referenced, octave-spanning frequency comb. The temperature of the optical cavity is actively regulated at the mK level, and the pressure measurements are SI-traceable. The sample is a NIST calibrated reference mixture of 11.98575(95)% CO in N_2. The absorption spectra are modeled using the Hartmann-Tran profile (HTP). The SNR in these spectra may exceed 10,000:1, which necessitates including the effects of speed dependence, collisional narrowing, and correlation between velocity-changing and dephasing collisions.

The relative uncertainties of the line strengths calculated in this study are better than 0.1%. There are systematic differences on the 1% level for ^{12}CO against both HITRAN [1] and the previous work by Wojtewitz et al [2]. The measurement uncertainties are nearly an order of magnitude lower than previous results. Additionally, the relative uncertainties in the integrated areas of selected ^{12}CO and ^{13}CO transitions are less than 0.006% and 0.02%, respectively, providing an excellent test case for determination of isotope ratios by direct use of theoretical line intensity calculations.

[1] Wojtewicz, S., et al., J Quant Spect and Rad Trans,2013. 130: p.191-200.
[2]Rothman, L.S., et al., Journal of Quant Spect and Rad Trans, 2013. 130: p. 4-50.

WJ05

RELAXATION MATRICES OF THE NH_3 MOLECULE IN PARALLEL AND PERPENDICUAR BANDS

QIANCHENG MA, *Applied Physis and Applied Mathematics, Columbia University, New York, NY, USA*; C. BOULET, *Institut des Sciences Moléculaires d'Orsay, Université Paris-Sud, Orsay, France*; RICHARD TIP-PING, *Physics and Astronomy, University of Alabama, Tuscaloosa, AL, USA*.

The phenomenon of collisional transfer of intensity due to line mixing has an increasing importance for atmospheric monitoring. From a theoretical point of view, all relevant information about the collisional processes is contained in the relaxation matrix W where the diagonal elements give half-widths and shifts, and the off-diagonal elements correspond to line interferences. For simple systems such as diatom-atom and diatom-diatom, fully quantum calculations are feasible, but become unrealistic for more complex systems. Meanwhile, the semi-classical Robert-Bonamy (RB) formalism widely used to calculate half-widths and shifts completely fails in calculating the off-diagonal elements of W resulting from applying the isolated line approximation. Recently, we have developed a new semi-classical formalism without this approximation that enables one not only to reduce uncertainties for calculated half-widths and shifts, but also to calculate the off-diagonal elements. This implies that we can address line mixing based on interaction potentials between molecular absorber and molecular perturber. In the present study, we have applied this method to calculate the relaxation matrices for self-broadened NH_3 lines in the parallel pure-rotational, ν_1, ν_2, and $2\nu_2$ bands and also in the perpendicular ν_4 band. Our studies have exhibited a significant off-diagonality of W in the pure-rotational, ν_1, and ν_4 bands. For the ν_2 and $2\nu_2$ bands, the off-diagonality is much less and even becomes completely absent. Given the fact that the inversion doublet splitting is the main source responsible for the off-diagonality of W and its value in these bands dramatically increases from less than 1 cm^{-1}, to 36 cm^{-1}, and further to 284 cm^{-1}, it is easy to understand these results. By comparing with half-widths derived from the RB formalism, our values in the pure-rotational, ν_1, and ν_4 bands are significantly reduced and match measurements very well. We have also compare calculated off-diagonal elements of W and Rosenkranz line mixing coefficients with measured results. In addition, we have compared the calculated profiles, including line mixing effects with the observed ones in various cases: a good agreement is obtained in the ^PP doublets of the ν_4 band as well as in the Q branch and the R(3,k) manifold in the ν_1 band. For some other measurements reported in literature, very large discrepancies (up to two orders) have been found and our comments on these measurements are presented.

WJ06

SATURATION DIP MEASUREMENTS OF HIGH-J TRANSITIONS IN THE $v_1 + v_3$ BAND OF C_2H_2: ABSOLUTE FREQUENCIES AND SELF-BROADENING

TREVOR SEARS[a], SYLVESTRE TWAGIRAYEZU[b], GREGORY HALL, *Division of Chemistry, Department of Energy and Photon Sciences, Brookhaven National Laboratory, Upton, NY, USA*.

Saturation dip spectra of acetylene in the $v_1 + v_3$ band have been obtained for rotational lines with $J = 31 - 37$ inclusive, using a diode laser referenced to a frequency comb. The estimated accuracy and precision of the measurements is better than 10 kHz in 194 THz. Data were obtained as a function of sample pressure to investigate the broadening of the saturation features. The observed line shapes are well modeled by convolution of a fixed Gaussian transit-time and varying Lorentzian lifetime broadening, *i.e.* a Voigt-type profile. The lines exhibit a significantly larger collisional (lifetime) broadening than has been measured in conventional Doppler and pressure-broadened samples at ambient temperatures. The figure shows the fitted Lorentzian width versus sample pressure for P(31). The

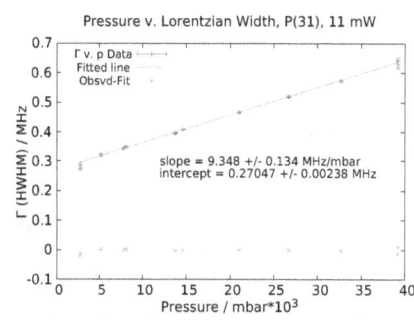

slope of this plot gives the pressure broadening coefficient, $\gamma_{self} = 9.35(13)$ MHz/mbar. For comparison, the coefficient derived from conventional Doppler and pressure broadened spectra for this transition is 2.7 MHz/mbar[c]. The sub-Doppler broadening coefficients are all significantly larger than the conventionally measured ones, due to the increased importance of velocity-changing collisions. The measurements therefore give information on the balance between hard phase- or state-changing and large cross-section velocity-changing collisions.

Acknowledgments: Work at Brookhaven National Laboratory was carried out under Contract No. DE-SC0012704 with the U.S. Department of Energy, Office of Science, and supported by its Division of Chemical Sciences, Geosciences and Biosciences within the Office of Basic Energy Sciences.

---

[a]also: *Chemistry Department, Stony Brook University, Stony Brook, New York 11794*
[b]now at: *Department of Chemistry, Lamar University, Beaumont, TX 77710*
[c]J. Molec. Spectrosc. **209**, 216-227 (2001) and J. Quant. Spectrosc. Rad. Transf. **76**, 237-267 (2003)

WJ07   3:27 – 3:42

TIME- AND FREQUENCY-DOMAIN SIGNATURES OF VELOCITY CHANGING COLLISIONS IN SUB-DOPPLER SATURATION SPECTRA AND PRESSURE BROADENING

GREGORY HALL, HONG XU, DAMIEN FORTHOMME, *Division of Chemistry, Department of Energy and Photon Sciences, Brookhaven National Laboratory, Upton, NY, USA*; PAUL DAGDIGIAN, *Department of Chemistry, Johns Hopkins University, Baltimore, MD, USA*; TREVOR SEARS[a], *Division of Chemistry, Department of Energy and Photon Sciences, Brookhaven National Laboratory, Upton, NY, USA*.

We have combined experimental and theoretical approaches to the competition between elastic and inelastic collisions of CN radicals with Ar, and how this competition influences time-resolved saturation spectra. Experimentally, we have measured transient, two-color sub-Doppler saturation spectra of CN radicals with an amplitude chopped saturation laser tuned to selected Doppler offsets within rotational lines of the A-X (2-0) band, while scanning a frequency modulated probe laser across the hyperfine-resolved saturation features of corresponding rotational lines of the A-X (1-0) band. A steady-state depletion spectrum includes off-resonant contributions ascribed to velocity diffusion, and the saturation recovery rates depend on the sub-Doppler detuning. The experimental results are compared with Monte Carlo solutions to the Boltzmann equation for the collisional evolution of the velocity distributions of CN radicals, combined with a pressure-dependent and speed-dependent lifetime broadening. Velocity changing collisions are included by appropriately sampling the energy resolved differential cross sections for elastic scattering of selected rotational states of CN (X). The velocity space diffusion of Doppler tagged molecules proceeds through a series of small-angle scattering events, eventually terminating in an inelastic collision that removes the molecule from the coherently driven ensemble of interest. Collision energy-dependent total cross sections and differential cross sections for elastic scattering of selected CN rotational states with Ar were computed with Hibridon quantum scattering calculations, and used for sampling in the Monte Carlo modeling.

Acknowledgments: Work at Brookhaven National Laboratory was carried out under Contract No. DE-SC0012704 with the U.S. Department of Energy, Office of Science, and supported by its Division of Chemical Sciences, Geosciences and Biosciences within the Office of Basic Energy Sciences.

[a]also: *Chemistry Department, Stony Brook University, Stony Brook, New York 11794*

Intermission

WJ08   4:01 – 4:16

RECENT PROGRESS ON LABFIT: A MULTISPECTRUM ANALYSIS PROGRAM FOR FITTING LINESHAPES INCLUDING THE HTP MODEL AND TEMPERATURE DEPENDENCE.

MATTHEW J. CICH, ALEXANDRE GUILLAUME, BRIAN DROUIN, *Jet Propulsion Laboratory, California Institute of Technology, Pasadena, CA, USA*; D. CHRIS BENNER, *Department of Physics, College of William and Mary, Williamsburg, VA, USA*.

Multispectrum analysis can be a challenge for a variety of reasons. It can be computationally intensive to fit a proper line shape model especially for high resolution experimental data. Band-wide analyses including many transitions along with interactions, across many pressures and temperatures are essential to accurately model, for example, atmospherically relevant systems. Labfit is a fast multispectrum analysis program originally developed by D. Chris Benner with a text-based interface. More recently at JPL a graphical user interface was developed with the goal of increasing the ease of use but also the number of potential users. The HTP lineshape model has been added to Labfit keeping it up-to-date with community standards. Recent analyses using labfit will be shown to demonstrate its ability to competently handle large experimental datasets, including high order lineshape effects, that are otherwise unmanageable.

WJ09 4:18 – 4:33

MULTISPECTRAL FITTING VALIDATION OF THE SPEED DEPENDENT VOIGT PROFILE AT UP TO 1300K IN WATER VAPOR WITH A DUAL FREQUENCY COMB SPECTROMETER

PAUL JAMES SCHROEDER, *Mechanical Engineering, University of Colorado Boulder, Boulder, CO, USA*; MATTHEW J. CICH, *Jet Propulsion Laboratory, California Institute of Technology, Pasadena, CA, USA*; JINYU YANG, *Mechanical Engineering, University of Colorado Boulder, Boulder, CO, USA*; BRIAN DROUIN, *Jet Propulsion Laboratory, California Institute of Technology, Pasadena, CA, USA*; GREG B RIEKER, *Mechanical Engineering, University of Colorado Boulder, Boulder, CO, USA.*

Using broadband, high resolution dual frequency comb spectroscopy, we test the power law temperature scaling relationship with Voigt, Rautian, and quadratic speed dependent Voigt profiles over a temperature range of 296-1300K for pure water vapor. The instrument covers the spectral range from 6800 cm^{-1} to 7200 cm^{-1} and samples the (101)-(000), (200)-(000), (021)-(000), (111)-(010), (210)-(010), and (031)-(010) vibrational bands of water. The data is sampled with a point spacing of 0.0033 cm^{-1} and absolute frequency accuracy of $<3.34e\text{-}6$ cm^{-1}. This region is of interest for detection and quantification of hot water and gas temperature within coal gasifiers and other high temperature systems. In order to extract water concentration and temperature, an extended range of lineshape parameters are needed. Lineshape parameters for pure and argon broadened water are obtained for 278 transitions using the multispectral fitting program Labfit, including self-broadening coefficients, power law temperature scaling exponents, and speed dependence coefficients. The extended temperature range of the data provides valuable insight into the application of the speed-dependence corrections of the line profiles, which are shown to have more reasonable line broadening temperature dependencies.

WJ10 4:35 – 4:50

HIGH PRECISION MEASUREMENTS OF LINE MIXING AND COLLISIONAL INDUCED ABSORPTION IN THE O_2 A-BAND

ERIN M. ADKINS, MÉLANIE GHYSELS, DAVID A. LONG, JOSEPH T. HODGES, *Chemical Sciences Division, National Institute of Standards and Technology, Gaithersburg, MD, USA.*

Molecular oxygen (O_2) has a well-known and uniform molar fraction within the Earth's atmosphere. Consequently, the O_2 A-band is commonly used in satellite and remote sensing measurements (GOSAT, OCO-2, TCCON) to determine the surface pressure-pathlength product for transmittance measurements that involve light propagation through the atmospheric column. For these missions, physics-based spectroscopic models and experimentally determined line-by-line parameters are used to predict the temperature- and pressure-dependence of the absorption cross-section as a function of wave number, pressure, temperature and water vapor concentration. At present, there remain airmass-dependent biases in retrievals of CO_2 which are linked to limitations in existing models of line mixing (LM) and collisional induced absorption (CIA) [a]. In order to better quantify these effects, we measured O_2 A-band spectra with a frequency-stabilized cavity ring-down spectroscopy (FS-CRDS) system. Because of the high molar fraction of O_2 in air samples, line cores and near wings of the dominant absorption transitions are heavily saturated, which makes it impossible to obtain continuous FS-CRDS spectra over the entire range of optical depth. Here, we focused on LM and CIA effects which dominate the valleys between strongly absorbing transitions. To this end, the FS-CRDS system employs a thresholding mechanism that avoids the optically thick regions and scans over the entire O_2 A-band and beyond the band head region. This approach provides high signal-to-noise ratio spectra that can be fit to yield LM and CIA parameters. These results are intended to provide strong constraints on multispectrum fits of continuous and broadband Fourier-transform-spectroscopy based O_2 A-band spectra.

---

[a] Long D.A and J.T. Hodges, *J. Geophys. Res.* 2012, 117: p. D12309.

184

COLLISON-INDUCED ABSORPTION OF OXYGEN MOLECULE AS STUDIED BY HIGH SENSITIVITY SPECTROSCOPY

<u>WATARU KASHIHARA</u>, ATSUSHI SHOJI, AKIO KAWAI, *Department of Chemistry, Tokyo Institute of Technology, Tokyo, Japan.*

Oxygen dimol is transiently generated when two oxygen molecules collide. At this short period, the electron clouds of molecules are distorted and some forbidden transition electronic transitions become partially allowed. This transition is called CIA (Collision-induced absorption). There are several CIA bands appearing in the spectral region from UV to near IR. Absorption of solar radiation by oxygen dimol is a small but significant part of the total budget of incoming shortwave radiation. However, a theory predicting the lineshape of CIA is still under developing. In this study, we measured CIA band around 630 nm that is assigned to optical transition, $a^1\Delta_g$(v=0):$a^1\Delta_g$(v=0)-$X^3\Sigma_g^-$(v=0):$X^3\Sigma_g^-$(v=0) of oxygen dimol. CRDS(Cavity Ring-down Spectroscopy) was employed to measure weak absorption CIA band of oxygen. Laser beam around 630 nm was generated by a dye laser that was pumped by a YAG Laser. Multiple reflection of the probe light was performed within a vacuum chamber that was equipped with two high reflective mirrors. We discuss the measured line shape of CIA on the basis of collision pair model.

THEORY OF COLLISION-INDUCED ABSORPTION FOR ELECTRONIC TRANSITIONS IN THE ATMOSPHERICALLY RELEVANT O_2-O_2 AND O_2-N_2 PAIRS.

<u>TIJS KARMAN</u>, AD VAN DER AVOIRD, GERRIT GROENENBOOM, *Institute for Molecules and Materials (IMM), Radboud University Nijmegen, Nijmegen, Netherlands.*

Collision-induced absorption of O_2-O_2 and O_2-N_2 pairs is observed in remote sensing of the Earth's atmosphere, and absorption by O_2-O_2 pairs has been put forward as a biomarker to be observed in exoplanetary transit spectra. The relevant electronic transitions, $X\,^3\Sigma_g^- \rightarrow a\,^1\Delta_g$ and $X\,^3\Sigma_g^- \rightarrow b\,^1\Sigma_g^+$, are electric-dipole forbidden by both spin and spatial selection rules, such that collision-induced absorption represents an important contribution to the absorption.

We present an *ab initio* study of collision-induced absorption for these electronic transitions using quantum-mechanical scattering calculations. Two mechanisms for breaking the spin-symmetry are taken into account: intramolecular spin-orbit coupling and intermolecular exchange interactions. We find and explain qualitative differences in the line shape and temperature dependence of the absorption due to these mechanisms. The contributions of these mechanisms furthermore explain qualitative differences between the atmospherically relevant O_2-O_2 and O_2-N_2 systems. Reasonable agreement with experimental data is obtained for various near-infrared transitions in both systems.

WK. Metal containing

Wednesday, June 21, 2017 – 1:45 PM

Room: 140 Burrill Hall

Chair: Leah C O'Brien, Southern Illinois University, Edwardsville, IL, USA

WK01   1:45 – 2:00

LASER INDUCED FLUORESCENCE SPECTROSCOPY OF JET-COOLED MgOMg

MICHAEL N. SULLIVAN, DANIEL J. FROHMAN, MICHAEL HEAVEN, *Department of Chemistry, Emory University, Atlanta, GA, USA*; WAFAA M FAWZY, *Department of Chemistry, Murray State University, Murray, KY, USA.*

The group IIA metals have stable hypermetallic oxides of the general form MOM. Theoretical interest in these species is associated with the multi-reference character of the ground states. It is now established that the ground states can be formally assigned to the $M^+O^{2-}M^+$ configuration, which leaves two electrons in orbitals that are primarily metal-centered ns orbitals. Hence the MOM species are diradicals with very small energy spacings between the lowest energy singlet and triplet states. Previously, we have characterized the lowest energy singlet transition ($^1\Sigma_u^+ \leftarrow ^1\Sigma_g^+$) of BeOBe. Preliminary data for the first electronic transition of the isovalent species, CaOCa, was presented previously (71st ISMS, talk RI10).

We now report the first electronic spectrum of MgOMg. Jet-cooled laser induced fluorescence spectra were recorded for multiple bands that occurred within the 21,000 - 24,000 cm^{-1} range. Most of the bands exhibited simple P/R branch rotational line patterns that were blue-shaded. Only even rotational levels were observed, consistent with the expected X $^1\Sigma_g^+$ symmetry of the ground state (^{24}Mg has zero nuclear spin). Molecular constants were extracted from the rovibronic bands using PGOPHER. The experimental results and interpretation of the spectrum, which was guided by the predictions of electronic structure calculation, will be presented.

WK02   2:02 – 2:17

HIGH RESOLUTION LASER SPECTROSCOPY OF THE [15.45]0 – a$^3\Delta_1$ TRANSITION OF TANTALUM MONONITRIDE, TaN

COLAN LINTON, *Department of Physics, University of New Brunswick, Fredericton, NB, Canada*; TIMOTHY STEIMLE[a], DAMIAN L KOKKIN, *School of Molecular Sciences, Arizona State University, Tempe, AZ, USA.*

Heavy polar molecules have been used for some time in experiments designed to measure the permanent molecular electric dipole moment, EDM, that is induced by the electron electric dipole moment, eEDM. More recently, ^{181}TaN has been proposed[b] as a prime candidate for experiments to measure the EDM induced by the magnetic quadrupole moment, MQM, of the nucleus. There have been a number of calculations to predict these quantities and hyperfine structure parameters are useful indicators of the quality of the electronic wavefunctions used in these calculations with the low-lying a$^3\Delta_1$ state being of particular interest[cd]. High resolution Laser Induced Fluorescence (LIF) spectra of the of the 0-0 band of the [15.45]0 – a$^3\Delta_1$ transition of TaN have been obtained using the laser ablation source at Arizona State University. Tantalum hyperfine structure was completely resolved and magnetic and quadrupole hyperfine parameters were determined and, where available, have been compared with predicted values. Calculations of the molecular hyperfine constants using atomic hyperfine parameters have been used to determine the nature and configurations of the electronic states.

---

[a]This research has been supported by the National Science Foundation, Division of Chemistry, CHE-1265885 (ASU)

[b]V.V. Flambaum, D. DeMille, M.G. Kozlov, Phys Rev Lett 113 (2014) 103003.

[c]V.Skripnikov, A.N. Petrov, N.S. Mosyagin, A.V. Titov, V.V. Flambaum, Phys. Rev. A: At., Mol., Opt. Phys. 92 (2015) 012521/1.

[d]T. Fleig, M.K. Nayak, M.G. Kozlov, Physical Review A 93 (2016) 012505.

186

WK03 2:19 – 2:34

ELECTRONIC TRANSITIONS OF TUNGSTEN MONOSULFIDE

L. F. TSANG, *Chemistry, The Chinese University of Hong Kong, Hong Kong, Hong Kong, China*; MAN-CHOR CHAN, *Department of Chemistry, The University of Hong Kong, Hong Kong, Hong Kong*; WENLI ZOU, *Institute of Modern Physics, Northwest University, Xi'an, China*; ALLAN S.C. CHEUNG, *Department of Chemistry, The University of Hong Kong, Hong Kong, Hong Kong*.

Electronic transition spectrum of the tungsten monosulfide (WS) molecule in the near infrared region between 725 nm and 885 nm has been recorded using laser ablation/reaction free-jet expansion and laser induced fluorescence spectroscopy. The WS molecule was produced by reacting laser - ablated tungsten atoms with 1% CS_2 seeded in argon. Fifteen vibrational bands with resolved rotational structure have been recorded and analyzed, which were organized into seven electronic transition systems. The ground state has been identified to be the $X^3\Sigma^-(0^+)$ state, and the determined vibrational frequency, $\Delta G_{1/2}$ and bond length, r_0, are respectively 556.7 cm^{-1} and 2.0676 Å. In addition, vibrational bands belong to another transition system involving lower state with $\Omega = 1$ component have also been analyzed. Least-squares fit of the measured line positions yielded molecular constants for the electronic states involved.

The low-lying Λ-S states and Ω sub-states of WS have been calculated using state-averaged complete active space self-consistent field (SA-CASSCF) and followed by MRCISD+Q (internally contracted multi-reference configuration interaction with singles and doubles plus Davidson's cluster correction). The active space consists of 10 electrons in 9 orbitals corresponding to the W 5d6s and S 3p shells. The lower molecular orbitals from W 5s5p and S 3s are inactive but are also correlated, and relativistic effective core potential (RECPs) are adopted to replace the core orbitals with 60 (W) and 10 (S) core electrons, respectively. Spin-orbit coupling (SOC) is calculated via the state-interaction (SI) approach with RECP spin-orbit operators using SA-CASSCF wavefunctions, where the diagonal elements in the SOC matrix are replaced by the corresponding MRCISD+Q energies calculated above. Spectroscopic constants and potential energy curves of the ground and many low-lying Λ-S states and Ω sub-states of the WS molecule are obtained. The calculated spectroscopic constants of the ground and low-lying states are generally in good agreement with our experimental determination. This work represents the first experimental investigation of the electronic and molecular structure of the WS molecule.

WK04 2:36 – 2:51

RE-VISITING THE ELECTRONIC ENERGY MAP OF THE COPPER DIMER BY DOUBLE-RESONANT FOUR-WAVE MIXING

BRADLEY VISSER, PETER BORNHAUSER, MARTIN BECK, GREGOR KNOPP, *Photonics, Paul Scherrer Institute, Villigen, Switzerland*; ROBERTO MARQUARDT, CHRISTOPHE GOURLAOUEN, *Laboratoire de Chimie Quantique, Institut de Chimie, Université de Strasbourg, 67008 Strasbourg, France*; JEROEN A. VAN BOKHOVEN, *Energy and Environment, Paul Scherrer Institute, Villigen, Switzerland*; PETER RADI, *Photonics, Paul Scherrer Institute, Villigen, Switzerland*.

The copper dimer is one of the most studied transition metal (TM) diatomics due to its alkali-metal like electronic shell structure, strongly bound ground state and chemical reactivity. The high electronic promotion energy in the copper atom yields numerous low-lying electronic states compared to TM dimers with (d)-hole electronic configurations. Thus, through extensive study the excited electronic structure of Cu_2 is relatively well known, however in practice few excited states have been investigated with rotational resolution or even assigned term symbols or dissociation limits.

The spectroscopic methods that have been used to investigate the copper dimer until now have not possessed sufficient *spectral* selectivity, which has complicated the analysis of the often overlapping transitions. Resonant four-wave mixing is a non-linear absorption based spectroscopic method. In favorable cases, the two-color version (TC-RFWM) enables purely optical mass selective spectral measurements in a mixed molecular beam. Additionally, by labelling individual rotational levels in the common intermediate state the spectra are dramatically simplified.

In this work, we report on the rotationally resolved characterization of low-lying electronic states of dicopper. Several term symbols have been assigned unambiguously. De-perturbation studies performed shed light on the complex electronic structure of the molecule. Furthermore, a new low-lying electronic state of Cu_2 is discovered and has important implications for the high-level theoretical structure calculations performed in parallel. In fact, the *ab initio* methods applied yield relative energies among the electronic levels that are almost quantitative and allow assignment of the newly observed state that is governed by spin-orbit interacting levels.

THE LOW-LYING ELECTRONIC STATES OF SCANDIUM MONOCARBIDE, ScC

CHIAO-WEI CHEN, <u>ANTHONY MERER</u>, YEN-CHU HSU, *Institute of Atomic and Molecular Sciences, Academia Sinica, Taipei, Taiwan.*

Extensive wavelength-resolved fluorescence studies have been carried out for the electronic bands of ScC and Sc^{13}C lying in the range 14000 - 16000 cm^{-1}. Taken together with detailed rotational analyses of these bands, these studies have clarified the natures of the low-lying electronic states. The ground state is an Ω = 3/2 state, with a vibrational frequency of 648 cm^{-1}, and the first excited electronic state is an Ω = 5/2 state, with a frequency of 712 cm^{-1}, lying 155.54 cm^{-1} higher. These states are assigned as the lowest spin-orbit components of X$^2\Pi_i$ and a$^4\Pi_i$, respectively. The quartet nature of the a state is confirmed by the observation of the $^4\Pi_{3/2}$ component, 18.71 cm^{-1} above the $^4\Pi_{5/2}$ component. The strongest bands in the region studied are two $^4\Delta_{7/2}$ - $^4\Pi_{5/2}$ transitions, where the upper states lie 14355 and 15445 cm^{-1} above X$^2\Pi_{3/2}$. Extensive doublet-quartet mixing occurs, which results in some complicated emission patterns. The energy order, a$^4\Pi$ above X$^2\Pi$, is consistent with the ab initio calculations of Kalemos et al.,[a] but differs from that found by Simard et al in the isoelectronic YC molecule.[b]

[a] A. Kalemos, A. Mavridis and J.F. Harrison, J. Phys. Chem. **A155**, 755 (2001).
[b] B. Simard, P.A. Hackett and W.J. Balfour, Chem. Phys. Lett., **230**, 103 (1994).

Intermission

THRESHOLD IONIZATION AND SPIN-ORBIT COUPLING OF CERIUM MONOXIDE

<u>WENJIN CAO</u>, YUCHEN ZHANG, LU WU, DONG-SHENG YANG, *Department of Chemistry, University of Kentucky, Lexington, KY, USA.*

Cerium oxides are widely used in heterogeneous catalysis due to their ability to switch between different oxidation states. We report here the mass-analyzed threshold ionization (MATI) spectroscopy of cerium monoxide (CeO) produced by laser ablating a Ce rod in a molecular beam source. The MATI spectrum in the range of 40000-45000 cm^{-1} exhibits several band systems with similar vibrational progressions. The strongest band is at 43015 (5) cm^{-1}, which can be assigned as the adiabatic ionization energy of the neutral species. The spectrum also shows Ce-O stretching frequencies of 817 and 890 cm^{-1} in the neutral and ion states, respectively. By comparing with spin-orbit coupled multireference quasi-degenerate perturbation theory (SO-MCQDPT) calculations, the observed band systems are assigned to transitions from various low-energy spin-orbit levels of the neutral oxide to the two lowest spin-orbit levels of the corresponding ion. The current work will also be compared with previous experimental and computational studies on the neutral species.

WK07

A DATABASE FOR TRANSITION METAL DIATOMICS

CORINNE DUPERROUZEL, NIKESH S. DATTANI, PAUL W. AYERS, *Department of Chemistry and Chemical Biology, McMaster University, Hamilton, Canada.*

Molecules containing transition metal elements are often notoriously difficult for most *ab initio* methods, especially single-reference methods. Transition metal diatomics are often treated using methods that depend on the choice of an active space (e.g. CASSCF, RASSCF, GASSCF) or on the choice of multiple references (e.g. MRCI, MRACPF, MRAQCC) and since the choice of active space or references needs to be done very carefully, these approaches cannot always be used in a black-box manner. Furthermore, spectroscopic data is usually extremely sparse or completely absent for transition metal dimers, especially when compared to dimers containing alkali, alkaline earth, halogen, and first or second row elements. For example, the chromium dimer is one of the most widely studied molecules by theorists interested in multi-reference or strongly correlated systems, but experimentally, only a few vibrational levels have been observed at the bottom of the potential energy curve, and a few more closer to the middle, making it unclear whether or not the potential has a double well or just one minimum and a kink. The isoelectronic molecule VMn, formed from Cr_2's direct neighbors, is even less well characterized, and an empirical ionization energy remains unknown. For Co_2 it is not even clear whether the ground state should be characterized with Δ or Σ symmetry, and the sparsity of experimental information is greater for dimers containing elements beyond the first row transition metals.

In this work we summarize the experimental data found for all transition metal diatomics, then for the first-row homonuclear transition metal dimers, we compare theoretical potential energy curves calculated with a wide range of *ab initio* methods using the largest basis sets openly available for those atoms: aug-cc-pwCV5Z. From the potential energy curves calculated with various theoretical approaches, we extract spectroscopic constants and compare them with experiments where available.

WK08

REANALYSIS OF THE $a\,^4\Sigma^-$ - $X\,^2\Pi_r$ TRANSITION OF GeH USING INTRACAVITY LASER SPECTROSCOPY

JACK C HARMS, *Chemistry and Biochemistry, University of Missouri, St. Louis, MO, USA*; LEAH C O'BRIEN, *Department of Chemistry, Southern Illinois University, Edwardsville, IL, USA*; JAMES J O'BRIEN, *Chemistry and Biochemistry, University of Missouri, St. Louis, MO, USA.*

The spin-forbidden $a\,^4\Sigma^-$ - $X\,^2\Pi_r$ transition of germanium hydride, GeH, was reported in emission in 1953 by Kleman and Werhagen. In our study, Intracavity Laser Spectroscopy, ILS, was used to obtain the first high resolution spectrum of this transition between 15,000 cm^{-1} and 16,500 cm^{-1}. The GeH molecules were produced in the plasma discharge of an Al-plate electrode, using 800 mTorr H_2 and 600 mTorr of GeH_4. The plasma was formed within the cavity of a tunable dye laser system, and the molecular absorption features are enhanced during an initial generation time prior to detection. The cathode length was 150 mm, the laser cavity was 1.15 m long, and a generation time of 180 μsec was used, resulting in an effective pathlength of 7 km. The spectra were collected intermittently with those from an external I_2 cell, and the spectra were calibrated using PGOHPER and the Doppler-limited I_2 spectrum of Salami and Ross. The obtained line positions were fit using PGOPHER. Results of the analysis will be presented.

WK09

A REEXAMINATION OF THE RED BAND OF CuO: ANALYSIS OF THE [16.5] $^2\Sigma^-$ - $X\,^2\Pi_i$ TRANSITION OF ^{63}CuO and ^{65}CuO

JACK C HARMS, ETHAN M GRAMES, SIRKHOO YUN, BUSHRA AHMED, *Chemistry and Biochemistry, University of Missouri, St. Louis, MO, USA*; LEAH C O'BRIEN, *Department of Chemistry, Southern Illinois University, Edwardsville, IL, USA*; JAMES J O'BRIEN, *Chemistry and Biochemistry, University of Missouri, St. Louis, MO, USA.*

The red band of CuO has been observed at high resolution using Intracavity Laser Spectroscopy (ILS). The red band was rotationally analyzed in 1974 by Appelblad and Lagerqvist and a portion of the band structure was assigned as the spectrum of the [16.5] $A\,^2\Sigma^+$ - $X\,^2\Pi_i$ transition. Subsequent analyses of CuO showed that the character of the A state was $^2\Sigma^-$ in character, and thus the Λ-doubling parameter, p, was inverted, and the *e*/*f* parity assignments were reversed. In this study, the spectrum of CuO was recorded in the in the regions 16,150 cm^{-1}– 16,270 cm^{-1} and 16,405 cm^{-1}- 16,545 cm^{-1}. The CuO molecules were produced in the plasma discharge of a copper hollow cathode within the cavity of a tunable dye laser, using 0.6 torr of argon as the sputter gas and a trace amount of O_2 as the source of oxygen. The plasma spectra were recorded intermittently with spectra from an external I_2 cell, and line positions from the widely used Iodine Atlas were used for calibration. In uncongested regions of the spectrum, both ^{63}CuO and ^{65}CuO were observed with appreciable intensity. The resulting spectra were rotationally analyzed for both isotopologues, fitting the data as a $^2\Sigma^-$ - $^2\Pi_i$ transition using PGOPHER. Line positions from the millimeter wave and FTIR studies of ^{63}CuO performed in the late 1990s were included in the fit to overcome potential complications due to the ambiguous parity assignments prevalent in the CuO literature. Previously unreported molecular constants were obtained from the fit for ^{65}CuO, and the constants of ^{63}CuO are determined to at least an order of magnitude greater than the results of Appelblad and Lagerqvist. Results of this analysis will be presented.

WK10

ANALYSIS OF SOME NEW ELECTRONIC TRANSITIONS OBSERVED USING INTRACAVITY LASER SPECTROSCOPY (ILS): POSSIBLE IDENTIFICATION OF HCuN

JACK C HARMS, ETHAN M GRAMES, *Chemistry and Biochemistry, University of Missouri, St. Louis, MO, USA*; LEAH C O'BRIEN, *Department of Chemistry, Southern Illinois University, Edwardsville, IL, USA*; JAMES J O'BRIEN, *Chemistry and Biochemistry, University of Missouri, St. Louis, MO, USA.*

Four new electronic transitions with blue-degraded bandheads were observed in the orange-red region of the visible spectrum. The transitions were observed in the plasma discharge of a hollow copper cathode placed within the cavity of a tunable dye laser system, allowing molecular absorbance to be enhanced upon laser amplification. To produce the molecules, the surface of the copper cathode was soaked in a dilute ammonia solution prior to installation, and 1 torr of H_2 was used as the sputter gas in the dc plasma discharge. The bandheads were observed at 16,560 cm^{-1}, 16,485 cm^{-1}, 16,027 cm^{-1}, and 15,960 cm^{-1}. Using 1.5 torr of D_2 as the sputter gas resulted in a -3 cm^{-1} shift in origin for the bands in the 16,000 cm^{-1} region. Four rotational branches have been identified in each transition, and the transitions have been fit to independent $^2\Sigma$ - $^2\Pi$ transitions using PGOPHER, with spin-orbit splittings in the Hund's case (a) Π-states of -71.2 cm^{-1} and -65.4 cm^{-1}. The transitions have tentatively been assigned to HCuN. Results of this analysis will be presented.

WK11

THE PURE ROTATIONAL SPECTRUM OF KO

<u>MARK BURTON</u>, *Department of Chemistry and Biochemistry, University of Arizona, Tucson, AZ, USA*; BEN-JAMIN RUSS, PHILLIP M. SHERIDAN, *Department of Chemistry and Biochemistry, Canisius College, Buffalo, NY, USA*; MATTHEW BUCCHINO, LUCY M. ZIURYS, *Department of Chemistry and Biochemistry, University of Arizona, Tucson, AZ, USA*.

The pure rotational spectrum of potassium monoxide (KO) has been recorded using millimeter-wave direct absorption spectroscopy. KO was synthesized by the reaction of potassium vapor, produced in a Broida-type oven, with nitrous oxide. No DC discharge was necessary. Eleven rotational transitions belonging to the $^2\Pi_{3/2}$ spin-orbit component have been measured and have been fit successfully to a case (c) Hamiltonian. Rotational and lambda-doubling constants for this spin-orbit component have been determined. It has been suggested that the ground electronic state of KO is either $^2\Pi$ (as for LiO and NaO) or $^2\Sigma$ (as for RbO and CsO), both of which lie close in energy. Recent computational studies favor a $^2\Sigma$ ground state. Further measurements of the rotational transitions of the $^2\Pi_{1/2}$ spin-orbit component and the $^2\Sigma$ state are currently in progress, as well as the potassium hyperfine structure.

RA. Plenary
Thursday, June 22, 2017 – 8:30 AM
Room: Foellinger Auditorium

Chair: Anthony Remijan, NRAO, Charlottesville, VA, USA

RA01   8:30 – 9:10

PRECISION SPECTROSCOPY OF MOLECULAR HYDROGEN AND THE SEARCH FOR NEW PHYSICS

WIM UBACHS, *Department of Physics and Astronomy, VU University , Amsterdam, Netherlands.*

The hydrogen molecule is the smallest neutral chemical entity and a benchmark system of molecular spectroscopy. The comparison between highly accurate measurements of transition frequencies and level energies with quantum calculations including all known phenomena (relativistic, vacuum polarization and self energy) provides a tool to search for physical phenomena in the realm of the unknown: are there forces beyond the three included in the Standard Model of physics plus gravity [1], are there extra dimensions beyond the 3+1 describing space time [2] ? Comparison of laboratory wavelengths of transitions in hydrogen may be compared with the lines observed during the epoch of the early Universe to verify whether fundamental constants of Nature have varied over cosmological time [3]. These concepts, as well as the precision laboratory experiments and the astronomical observations used for such searches of new physics [4] will be discussed.

[1] E.J. Salumbides, J.C.J. Koelemeij, J. Komasa, K. Pachucki, K.S.E. Eikema, W. Ubachs, *Bounds on fifth forces from precision measurements on molecules*, Phys. Rev. D87, 112008 (2013).

[2] E.J. Salumbides, A.N. Schellekens, B. Gato-Rivera, W. Ubachs *Constraints on extra dimensions from molecular spectroscopy*, New. J. Phys. 17, 033015 (2015).

[3] W. Ubachs, J. Bagdonaite, E.J. Salumbides, M.T. Murphy, L. Kaper, *Search for a drifting proton-electron mass ratio from H_2*, Rev. Mod. Phys. 88, 021003 (2016).

[4] W. Ubachs, J.C.J. Koelemeij, K.S.E. Eikema, E.J. Salumbides, *Physics beyond the Standard Model from hydrogen spectroscopy*, J. Mol. Spectr. 320, 1 (2016).

RA02   9:15 – 9:55

EXPLORING THE DETAILS OF INTERMOLECULAR INTERACTIONS VIA A SYSTEMATIC CHARACTERIZATION OF THE STRUCTURES OF THE BIMOLECULAR HETERODIMERS FORMED BETWEEN PROTIC ACIDS AND HALOETHYLENES

HELEN O. LEUNG, *Chemistry Department, Amherst College, Amherst, MA, USA.*

In the early 2000's, the work of Cole and Legon,[a,b] combined with that done earlier by Kisiel, Fowler, and Legon,[c] demonstrated that comparisons among the complexes of HF, HCl, and HCCH each with vinyl fluoride could provide information concerning the strength of intermolecular interactions. Specifically, that the length of the hydrogen bond and its deviation from linearity as a result of a secondary interaction with the nucleophilic portion of the protic acid could be correlated with the hydrogen bond strength. Building on this foundation, we undertook a systematic characterization of the molecular structures of complexes formed between these three acids and the remaining polar fluoroethylenes, seeking to unravel the nature of their intermolecular interactions. What started out as a simple confirmation of chemical intuition regarding relative interaction strengths developed into a fuller appreciation of the competition between electrostatic and steric forces in determining the lowest energy configuration for the heterodimer.

Additional surprises were in store for us as we expanded the study to chlorofluoroethylenes. Although the first few examples again served to confirm earlier conclusions, subsequent complexes provided unexpected results that signaled an increasing importance of the dispersion interaction in determining the geometry of the complex as well as the fundamental differences in the electron distributions surrounding the halogens in a C–F versus C–Cl bond.

Our work with these species has not only allowed us to investigate fundamental questions regarding intermolecular interactions, but obtaining and analyzing the spectra of these complexes along with those of the various haloethylene monomers and their complexes with the argon atom have provided an introduction to molecular spectroscopy and structure determination for many undergraduate students.

---

[a] G.C. Cole and A.C. Legon, *Chem. Phys. Lett.* 369, 31-40 (2003).
[b] G.C. Cole and A.C. Legon, *Chem. Phys. Lett.* 400, 414-424 (2004).
[c] Z. Kisiel, P.W. Fowler, and A.C. Legon, *J. Chem. Phys.* 93, 3054-3062 (1990).

Intermission

Presentation of Awards by Gary Douberly, University of Georgia

2016 Rao Award Winners
David Grimes, Massachusetts Institute of Technology
Kasper Mackeprang, University of Copenhagen
Josey E. Topolski, Indiana University Bloomington

MILLER PRIZE   **10:45**
Introduction by Susanna Widicus Weaver, Emory University

RA03   ***Miller Prize Lecture***   **10:50 – 11:05**

AUTOMATED MICROWAVE DOUBLE RESONANCE SPECTROSCOPY: A TOOL TO IDENTIFY AND CHARACTER-
IZE CHEMICAL COMPOUNDS

MARIE-ALINE MARTIN-DRUMEL, *CNRS, Institut des Sciences Moleculaires d'Orsay, Orsay, France*;
MICHAEL C McCARTHY, *Atomic and Molecular Physics, Harvard-Smithsonian Center for Astrophysics, Cambridge, MA, USA*; DAVID PATTERSON, *Department of Physics, Harvard University, Cambridge, MA, USA*;
BRETT A. McGUIRE, *NAASC, National Radio Astronomy Observatory, Charlottesville, VA, USA*; KYLE N.
CRABTREE, *Department of Chemistry, The University of California, Davis, CA, USA.*

Owing to its unparalleled structural specificity, rotational spectroscopy is a powerful technique to unambiguously identify and characterize polar molecules. We present here an experimental approach, automated microwave double resonance (AMDOR) spectroscopy, that allows to rapidly determine the rotational constants of such compounds without any a priori knowledge of elemental composition or molecular structure. This task is achieved by acquiring the classical (frequency vs. intensity) broadband spectrum of a molecule using chirped-pulse Fourier transform microwave (FTMW) spectroscopy, and subsequently analyzing it in near-real time using complementary cavity FTMW detection and double resonance. AMDOR measurements provide a unique "barcode" for each compound from which rotational constants can be extracted. Results obtained on the characterization of individual compounds and mixtures will be described.

COBLENTZ AWARD   **11:10**
Presentation of Award by Linda Kidder, Coblentz Society

RA04   ***Coblentz Society Award Lecture***   **11:15 – 11:55**

BIOORTHOGONAL CHEMICAL IMAGING FOR BIOMEDICINE

WEI MIN, *Chemistry, Columbia University, New York, NY, USA.*

Innovations in light microscopy have tremendously revolutionized the way researchers study biological systems with subcellular resolution. Although fluorescence microscopy is currently the method of choice for cellular imaging, it faces fundamental limitations for studying the vast number of small biomolecules. This is because relatively bulky fluorescent labels could introduce considerable perturbation to or even completely alter the native functions of vital small biomolecules. Hence, despite their immense functional importance, these small biomolecules remain largely undetectable by fluorescence microscopy. To address this challenge, we have developed a bioorthogonal chemical imaging platform. By coupling stimulated Raman scattering (SRS) microscopy, an emerging nonlinear Raman microscopy technique, with tiny and Raman-active vibrational probes (e.g., alkynes, nitriles and stable isotopes including 2H and 13C), bioorthogonal chemical imaging exhibits superb sensitivity, specificity, multiplicity and biocompatibility for imaging small biomolecules in live systems including tissues and organisms. Exciting biomedical applications such as imaging fatty acid metabolism related to lipotoxicity, glucose uptake and metabolism, drug trafficking, protein synthesis, DNA replication, protein degradation, RNA synthesis and tumor metabolism will be presented. This bioorthogonal chemical imaging platform is compatible with live-cell biology, thus allowing real-time imaging of small-molecule dynamics. Moreover, further chemical and spectroscopic strategies allow for multicolor bioorthogonal chemical imaging, a valuable technique in the era of "omics". We envision that the coupling of SRS microscopy with vibrational probes would do for small biomolecules what fluorescence microscopy of fluorophores has done for larger molecular species, bringing small molecules under the illumination of modern light microscopy.

RF. Mini-symposium: ALMA's Molecular View
Thursday, June 22, 2017 – 1:45 PM
Room: 274 Medical Sciences Building

Chair: Marie-Aline Martin-Drumel, Institut des Sciences Moléculaires d'Orsay, Orsay, France

---

RF01   ***INVITED TALK***   **1:45 – 2:15**

FEEDING, FEEDBACK AND THE GROWTH OF GALAXIES – MOLECULES AS TOOLS FOR PROBING GALAXY EVOLUTION

SUSANNE AALTO, *Department of Earth and Space Sciences, Onsala Space Observatory, Chalmers University of Technology, Onsala, Sweden.*

Cold gas plays a central role in feeding and regulating star formation and growth of supermassive black holes (SMBH) in galaxy nuclei. Particularly powerful activity occurs when interactions of gas-rich galaxies funnel large amounts of gas and dust into nuclei of luminous and ultra luminous infrared galaxies (LIRGs/ULIRGs). These dusty objects are of key importance to galaxy mass assembly over cosmic time. Some (U)LIRGS have deeply embedded galaxy nuclei that harbour a very active evolutionary stage of AGNs and/or starbursts. The nuclear activity will often drive mechanical feedback in the form of molecular winds, jets and outflows. This feedback can for example remove baryons from low-mass galaxies, prevent overgrowth of galaxies, be linked to the M_{BH}-σ relation, and explain "red-and dead" properties of local ellipticals.

With the ALMA and NOEMA telescopes we can use molecules as diagnostic tools to probe the properties of dust-enshrouded galaxy nuclei and their associated cold winds and outflows. Their morphology, velocity structure, physical conditions and even chemistry can be studied at unprecedented sensitivity and resolution, opening new avenues to further our understanding of the growth of galaxies.

I will give a brief review of the ALMA/NOEMA view of AGN and starburst radiative and mechanical feedback, and how it is linked to the properties of the nuclear power source. I will discuss the use of molecules (e.g. H_2O, H_3O^+, HCN, HCO^+, H_2S) for studying dusty nuclei and the nature of the embedded activity. We can, for example, investigate ionization rates and the impact of cosmic ray-, X-ray- and PDR-chemistry and the onset of outflows and winds. Interestingly, in some deeply obscured nuclei the chemistry shows strong similarities to that of Galactic hot cores. Finally I will show peculiar molecular jets and very recent ALMA observations at resolutions of tens of milli-arcseconds (few pc) of vibrationally excited HCN in opaque nuclei. These regions offer both challenges and opportunities for IR and submm studies of the nature of the buried activity – which we suggest is a deeply dust-enshrouded SMBH in a high-accretion state, or an extreme, high-temperature, burst of star formation.

RF02   **2:19 – 2:34**

WATER EMISSION FROM EARLY UNIVERSE

SREEVANI JARUGULA, JOAQUIN VIEIRA, *Department of Astronomy, University of Illinois at Urbana-Champaign, Urbana, IL, USA.*

The study of dusty star forming galaxies (DSFGs) is important to understand galaxy assembly in early universe. A bulk of star formation at $z \sim 2 - 3$ takes place in DSFGs but are obscured by dust in optical/UV. However, they are extremely bright in far infrared (FIR) and submillimeter with infrared luminosities of $10^{11} - 10^{13} L_{\odot}$. ALMA, with its high spatial and spectral resolution, has opened up a new window to study molecular lines, which are vital to our understanding of the excitation and physical processes in the galaxy. Carbon monoxide (CO) being the second most abundant and bright molecule after hydrogen (H_2), is an important tracer of star forming potential. Besides CO, water (H_2O) is also abundant and it's line strength is comparable to high-J CO lines in high redshift Ultra Luminous Infrared Galaxies (ULIRGs). Studies have shown H_2O to directly trace the FIR field and hence the star forming regions. Moreover, L_{H_2O}/L_{IR} ratio is nearly constant for five of the most important water lines and does not depend on the presence of AGN implying that H_2O is one of the best tracers of star forming regions (SFRs). This incredible correlation holds for nearly five orders of magnitude in luminosity and observed in both local and high redshift luminous infrared galaxies.

In this talk, I will discuss the importance of H_2O in tracing FIR field and show the preliminary results of resolved water emission from three high-redshift gravitationally lensed South Pole Telescope (SPT) sources obtained from ALMA cycle 3 and cycle 4. These sources are among the first H_2O observations with resolved spatial scales ~ 1 kpc and will prove to be important for ALMA and galaxy evolution studies.

RF03

THE SPT+ALMA CO REDSHIFT SURVEY OF DUSTY GALAXIES

JOAQUIN VIEIRA, *Department of Astronomy, University of Illinois at Urbana-Champaign, Urbana, IL, USA.*

In a 2500 square degree cosmological survey, the South Pole Telescope has systematically identified a large number (100) of high-redshift strongly gravitationally lensed sub-millimeter galaxies (SMGs). We are conducting a unique spectroscopic redshift survey with ALMA, targeting carbon monoxide (CO) line emission in these sources, across the 3mm spectral window. To date, we have obtained spectroscopic redshifts for 54 sources from $1.8 < z < 6.9$, with a median of $z=3.9$. This sample comprises 70% of the total spectroscopically confirmed SMGs at $z > 4$ and extends into the epoch of re-ionization. Once we determine the redshift for these sources, we are able to obtain high-resolution CO, [CII], [NII], H_2O, OH, and HCN for these sources with ALMA, making this the largest and most well-studied samples of high-redshift starburst galaxies. We are undertaking a comprehensive and systematic followup campaign to use these "cosmic magnifying glasses" to study the physical conditions and chemical evolution of the dust-obscured universe in unprecedented detail. I will describe our team's method for obtaining and confirming spectroscopic redshifts, detail our current knowledge of the redshifts distribution of SMGs, present a method for selecting the highest redshift SMGs, describe our high-resolution imaging of molecular lines, and discuss future directions for obtaining large samples of mm-wave spectra.

RF04

PHOSPHORUS CHEMISTRY IN OXYGEN RICH STARS

JACOB BERNAL, *Department of Chemistry and Biochemistry, University of Arizona, Tucson, AZ, USA*; DEB-ORAH SCHMIDT, *Department of Astronomy, University of Arizona, Tucson, AZ, USA*; JULIE ANDERSON, LUCY M. ZIURYS, *Department of Chemistry and Biochemistry, University of Arizona, Tucson, AZ, USA.*

Observations of PO and PN have been carried out at the Arizona Radio Observatory at 1, 2, and 3 mm. Multiple transitions of PO and PN have been detected towards the O-rich AGB stars TX Cam and RCas. Data obtained toward supergiant stars VY Canis Majoris and NML Cyg have also been analyzed. Abundances were obtained for these molecules in all four objects using the radiative transfer code ESCAPADE, which is suitable for symmetric and asymmetric stellar outflows. The abundances of PN and PO were found to be in the range 10^{-8} - 10^{-7} relative to H_2. While PN appears to be a parent molecule formed by LTE chemistry near the stellar photosphere, PO appears to be created further out from the star at r $> 400\,R_*$.

RF05

THE HIGH RESOLUTION VIBRATION-ROTATION SPECTRUM OF SiH^+

JOSE LUIS DOMENECH, *Molecular Physics, Instituto de Estructura de la Materia (IEM-CSIC), Madrid, Spain.*

Silicon bearing molecules account for $\sim 10\%$ of the identified molecules in space. Among those containing hydrogen, SiH and SiH_4 have been identified in the solar spectrum, in some cold stars, and SiH_4 in IRC+1026. However the simple SiH^+ cation (silylidinium) has only been observed in the solar photosphere and it remains undetected in interstellar space. Most of the spectroscopic information on SiH^+ comes from the analisis of its vis-UV spectrum[a,b], and from a diode laser spectrum combined with velocity modulation of the v=1-0 band[c]. The latter contained just eight lines measured with an estimated accuracy of 0.001 cm^{-1}. We present the results obtained with a difference frequency laser spectrometer coupled to a hollow cathode discharge, with an increased number of lines and improved accuracy (1×10^{-4} cm^{-1}), allowing for an accurate prediction of the pure rotational transitions. These will be searched for in the Cologne Center for Terahertz Spectroscopy (CCTS). These data will be of use in future searches for this molecule in different astronomical environments.

[a] A. E. Douglas & B. Lutz, Can. J. Phys. 48 (1970) 247
[b] T. Carlson et al. Astron. & Astrophys. 83 (1980) 238
[c] P. B. Davies, P. M. Martineau, J. Chem. Phys. 88 (1985) 485

Intermission

RF06 *Journal of Molecular Spectroscopy Review Lecture* 3:44 – 4:14

BUILDING BLOCKS OF DUST AND LARGE ORGANIC MOLECULES: A COORDINATED LABORATORY AND ASTRONOMICAL STUDY OF AGB STARS

MICHAEL C McCARTHY, *Atomic and Molecular Physics, Harvard-Smithsonian Center for Astrophysics, Cambridge, MA, USA*; CARL A GOTTLIEB, *Radio and Geoastronomy Division, Harvard-Smithsonian Center for Astrophysics, Cambridge, MA, USA*; JOSE CERNICHARO, *Departamento de Astrofísica, Centro de Astrobiología CAB, CSIC-INTA, Madrid, Spain.*

The increased sensitivity and angular resolution of high-altitude ground-based interferometers in the sub-millimeter band has enabled the physics and chemistry of carbon- and oxygen-rich evolved stars to be re-examined at an unprecedented level of detail. Observations of rotational lines in the inner envelope — the region within a few stellar radii of the central star where the molecular seeds of dust are formed — allows one to critically assess models of dust growth. Interferometric observations of the outer envelope provide stringent tests of neutral and ionized molecule formation. All of the astronomical studies are crucially dependent on precise laboratory measurements of the rotational spectra of new species and of vibrationally excited levels of known molecules and their rare isotopic species. By means of a closely coordinated laboratory and astronomical program, a number of exotic species including the disilicon carbide SiCSi, titanium oxides TiO and TiO_2, and

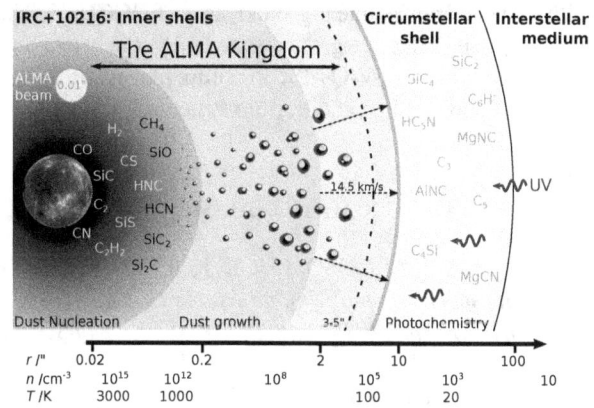

carbon chain anions ranging from CN^- to C_8H^- have recently been observed in evolved stars. This talk will provide overview of these findings, and how they impact current models of the "chemical laboratories" of evolved stars. Ongoing laboratory studies of small silicon-bearing molecules such as H_2SiO_2 and vibrationally excited SiC_2 will be highlighted.

RF07 4:18 – 4:33

THE (SUB-)MILLIMETER-WAVE SPECTRUM OF PROPANAL

OLIVER ZINGSHEIM, HOLGER S. P. MÜLLER, FRANK LEWEN, STEPHAN SCHLEMMER, *I. Physikalisches Institut, Universität zu Köln, Köln, Germany.*

The microwave spectrum of propanal, also known as propionaldehyde, CH_3CH_2CHO, has been investigated in the laboratory already since 1964[1] and has also been detected in space[2]. Recently, propanal was detected with the Atacama Large Millimeter/submillimeter Array (ALMA), Protostellar Interferometric Line Survey (PILS)[3]. The high sensitivity and resolution of ALMA indicated small discrepancies between observed and predicted rotational spectra of propanal. As higher accuracies are desired the spectrum of propanal was measured up to 500 GHz with the Cologne (Sub-)Millimeter spectrometer. Propanal has two stable conformers, *syn* and *gauche*, which differ mainly in the rotation of the aldehyd group with respect to the rigid C-atom framework of the molecule. We extensively studied both of them. The lower *syn*-conformer shows small splittings caused by the internal rotation of the methyl group, whereas the spectrum of *gauche*-propanal is complicated due to the tunneling rotation interaction from two stable degenerate conformers. Additionally, we analyzed vibrationally excited states.

[1] Butcher et al., *J. Chem. Phys.* **40** 6 (1964)
[2] Hollis et al., *Astrophys. J.* **610** L21 (2004)
[3] Lykke et al., *A&A* **597** A53 (2017)

RF08   4:35 – 4:50

ROTATIONAL SPECTRA AND STRUCTURAL DETERMINATION OF HCCNCS

WENHAO SUN, REBECCA DAVIS, JENNIFER VAN WIJNGAARDEN, *Department of Chemistry, University of Manitoba, Winnipeg, MB, Canada.*

The ground state of HCCNCS, prepared by high voltage electric discharge of a gas mixture of acetylene and CH_3NCS in neon during supersonic expansion, was studied using both chirped pulse Fourier transform microwave (cp-FTMW) and Balle Flygare FTMW spectrometers. The pure rotational spectra were measured for the parent, ^{34}S, and three ^{13}C isotopologues in natural abundance and the ^{14}N nuclear quadrupole hyperfine structure was resolved. The observed spectra are consistent with a linear or quasilinear ground state of HCCNCS. The corresponding rotational constants were used to derive the substitution (r_s) and effective ground state (r_0) geometries. Supporting calculations at the MP2/cc-pVQZ and CCSD(T)/cc-pVQZ (expanded basis cc-pV(Q+d)Z for sulfur) levels of theory reveal that the potential energy surface is virtually flat around the minimum and yield an equilibrium structure (r_e) that is consistent with experiment.

RF09   4:52 – 5:07

THE JET-COOLED ROTATIONAL SPECTRUM OF GLYCINAMIDE, AN AMINOACID PRECURSOR

ELENA R. ALONSO, LUCIE KOLESNIKOVÁ, *Grupo de Espectroscopia Molecular, Lab. de Espectroscopia y Bioespectroscopia, Unidad Asociada CSIC, Universidad de Valladolid, Valladolid, Spain*; ZBIGNIEW KISIEL, *ON2, Institute of Physics, Polish Academy of Sciences, Warszawa, Poland*; J.-C. GUILLEMIN, *UMR 6226 CNRS - ENSCR, Institut des Sciences Chimiques de Rennes, Rennes, France*; JOSÉ L. ALONSO, *Grupo de Espectroscopia Molecular, Lab. de Espectroscopia y Bioespectroscopia, Unidad Asociada CSIC, Universidad de Valladolid, Valladolid, Spain.*

The glycinamide $H_2NCH_2CONH_2$, considered as one of the possible precursors of glycine, has been generated in the gas phase via laser ablation of glycinamide hydrochloride. The vaporized products were seeded in neon, expanded adiabatically into the vacuum chamber of the spectrometer and probed by broadband chirped pulsed Fourier transform microwave spectroscopy. The most stable conformer is stabilized by an intramolecular hydrogen bonding interaction between the lone pair on the nitrogen in the amine group and the H-N bond in the amide group was observed in accordance with the previous millimeter wave study [a]. Glycinamide possesses two ^{14}N nuclei with a nuclear quadrupole moment I=1, which give rise to a complex hyperfine structure. We took advantage of the higher resolution of our narrowband LA-MB-FTMW spectrometer [b] to fully resolve the nuclear quadrupole hyperfine structure. More than 90 nuclear quadrupole hyperfine components belonging to 5 different rotational transitions were analyzed. This provides a definitive evidence to establish the most stable observed conformer.

[a] Z.Kisiel, E.Białkowska-Jaworska,L.Pszczółkowski,J.C.Guillemin, 21st HRMS, Poznań, 2010.
[b] C.Bermudez, S. Mata, C.Cabezas, and J.L.Alonso, Angew. Chem. 2014, 126, 11195–11198.

RF10   5:09 – 5:24

THE MICROWAVE SPECTRUM OF LACTALDEHYDE, THE SIMPLEST CHIRAL SUGAR.

ELENA R. ALONSO, LUCIE KOLESNIKOVÁ, CARLOS CABEZAS, SANTIAGO MATA, *Grupo de Espectroscopia Molecular, Lab. de Espectroscopia y Bioespectroscopia, Unidad Asociada CSIC, Universidad de Valladolid, Valladolid, Spain*; J.-C. GUILLEMIN, *UMR 6226 CNRS - ENSCR, Institut des Sciences Chimiques de Rennes, Rennes, France*; JOSÉ L. ALONSO, *Grupo de Espectroscopia Molecular, Lab. de Espectroscopia y Bioespectroscopia, Unidad Asociada CSIC, Universidad de Valladolid, Valladolid, Spain.*

Among the sugar compounds whose conformations have been determined by different spectroscopic techniques the structure of the lactaldehyde ($CH_3CH(OH)CHO$), the simplest chiral sugar, is conspicuously absent. It is of great interest in the field of astrophysics, where the ongoing search in the interstellar medium (ISM) has been able to detect, based on the rotational spectra identification, the simplest C_2 sugar glycoaldehyde [a, b] Lactaldehyde is a solid with high melting point and low vapor pressure, preventing easy measurements of its gas-phase spectra. Herein, crystalline DL-lactaldehyde samples have been vaporized by laser ablation (LA) and the monomer and the non-centrosymmetric hemiacetal dimer have been revealed in a supersonic expansion by broadband Fourier transform microwave (CP-FTMW) spectroscopy. This rotational study enables the search of the lactaldehyde in the ISM.

[a] Hollis JM; Lovas FJ; Jewell. Astrophys J. (2000), 540(2):L107–L110
[b] Hollis JM; Jewell PR; Lovas FJ; Remijan A. Astrophys J. (2004), 613(1):L45–L48.

EXTENDING THE MILLIMETER-SUBMILLIMETER SPECTRUM OF PROTONATED FORMALDEHYDE

KEVIN ROENITZ, LUYAO ZOU, SUSANNA L. WIDICUS WEAVER, *Department of Chemistry, Emory University, Atlanta, GA, USA.*

Protonated formaldehyde has been detected in the interstellar medium, where it participates in the formation and destruction of methanol. The rotational spectrum for protonated formaldehyde has been previously recorded by Amano and coworkers from 120–385 GHz using a hollow cathode discharge source for ion production. Additionally, protonated formaldehyde was produced in a supersonic expansion discharge source by Duncan and coworkers, but it was detected using time-of-flight mass spectrometry. Higher frequency spectra would help to guide additional observational studies of protonated formaldehyde using instruments such as the ALMA and SOFIA observatories. As such, we have used a supersonic expansion discharge source to produce protonated formaldehyde, and recorded its spectrum using millimeter-submillimeter direct absorption spectroscopy. The rotational spectrum was recorded from 350–1000 GHz. Here we will present the experimental design, specifically focusing on the optimization of the source for production of organic ions. We will also present the spectroscopic results for protonated formaldehyde and a spectral analysis with associated prediction that can be extended to frequencies above 1 THz.

RG. Mini-symposium: Chirality-Sensitive Spectroscopy
Thursday, June 22, 2017 – 1:45 PM
Room: 116 Roger Adams Lab

Chair: Yunjie Xu, University of Alberta, Edmonton, AB, Canada

---

RG01   ***INVITED TALK***   1:45 – 2:15

WHAT CAN WE LEARN ON GAS PHASE CHIRAL COMPOUNDS BY PHOTOELECTRON CIRCULAR DICHROISM ?

LAURENT NAHON, *DESIRS beamline, Synchrotron SOLEIL, Gif-sur-Yvette, France.*

Since 15 years, a new type of chiroptical effect has been the subject of a large array of both theoretical and experimental studies: Photoelectron Circular Dichroism (PECD) in the angular distribution of photoelectrons produced by CPL-ionization of pure enantiomers in the gas phase observed as a very intense (up to 35 %) forward/backward asymmetry with respect to the photon axis and which reveals the chirality of the molecule (configuration).

PECD happens to be an orbital-specific, photon energy dependent effect and is a very subtle probe of the molecular potential being very sensitive to static molecular structures such as conformers, chemical substitution, clusters, as well as to vibrational motion, much more so than other observables in photoionization such as the cross section or the β asymmetry parameter (for a recent review see L. Nahon, G. A. Garcia, and I. Powis, J. Elec. Spec. Rel. Phen. 204, 322 (2015)). Therefore PECD studies have both a fundamental interest as well and analytical interest, especially since chiral species are ubiquitous in the biosphere, food and medical industry. This last aspect is probably the driving force for the recent extension of PECD studies by the laser community using UV REMPI schemes.

After a large introduction to the PECD process itself, and a description of our double imaging electron/ion coincidence set-up, several recent results on one-photon VUV PECD will be presented, including:

- Sensitivity to chemical substitutions, isomerism and conformation
- Case of floppy biomolecules such as amino acids alanine and proline with a conformer analysis and possible consequences for the origin of life's homochirality
- Analytical capabilities in terms of enantiomeric excess determination on a pure molecule as well as on a mixture of compounds.

Future trends for PECD studies will be given regarding the case of more complex/structured chiral systems as well as opportunities for time-resolved PECD opened by the recent first performance of PECD with fs HHG pulses and REMPI time-resolved PECD.

RG02   2:19 – 2:34

INTERNAL DYNAMICS AND CHIRAL ANALYSIS OF PULEGONE, USING MICROWAVE BROADBAND SPECTROSCOPY

ANNA KRIN, CRISTOBAL PEREZ, MELANIE SCHNELL, *CoCoMol, Max-Planck-Institut für Struktur und Dynamik der Materie, Hamburg, Germany*; MARÍA DEL MAR QUESADA-MORENO, JUAN JESÚS LÓPEZ-GONZÁLEZ, JUAN RAMÓN AVILÉS-MORENO, *Department of Physical and Analytical Chemistry, University of Jaén, Jaén, Spain*; PABLO PINACHO, *Departamento de Química Física y Química Inorgánica, Universidad de Valladolid, Valladolid, Spain*; SUSANA BLANCO, JUAN CARLOS LOPEZ, *Departamento de Química Física y Química Inorgánica / Grupo de Espectroscopía Molecular, Universidad de Valladolid, Valladolid, Spain.*

Essential oils, such as peppermint or pennyroyal oil, are widely used in medicine, pharmacology and cosmetics. Their major constituents, terpenes, are mostly chiral molecules and thus may exhibit different biological functionality with respect to their enantiomers. Here, we present recent results on the enantiomers of pulegone, one of the components of the peppermint (*Mentha piperita L.*) and pennyroyal (*Mentha pulegium*) essential oils, using the microwave three-wave mixing (M3WM) technique.

M3WM relies on the fact that the scalar triple product of the dipole moment components μ_a, μ_b and μ_c differs in sign between the enantiomers. A loop of three dipole-allowed rotational transitions is required for the analysis of a chiral molecule. Since the recorded signal will be exactly out of phase for the two enantiomers, an unambiguous differentiation between them is possible, even in complex mixtures.

In addition to the chiral analysis of pulegone, its internal dynamics, resulting from the independent rotation of two of its three methyl groups, will be discussed. Moreover, a cluster of pulegone with one water molecule will be presented.

RG03 2:36 – 2:51

A CHIRAL TAGGING STRATEGY FOR DETERMINING ABSOLUTE CONFIGURATION AND ENANTIOMERIC EX-CESS BY MOLECULAR ROTATIONAL SPECTROSCOPY

LUCA EVANGELISTI, WALTHER CAMINATI, *Dipartimento di Chimica G. Ciamician, Università di Bologna, Bologna, Italy*; DAVID PATTERSON, *Department of Physics, Harvard University, Cambridge, MA, USA*; JAVIX THOMAS, YUNJIE XU, *Department of Chemistry, University of Alberta, Edmonton, AB, Canada*; CHANNING WEST, <u>BROOKS PATE</u>, *Department of Chemistry, The University of Virginia, Charlottesville, VA, USA.*

The introduction of three wave mixing rotational spectroscopy by Patterson, Schnell, and Doyle [1,2] has expanded applications of molecular rotational spectroscopy into the field of chiral analysis. Chiral analysis of a molecule is the quantitative measurement of the relative abundances of all stereoisomers of the molecule and these include both diastereomers (with distinct molecular rotational spectra) and enantiomers (with equivalent molecular rotational spectra). This work adapts a common strategy in chiral analysis of enantiomers to molecular rotational spectroscopy. A "chiral tag" is attached to the molecule of interest by making a weakly bound complex in a pulsed jet expansion. When this tag molecule is enantiopure, it will create diastereomeric complexes with the two enantiomers of the molecule being analyzed and these can be differentiated by molecule rotational spectroscopy. Identifying the structure of this complex, with knowledge of the absolute configuration of the tag, establishes the absolute configuration of the molecule of interest. Furthermore, the diastereomer complex spectra can be used to determine the enantiomeric excess of the sample. The ability to perform chiral analysis will be illustrated by a study of solketal using propylene oxide as the tag. The possibility of using current methods of quantum chemistry to assign a specific structure to the chiral tag complex will be discussed. Finally, chiral tag rotational spectroscopy offers a "gold standard" method for determining the absolute configuration of the molecule through determination of the substitution structure of the complex. When this measurement is possible, rotational spectroscopy can deliver a quantitative three dimensional structure of the molecule with correct stereochemistry as the analysis output.

[1] David Patterson, Melanie Schnell, John M. Doyle, Nature 497, 475 (2013). [2] David Patterson, John M. Doyle, Phys. Rev. Lett. 111, 023008 (2013).

RG04 2:53 – 3:08

HIGH SENSITIVITY 1-D AND 2-D MICROWAVE SPECTROSCOPY VIA CRYOGENIC BUFFER GAS COOLING

<u>DAVID PATTERSON</u>, SANDRA EIBENBERGER, *Department of Physics, Harvard University, Cambridge, MA, USA.*

All rotationally resolved spectroscopic methods rely on sources of cold molecules. For the last three decades, the workhorse technique for producing highly supersaturated samples of cold molecules has been the pulsed supersonic jet. We present here progress on our alternative method, cryogenic buffer gas cooling. Our high density, continuous source, and low noise temperature allow us to record microwave spectra at unprecedented sensitivity, with a dynamic range in excess of 10^6 achievable in a few minutes of integration time. This high sensitivity enables new protocols in both 1-D and 2-D microwave spectroscopy, including sensitive chiral analysis via nonlinear three wave mixing and applications as an analytical chemistry tool

RG05   3:10 – 3:25

NATURAL OPTICAL ACTIVITY OF CHIRAL EPOXIDES: THE INFLUENCE OF STRUCTURE AND ENVIRONMENT ON THE INTRINSIC CHIROPTICAL RESPONSE

PAUL M LEMLER, CLAYTON L. CRAFT, PATRICK VACCARO, *Department of Chemistry, Yale University, New Haven, CT, USA.*

Chiral epoxides built upon nominally rigid frameworks that incorporate aryl substituents have been shown to provide versatile backbones for asymmetric syntheses designed to generate novel pharmaceutical and catalytic agents. The ubiquity of these species has motivated the present studies of their intrinsic (solvent-free) circular birefringence (CB), the measurement of which serves as a benchmark for quantum-chemical predictions of non-resonant chiroptical behavior and as a beachhead for understanding the often-pronounced mediation of such properties by environmental perturbations (e.g., solvation). The optical rotatory dispersion (or wavelength-resolved CB) of (R)-styrene oxide (R-SO) and (S,S)-phenylpropylene oxide (S-PPO) have been interrogated under ambient solvated and isolated conditions, where the latter efforts exploited the ultrasensitive techniques of cavity ring-down polarimetry. Both of the targeted systems display marked solvation effects as evinced by changes the magnitude and (in the case of R-SO) the sign of the extracted specific optical rotation, with the anomalously large response evoked from S-PPO distinguishing it from other members of the epoxide family. Linear-response calculations of dispersive optical activity have been performed at both density-functional and coupled-cluster levels of theory to unravel the structural and electronic origins of experimental findings, thereby suggesting the possible involvement of hindered torsional motion along dihedral coordinates adjoining phenyl and epoxide moieties.

RG06   3:27 – 3:42

CHARACTERIZATION OF INTERMOLECULAR INTERACTIONS AT PLAY IN THE 2,2,2-TRIFLUOROETHANOL TRIMERS USING CAVITY AND CHIRPED-PULSE MICROWAVE SPECTROSCOPY

NATHAN A SEIFERT, JAVIX THOMAS, WOLFGANG JÄGER, YUNJIE XU, *Department of Chemistry, University of Alberta, Edmonton, AB, Canada.*

2,2,2-trifluoroethanol (TFE) is a common aqueous co-solvent in biological chemistry which may induce or destabilize secondary structures of proteins and polypeptides, thanks to its diverse intermolecular linkages originating from the hydrogen bonding potential of both the hydroxyl and perfluoro groups.[a] Theoretically, the TFE monomer is predicted to have two stable *gauche* (*gauche*[+]/*gauche*[−]) conformations whereas the *trans* form is unstable or is supported only by a very shallow potential. Only the gauche conformers have been identified in the gas phase, whereas liquid phase studies suggest a *trans:gauche* ratio of 2:3.[b] The question at which sample (cluster) size the *trans* form of TFE would appear was one major motivation for our study.

Here, we report the detection of three trimers of TFE using Balle-Flygare cavity and chirped-pulse Fourier transform microwave spectroscopy (CP-FTMW) techniques. The most stable observed trimer features one *trans*- and two *gauche*-TFE subunits. The other two trimers, observed using a newly constructed 2-6 GHz CP-FTMW spectrometer, consist of only the two *gauche* conformers of TFE. Quantum Theory of Atoms in Molecules (QTAIM)[c] and non-covalent interactions (NCI)[d] analyses give detailed insights into which intermolecular interactions are at play to stabilize the *trans* form of TFE in the most stable trimer.

[a]M. Buck, Q. Rev. Biophys. 1998, 31, 297-335.
[b]I. Bakó, T. Radnai, M. Claire, B. Funel, J. Chem. Phys. 2004, 121, 12472-12480.
[c]R. F. W. Bader, Chem. Rev. 1991, 91, 893-928.
[d]E. R. Johnson, S. Keinan, P. Mori-Sánchez, J. Contreras-Garcia, A. J. Cohen, W. Yang, J. Am. Chem. Soc., 2010, 132, 6498-6506.

Intermission

RG07   *INVITED TALK*   4:01 – 4:31

ADVANCED APPLICATIONS OF VIBRATIONAL CIRCULAR DICHROISM: FROM SMALL CHIRAL MOLECULES TO FIBRILS

RINA K. DUKOR, *R&D, BioTools, Inc., Jupiter, FL, USA.*

Vibrational Circular Dichroism (VCD), first discovered in the early 1970s, and commercialized in the late 1990's, is finally coming of age! No longer a curiosity of the few selected academic groups, it is now used by all major pharmaceutical companies, regulatory agencies, government labs and academic institutions. The main application for the technology has been determination of absolute configuration of small pharmaceutical molecules. In more recent years, this has extended to more complicated molecules such as natural products with many chiral centers and conformational flexibility. Other applications include determination of enantiomeric purity, chiral polymers, and characterization of other biological molecules such as proteins, carohydrates and nucleic acids.

One of the most fascinating discoveries in the VCD field has been been unusual enhancement in intensity for proteins that form fibrils. We have demonstrated sensitivity of VCD to in situ solution-phase probe of the process of fibrillogenesis and subsequent development that currently can only be studied in detail with dried samples by such techniques as scanning electron microscopy or atomic force microscopy. We have further shown that several different proteins, that in their native state have different secondary structures, have a very similar unique signature of mature fibrils.

In this presentation, we will discuss fundamentals of VCD, demonstrate a few examples of different applications and showcase the sensitivity to structure of fibrils, including new results on micro-sampling.

RG08   4:35 – 4:50

THE MICROWAVE SPECTRA AND MOLECULAR STRUCTURES OF 2-(TRIFLUOROMETHYL)-OXIRANE AND 2-VINYLOXIRANE, TWO CANDIDATES FOR CHIRAL ANALYSIS VIA NONCOVALENT CHIRAL TAGGING

MARK D. MARSHALL, HELEN O. LEUNG, DESMOND ACHA, KEVIN WANG, *Chemistry Department, Amherst College, Amherst, MA, USA.*

The conversion of enantiomeric molecules into spectroscopically distinct diasteromeric complexes has been proposed as a promising new means for chiral analysis. The success of this method requires the characterization of potential chiral tags as well as demonstrations of the feasibility and power of the technique. 2-(trifluoromethyl)- and 2-vinyloxirane are chiral molecules with simple, hyperfine-free spectra. They are high vapor pressure liquids that can easily be incorporated into a free jet expansion for complex formation and spectroscopic analysis, and they are commercially available in enantiomerically pure forms as well as racemic mixtures. The microwave spectra of these two molecules and their carbon and oxygen atom substitution structures are obtained as well as the spectrum and structure of the 2-(trifluoromethyl)-oxirane-argon complex.

RG09

INTRINSIC OPTICAL ACTIVITY AND CONFORMATIONAL FLEXIBILITY: NEW INSIGHTS ON THE ROLE OF RING MORPHOLOGY FROM CYCLIC AMINES

CLAYTON L. CRAFT, PAUL M LEMLER, PATRICK VACCARO, *Department of Chemistry, Yale University, New Haven, CT, USA.*

Electronic circular birefringence (ECB), which causes rotation of the linear-polarization state for non-resonant light traversing an isotropic sample of chiral molecules, long has served as a robust means for assessing enantiomeric purity, but quantitative studies of this important property historically have been restricted to condensed phases where environmental effects (e.g., solvent-solute interactions or crystal-packing forces) can alter the magnitude and even the sign of the intrinsic behavior. As part of a continuing effort to elucidate the structural and electronic origins of such chiroptical phenomena, the dependence of optical rotatory dispersion (or wavelength-resolved ECB) on ring morphology has been explored for two saturated monocyclic amines, (R)-2-methylpyrrolidine and (S)-2-methylpiperidine. To assess the putative role of extrinsic perturbations, ambient measurements of specific optical rotation were performed under both solvated and isolated conditions, where the latter gas-phase work involved use of ultrasensitive cavity ring-down polarimetry. Each of the targeted compounds support active conformational degrees of freedom in the form of large-amplitude puckering motion of the heterocyclic ring combined with internal rotation of methyl substituents, with the antagonistic chiroptical properties exhibited by the resulting conformers combining to yield the overall response observed from a thermally equilibrated ensemble of molecules. Experimental ECB findings will be contrasted with those reported previously for ketones built upon comparable carbocyclic frameworks, and interpreted, in part, by reference to electronic-structure and linear-response calculations performed at various levels of quantum-chemical theory.

RG10

ABSOLUTE CONFIGURATION OF 3-METHYLCYCLOHEXANONE BY CHIRAL TAG ROTATIONAL SPECTROSCOPY AND VIBRATIONAL CIRCULAR DICHROISM

LUCA EVANGELISTI, *Dipartimento di Chimica G. Ciamician, Università di Bologna, Bologna, Italy*; MARTIN S. HOLDREN, KEVIN J MAYER, TAYLOR SMART, CHANNING WEST, BROOKS PATE, *Department of Chemistry, The University of Virginia, Charlottesville, VA, USA.*

The absolute configuration of 3-methylcyclohexanone was established by chiral tag rotational spectroscopy measurements using 3-butyn-2-ol as the tag partner. This molecule was chosen because it is a benchmark measurement for vibrational circular dichroism (VCD). A comparison of the analysis approaches of chiral tag rotational spectroscopy and VCD will be presented. One important issue in chiral analysis by both methods is the conformational flexibility of the molecule being analyzed. The analysis of conformational composition of samples will be illustrated. In this case, the high spectral resolution of molecular rotational spectroscopy and potential for spectral simplification by conformational cooling in the pulsed jet expansion are advantages for chiral tag spectroscopy. The computational chemistry requirements for the two methods will also be discussed. In this case, the need to perform conformer searches for weakly bound complexes and to perform reasonably high level quantum chemistry geometry optimizations on these complexes makes the computational time requirements less favorable for chiral tag rotational spectroscopy. Finally, the issue of reliability of the determination of the absolute configuration will be considered. In this case, rotational spectroscopy offers a "gold standard" analysis method through the determination of the ^{13}C-subsitution structure of the complex between 3-methylcyclohexanone and an enantiopure sample of the 3-butyn-2-ol tag.

RG11   *Post-Deadline Abstract*   **5:26 – 5:41**

CHIRALITY RECOGNITION IN CAMPHOR - 1,2-PROPANEDIOL COMPLEXES

CRISTOBAL PEREZ, MARIYAM FATIMA, ANNA KRIN, <u>MELANIE SCHNELL</u>, *CoCoMol, Max-Planck-Institut für Struktur und Dynamik der Materie, Hamburg, Germany.*

The molecular interactions in complexes involving chiral molecules are of particular interest, because the interactions change in a subtle way upon replacing one of the partners by its mirror image. This is based on the fact that chiral molecules are sensitive probes for other chiral objects and chiral interactions. In this particular case, we will concentrate on molecule-molecule interactions and investigate them with broadband rotational spectroscopy. When two chiral molecules form complexes, the homochiral and heterochiral forms have different structures (and thus rotational constants and spectra) and different energies. They are diastereomers, which can easily be differentiated, for example via molecular spectroscopy. This is often exploited in chemical synthesis for identifying and separating enantiomers. The phenomena involving chirality recognition are relevant in the biosphere, in organic synthesis and in polymer design.

We use chirped-pulse Fourier transform microwave (CP-FTMW) spectroscopy to study the structures and the underlying interactions of camphor-1,2-propanediol complexes. This system is also interesting because the complex formation can be expected to be ruled by an interplay between hydrogen bonding to the polar carbonyl group in camphor and dispersion interactions. The spectra are extremely rich because of the high number of conformers for 1,2-propanediol. We started out with racemic mixtures of both camphor and 1,2-propanediol. Using enantiopure samples of different handedness of the two partners nicely simplifies the spectra and guides the assignment. In the talk, we will report on the latest results for this chiral complex.

RG12   **5:43 – 5:58**

THE COMPLETE HEAVY-ATOM STRUCTURE OF A CP-FTMW CHIRAL TAG PRECURSOR, VERBENONE

<u>FRANK E MARSHALL</u>, *Department of Chemistry, Missouri University of Science and Technology, Rolla, MO, USA*; CHANNING WEST, GALEN SEDO, *Department of Natural Sciences, University of Virginia's College at Wise, Wise, VA, USA*; BROOKS PATE, *Department of Chemistry, The University of Virginia, Charlottesville, VA, USA*; G. S. GRUBBS II, *Department of Chemistry, Missouri University of Science and Technology, Rolla, MO, USA.*

The microwave spectrum of the chiral molecule verbenone has been recorded from 2-18 GHz using two CP-FTMW spectrometers. 2-8 GHz data has been acquired on a 2-8 GHz CP-FTMW located at the University of Virginia and 8-18 data has been acquired on a 6-18 GHz spectrometer located at Missouri S&T. From the experiments the authors were able to assign and fit isotopologues corresponding to each heavy atom position (either ^{13}C or ^{18}O), providing for the heavy-atom structure. Previous studies by Evans and coworkers have been added to these measurements in a global fit of the parent species.[a,b] The measurement and assignment of these transitions provide preliminary information needed for enatiomeric excess experiments using CP-FTMW van der Waals-type chiral tagging processes already being performed at UVa. Details of the experiment, fits, and structure will be discussed.

[a]C. J. Evans, S. M. Allpress, P. D. Godfrey, D. McNaughton, *67th International Symposium on Molecular Spectroscopy*, 2012, **RH13**

[b]S. M. Allpress, *Spectroscopic and Computational Chemistry Studies on Terpene Related Compounds*, University of Leicester, 2015, Chapter 6: Microwave Spectroscopy of Verbenone

RH. Clusters/Complexes
Thursday, June 22, 2017 – 1:45 PM
Room: 1024 Chemistry Annex

Chair: Jer-Lai Kuo, Academia Sinica, Taipei, Taiwan

RH01 1:45–2:00

MILLIMETER-WAVE SPECTROSCOPY OF He-HCN AND He-DCN: ENERGY LEVELS NEAR THE DISSOCIATION LIMIT.

__KENSUKE HARADA__, KEIICHI TANAKA, *Department of Chemistry, Kyushu University, Fukuoka, Japan.*

The He-HCN complex is a weakly bound complex with binding energy of about 9 cm^{-1}. We have measured the the $j = 1 \leftarrow 0$ internal rotation fundamental band of the He-HCN complex by millimeter-wave absorption spectroscopy and reported the potential energy surface (PES) to reproduce the observed transition frequencies.[a]

In the present study, we have extended the measurement to the $j = 2 \leftarrow 1$ internal rotation hot bands of the He-HCN and He-DCN complexes. In the analysis, the upper state of several observed transitions are found to be located above the "dissociation limit" (D_0). The rovibrational levels with e label dissociate to the HCN molecule with $j = 0$ and the He atom (D_0), while those with f label, due to the parity conservation, to the HCN molecule with $j = 1$ and the He atom which is higher in energy by about 2.96 cm^{-1} ($2B_{\mathrm{HCN}}$) than D_0. The f levels are bound up to $D_0 + 2B_{\mathrm{HCN}}$.

The revised PES of He-HCN has a global minimum in the linear He–HCN configuration with a depth of 29.9 cm^{-1} and has a saddle point at the anti-linear He–NCH configuration with a depth of 20.9 cm^{-1}. The ν_s intermolecular stretching first excited state and the $j = 2$ internal rotation second excited state are determined to be located 9.1405 and 9.0530 cm^{-1} above the ground state and very close to the calculated dissociation limit (D_0) of 9.32 cm^{-1}. Life times of several quasi-bound levels (both of e and f labels) and line widths of the related transitions are predicted for He-HCN and He-DCN from the revised PESs.

---

[a]K. Harada, K. Tanaka, T. Tanaka, S. Nanbu, and M. Aoyagi, J. Chem. Phys. **117**, 7041 (2002).

RH02 2:02–2:17

THEORETICAL STUDY OF GROUP 14 M$^+$(^2P_J)-RG COMPLEXES (M$^+$ = C$^+$, Si$^+$; RG = He - Ar)

__WILLIAM DUNCAN TUTTLE__, REBECCA L. THORINGTON, TIMOTHY G. WRIGHT, *School of Chemistry, University of Nottingham, Nottingham, United Kingdom*; LARRY A. VIEHLAND, *Science Department, Chatham University, Pittsburgh, USA.*

The light group 14 cations are found in a wide variety of environments, with, for example, C$^+$ ions thought to play a key role in the chemistry of the interstellar medium,[a] while Si$^+$ ions are an important component of the upper atmosphere of the Earth due to their presence in meteoroids.[b]

We calculate accurate interatomic potentials for a singly charged carbon cation[c,d] and a singly charged silicon cation[e] interacting with the rare gas atoms helium, neon and argon. The RCCSD(T) method is employed, with basis sets of quadruple-ζ and quintuple-ζ quality, and the energies counterpoise corrected and extrapolated to the basis set limit at each point. In all cases, we consider the lowest electronic states of the M$^+$ atom, (^2P_J), interacting with the ground electronic state of the RG atom, ($^1S_0$), and compute potentials corresponding to the molecular terms, $^2\Pi$ and $^2\Sigma^+$, as well as the spin-orbit levels which arise: $^2\Pi_{3/2}$, $^2\Pi_{1/2}$ and $^2\Sigma_{1/2}{}^+$. The potentials are employed to calculated spectroscopic constants and ion transport properties.

---

[a]S. Petrie and D. K. Bohme, *Mass Spec. Rev.*, **26**, 258 (2007).
[b]J. M. C. Plane, J. C. Gómez-Martin, W. Feng, and D. Janches, *J. Geophys. Res. Atmos.* **121**, 3718 (2016).
[c]W. D. Tuttle, R. L. Thorington, L. A. Viehland and T. G. Wright, *Mol. Phys.* **113**, 3767 (2015).
[d]W. D. Tuttle, R. L. Thorington, L. A. Viehland and T. G. Wright (in preparation).
[e]W. D. Tuttle, R. L. Thorington, L. A. Viehland and T. G. Wright, *Mol. Phys.* **115**, 437 (2017).

RH03

ROVIBRATIONAL SPECTRUM OF THE Ar-NO COMPLEX IN 5.3 μm REGION

CHUANXI DUAN, *College of Physical Science and Technology, Central China Normal University, Wuhan, China.*

The rovibrational spectrum of the open-shell complex Ar-NO was recorded in the 1870-1892 cm^{-1}range with a segmented rapid-scan pulsed supersonic jet infrared absorption spectrometer based on distributed-feedback quantum cascade lasers. Four b-type subbands were observed. The progress on the rotational analysis will be presented.

RH04

WEAK INTERACTIONS AND CO_2 MICROSOLVATION IN THE CIS-1,2-DIFLUOROETHYLENE...CO_2 COMPLEX

WILLIAM TRENDELL, REBECCA A. PEEBLES, SEAN A. PEEBLES, *Department of Chemistry, Eastern Illinois University, Charleston, IL, USA.*

The need for a deep understanding of CO_2 interactions is significant given the importance of supercritical CO_2 (sc-CO_2) as a green solvent. Fluorinated compounds often have higher solubility in sc-CO_2 than their hydrocarbon analogs, and the reasons for this are not well understood. Investigations of dimers of one CO_2 molecule with a simple fluorinated hydrocarbon provide an initial step towards understanding the complex balance of forces that is likely to be present as a larger solvation shell of sc-CO_2 is built.

The weakly bound dimer *cis*-1,2-difluoroethylene...CO_2 is the latest in a series of complexes of CO_2 with fluorinated ethylenes that has recently been studied using chirped-pulse (CP) Fourier-transform microwave spectroscopy. Unlike all previous members of the series, the observed structure of *cis*-1,2-difluoroethylene...CO_2 is nonplanar, with CO_2 sitting above the ethylene plane and crossed relative to the C=C bond. This nonplanar arrangement is consistent with predictions made using symmetry adapted perturbation theory (SAPT), where the dispersion energy of the nonplanar structure is significantly more favorable than for a structure where CO_2 lies in the same plane as the ethylene moiety. Observed transitions are doubled as a result of CO_2 tunneling between equivalent positions above and below the ethylene plane, leading to inversion of the μ_c dipole moment component. Observed transitions for the most abundant isotopologue have been fitted to a two state Hamiltonian to give an energy difference between tunneling states of $\Delta E \approx 333$ MHz, and analysis using Meyer's one dimensional model to determine the barrier to inversion is presently in progress.

RH05

MICROWAVE SPECTROSCOPIC STUDY OF THE ATMOSPHERIC OXIDATION PRODUCT *m*-TOLUIC ACID AND ITS MONOHYDRATE

MOHAMAD AL-JABIRI, ELIJAH G SCHNITZLER, NATHAN A SEIFERT, WOLFGANG JÄGER, *Department of Chemistry, University of Alberta, Edmonton, AB, Canada.*

m-Toluic acid is a photo-oxidation product of *m*-xylene, a chemical byproduct of the oil and gas industry, and is a common component of secondary atmospheric aerosol. Organic acids, such as *m*-toluic acid, are also thought to play an important role in the initial steps of aerosol formation, which involves formation of hydrogen bonded clusters with molecular species, such as water, ammonia, and sulfuric acid.

Somewhat surprisingly, the rotational spectrum of the *m*-toluic acid monomer has not been studied before. We have identified four stable conformers using ab initio calculations at the MP2/6-311++G(2df,2pd) level of theory. The two lowest energy conformers are rather close in energy and their rotational spectra were measured using a Balle-Flygare type microwave spectrometer. The structures and barriers to methyl internal rotation were determined.

We have identified four isomers of the monohydrate of *m*-toluic acid using ab initio calculations. Measurements of the microwave spectra of the two lowest energy isomers are underway with a newly constructed chirped pulse microwave Fourier transform spectrometer in the frequency range from 2 to 6 GHz. The spectra and analyses will be presented.

RH06

THE ETHANOL-CO_2 DIMER IS AN ELECTRON DONOR-ACCEPTOR COMPLEX

BRETT A. McGUIRE, *NAASC, National Radio Astronomy Observatory, Charlottesville, VA, USA*; MARIE-ALINE MARTIN-DRUMEL, *CNRS, Institut des Sciences Moleculaires d'Orsay, Orsay, France*; MICHAEL C McCARTHY, *Atomic and Molecular Physics, Harvard-Smithsonian Center for Astrophysics, Cambridge, MA, USA.*

Supercritical (sc) CO_2 is a common industrial solvent for the extraction of caffeine, nicotine, petrochemicals, and natural products. The ability of apolar scCO_2 to dissolve polar solutes is greatly enhanced by the addition of a polar co-solvent, often methanol or ethanol. Experimental and theoretical work show that methanol interactions in scCO_2 are predominantly hydrogen bonding, while the gas-phase complex is an electron donor-acceptor (EDA) configuration. Ethanol, meanwhile, is predicted to form EDA complexes both in scCO_2 and in the gas phase, but there have been no experimental measurements to support this conclusion. Here, we report a combined chirped-pulse and cavity FTMW study of the ethanol-CO_2 complex. Comparison with theory indicates the EDA complex is dominant under our experimental conditions. We confirm the structure with isotopic substitution, and derive a semi-experimental equilibrium structure. Our results are consistent with theoretical predictions that the linearity of the CO_2 subgroup is broken by the complexation interaction.

Intermission

RH07

MICROWAVE SPECTRUM AND STRUCTURE OF THE METHANE-PROPANE COMPLEX

KAREN I. PETERSON, *Chemistry and Biochemistry, San Diego State University, San Diego, CA, USA*; WEI LIN, *Chemistry, University of Texas Rio Grande Valley, Brownsville, TX, USA*; ERIC A. ARSENAULT, YOON JEONG CHOI, STEWART E. NOVICK, *Department of Chemistry, Wesleyan University, Middletown, CT, USA.*

Methane is exceptional in its solid-phase orientational disorder that persists down to 24 K. Only below that temperature does the structure become partially ordered, and full crystallinity requires even lower temperatures and high pressures. Not surprisingly, methane appears to freely rotate in most van der Waals complexes, although two notable exceptions are CH_4-HF and CH_4-C_5H_5N. Of interest to us is how alkane interactions affect the methane rotation. Except for CH_4-CH_4, rotationally-resolved spectra of alkane-alkane complexes have not been studied. To fill this void, we present the microwave spectrum of CH_4-C_3H_8 which is the smallest alkane complex with a practical dipole moment. The microwave spectrum of CH_4-C_3H_8 was measured using the Fourier Transform microwave spectrometer at Wesleyan University. In the region between 7100 and 25300 MHz, we observed approximately 70 transitions that could plausibly be attributed to the CH_4-C_3H_8 complex (requiring high power and the proper mixture of gases). Of these, 16 were assigned to the A-state (lowest internal rotor state of methane) and four to the F-state. The A-state transitions were fitted with a Watson Hamiltonian using nine spectroscopic constants of which A = 7553.8144(97) MHz, B = 2483.9183(35) MHz, and C = 2041.8630(21) MHz. The A rotational constant is only 1.5 MHz higher than that of Ar-C_3H_8 and, since the a-axis of the complex passes approximately through the centers of mass of the subunits, this indicates a similar relative orientation. Thus, we find that the CH_4 is located above the plane of the propane. The center-of-mass separation of the subunits in CH_4-C_3H_8 is calculated to be 3.993 Å, 0.16 Å longer than the Ar-C_3H_8 distance of 3.825 Å, a reasonable difference considering the larger van der Waals radius of CH_4. The four F-state lines, which were about twice as strong as the A-state lines, could be fitted to A, B, and C rotational constants, and further analysis is in progress.

RH08   4:01 – 4:16

NON-COVALENT INTERACTIONS AND INTERNAL DYNAMICS IN PYRIDINE-AMMONIA: A COMBINED QUANTUM-CHEMICAL AND MICROWAVE SPECTROSCOPY STUDY

LORENZO SPADA, NICOLA TASINATO, FANNY VAZART, VINCENZO BARONE, *Scuola Normale Superiore, Scuola Normale Superiore, Pisa, Italy*; WALTHER CAMINATI, CRISTINA PUZZARINI, *Dep. Chemistry 'Giacomo Ciamician', University of Bologna, Bologna, Italy.*

The 1:1 complex of ammonia with pyridine has been characterized by using state-of-the-art quantum-chemical computations combined with pulsed-jet Fourier-Transform microwave spectroscopy. The computed potential energy landscape pointed out the formation of a stable σ-type complex, which has been confirmed experimentally: the analysis of the rotational spectrum showed the presence of only one 1:1 pyridine – ammonia adduct. Each rotational transition is split into several components due to the internal rotation of NH_3 around its C_3 axis and to the hyperfine structure of both ^{14}N quadrupolar nuclei, thus providing the unequivocal proof that the two molecules form a σ-type complex involving both a N-H$\cdots$N and a C-H$\cdots$N hydrogen bond. The dissociation energy (BSSE and ZPE corrected) has been estimated to be 11.5 kJ$\cdot$mol^{-1}. This work represents the first application of an accurate, yet efficient computational scheme, designed for the investigation of small biomolecules, to a molecular cluster.

RH09   4:18 – 4:33

SPECTROSCOPIC CHARACTERIZATION OF N_2O_5 HALIDE CLUSTERS AND THE FORMATION OF HNO_3

JOANNA K. DENTON, PATRICK J KELLEHER, FABIAN MENGES, MARK JOHNSON, *Department of Chemistry, Yale University, New Haven, CT, USA.*

N_2O_5 is an atmospheric species which serves as night-time sink for NO_x species. Its reconversion to NO_x products occurs through solvation in atmospheric aerosols. Detection of N_2O_5 and NO_3^- fragmentation products in such aerosols has previously utilized chemical ionization featuring halides (of which chlorine is ubiquitous in sea-spray aerosols). We examine the solvation behavior of N_2O_5 and the critical number of water molecules to form HNO_3 from N_2O_5 and water. We have been able to generate and spectroscopically characterize N_2O_5-halide ions formed from halide-water clusters. We observe $X^--N_2O_5$ species whose spectra best correspond to a calculated $(O_2NX)(ONO_2^-)$ species.[a]

[a] Funding for this work was provided by the NSF's Center for Aerosol Impacts on Climate and the Environment.

RH10   4:35 – 4:50

ETHANOL DIMER: OBSERVATION OF THREE NEW CONFORMERS BY BROADBAND ROTATIONAL SPECTROSCOPY

DONATELLA LORU, ISABEL PEÑA, M. EUGENIA SANZ, *Department of Chemistry, King's College London, London, United Kingdom.*

The conformational behaviour of the hydrogen-bonded cluster ethanol dimer has been reinvestigated by chirped pulse Fourier transform microwave spectroscopy in the 2-8 GHz frequency region. Three new conformers (tt, $tg+$, and g-$g+$) have been identified together with the three ($g+g+$, g-t, and $g+t$) previously observed by Hearn et al. (J. Chem. Phys. 123, 134324, 2005) and their rotational and centrifugal distortion constants have been determined. By using different carrier gases in the supersonic expansion, the relative abundances of the observed conformers have been estimated. The monosubstituted ^{13}C species and some of the ^{18}O species of the most abundant conformers $g+g+$, g-t, and tt have been observed in their natural abundance, which led to the partial determination of their r_s structures, and the r_0 structure for the tt conformer. The six observed conformers are stabilized by the delicate interplay of primary O-H$\cdots$O and secondary C-H$\cdots$O hydrogen bonds, and dispersion interactions between the methyl groups. Density functional and ab initio methods with different basis sets are benchmarked against the experimental data.

Post-Deadline Abstract

BROADBAND FTMW SPECTROSCOPY OF THE UREA-ARGON AND THIOUREA-ARGON COMPLEXES

<u>CHRIS MEDCRAFT</u>, *School of Chemistry, Newcastle University, Newcastle-upon-Tyne, United Kingdom*; DROR M. BITTNER, *Department of Chemistry and Biochemistry, Old Dominion University, Norfolk, VA, USA*; GRAHAM A. COOPER, JOHN C MULLANEY, NICK WALKER, *School of Chemistry, Newcastle University, Newcastle-upon-Tyne, United Kingdom.*

The rotational spectra complexes of argon-urea, argon-thiourea and water-thiourea have been measured by chirped-pulse Fourier transform microwave spectroscopy from 2-18.5 GHz. The sample was produced via laser vaporisation of a rod containing copper and the organic sample as a stream of argon was passed over the surface and subsequently expanded into the vacuum chamber cooling the sample. Argon was found to bind to π system of the carbonyl bond for both the urea and thiourea complexes.

RI. Instrument/Technique Demonstration

Thursday, June 22, 2017 – 1:45 PM

Room: B102 Chemical and Life Sciences

Chair: Kyle N. Crabtree, University of California, Davis, CA, USA

RI01　　**1:45 – 2:00**

DOPPLER-FREE TWO-PHOTON ABSORPTION SPECTROSCOPY OF VIBRONIC EXCITED STATES OF NAPHTHALENE ASSISTED BY AN OPTICAL FREQUENCY COMB

AKIKO NISHIYAMA, *Department of Engineering Science, Graduate School of Informatics, The University of Electro-Communications, Tokyo, Japan*; KAZUKI NAKASHIMA, MASATOSHI MISONO, *Applied Physics, Fukuoka University, Fukuoka, Japan*; MASAAKI BABA, *Division of Chemistry, Graduate School of Science, Kyoto University, Kyoto, Japan.*

We observe Doppler-free two-photon absorption spectra of three bands of $S_1 \leftarrow S_0$ transition of naphthalene. We use an optical frequency comb stabilized to a GPS clock as a frequency reference of a scanning cw laser. The use of the optical frequency comb enables us to decide transition frequencies of rovibronic lines and their linewidths with uncertainties of several tens of kHz[a]. We discuss the interactions in vibronic excited states of naphthalene based on the dependences of frequency shifts and linewidths on vibrational and on rotational quantum numbers.

[a] A. Nishiyama, K. Nakashima, A. Matsuba, M. Misono, J. Mol. Spectrosc. 318, 40 (2015).

RI02　　**2:02 – 2:17**

TWO-PHOTON ABSORPTION SPECTROSCOPY OF RUBIDIUM WITH A DUAL-COMB TEQUNIQUE

AKIKO NISHIYAMA, SATORU YOSHIDA, TAKUYA HARIKI, YOSHIAKI NAKAJIMA, KAORU MINOSHIMA, *Department of Engineering Science, Graduate School of Informatics, The University of Electro-Communications, Tokyo, Japan.*

Dual-comb spectroscopies have great potential for high-resolution molecular and atomic spectroscopies, thanks to the broadband comb spectrum consisting of dense narrow modes[a]. In this study, we apply the dual-comb system to Doppler-free two-photon absorption spectroscopy. The outputs of two frequency combs excite several two-photon transitions of rubidium[b], and we obtained broadband Doppler-free spectra from dual-comb fluorescence signals. The fluorescence detection scheme circumvents the sensitivity limit which is effectively determined by the dynamic range of photodetectors in absorption-based dual-comb spectroscopies. Our system realized high-sensitive, Doppler-free high-resolution and broadband atomic spectroscopy.

A part of observed spectra of $5S_{1/2}$ - $5D_{5/2}$ transition is shown in the figure. The hyperfine structures of the $F" = 1$ - $F' = 3,2,1$ transitions are fully-resolved and the spectral widths are approximately 5 MHz. The absolute frequency axis is precisely calibrated from comb mode frequencies which were stabilized to a GPS-disciplined clock.

This work was supported by JST through the ERATO MINOSHIMA Intelligent Optical Synthesizer Project and Grant-in-Aid for JSPS Fellows (16J02345).

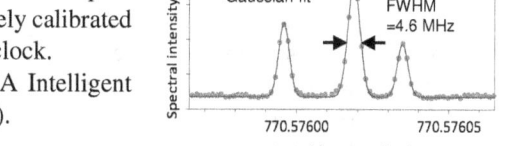

[a] A. Nishiyama, S. Yoshida, Y. Nakajima, H. Sasada, K. Nakagawa, A. Onae, K. and Minoshima, Opt. Express 24, 25894 (2016).
[b] A. Hipke, S. A. Meek, T. Ideguchi, T.W. Hänsch, and N. Picqué, Phys. Rev. A 90, 011805(R) (2014).

RI03   2:19–2:34

SPIN POLARIZATION SPECTROSCOPY OF ALKALI-NOBLE GAS INTERATOMIC POTENTIALS

ANDREY E. MIRONOV, WILLIAM GOLDSHLAG, J. GARY EDEN, *Department of Electrical and Computer Engineering, University of Illinois at Urbana-Champaign, Urbana, IL, USA.*

We report a new laser spectroscopic technique capable of detecting weak state-state interactions in diatomic molecules. Specifically, a weak interaction has been observed between the $6p\sigma$ antibonding orbital of the CsXe (B $^2\Sigma_{\frac{1}{2}}^+$) state and a $5d\sigma$ MO associated with a $5d\Lambda$ ($\Lambda = 0, 1$) state. Thermal Cs-rare gas collision pairs are photoexcited by a circularly-polarized optical field having a wavelength within the B $^2\Sigma_{\frac{1}{2}}^+ \longleftarrow$ X $^2\Sigma_{\frac{1}{2}}^+$ (free$\longleftarrow$free) continuum. Subsequent dissociation of the B $^2\Sigma_{\frac{1}{2}}^+$ transient diatomic selectively populates the $F = 4, 5$ hyperfine levels of the Cs 6p $^2P_{\frac{3}{2}}$ state, and circularly-polarized (σ^+) amplified spontaneous emission (ASE) is generated on the Cs D_2 line. The dependence of Cs 6p spin polarization on the Cs(6p)-Xe internuclear separation (R), clearly shows an interaction between the CsXe(B $^2\Sigma_{\frac{1}{2}}^+$) state and a $5d\Lambda$ ($\Lambda = 0, 1$) potential of the diatomic molecule.

RI04   2:36–2:51

MOLECULAR STRUCTURE AND DYNAMICS PROBED BY PHOTOIONIZATION OUT OF RYDBERG STATES

FEDOR RUDAKOV, *Department of Chemistry, University of Missouri - Kansas City, Kansas City, MO, USA.*

Probing the structure of a molecule as a chemical reaction unfolds has been a long standing goal in chemical physics. Most spectroscopic and diffraction techniques work well when the molecules are cold and thus vibrational motion is minimized. Yet, the very ability of a molecule to undergo structural changes implies that a significant amount of energy resides within the molecule. In order to probe structures of even medium sized molecules on an ultrafast time scale a technique that is sensitive to the molecular structure, yet insensitive to the vibrational motion is required.

In our research we demonstrated that Rydberg electrons are remarkably sensitive to the molecular structure. Photoionization of a molecule out of Rydberg states reveals a purely electronic spectrum which is largely insensitive to vibrational motion. The talk illustrates how Rydberg electrons can serve as a probe for ultrafast structural dynamics in polyatomic molecules. The talk also demonstrates that photoionization through Rydberg states can be utilized for non-intrusive detection of polyatomic combustion intermediates in flames.

RI05   2:53–3:08

LASER-MILLIMETER-WAVE TWO-PHOTON RABI OSCILLATIONS EN ROUTE TO COHERENT POPULATION TRANSFER

DAVID GRIMES, TIMOTHY J BARNUM, *Department of Chemistry, MIT, Cambridge, MA, USA*; YAN ZHOU, *JILA, National Institute of Standards and Technology and Univ. of Colorado Department of Physics, University of Colorado, Boulder, Boulder, CO, USA*; TONY COLOMBO, *Physical Chemistry, Sandia National Laboratories, Albuquerque, NM, USA*; ROBERT W FIELD, *Department of Chemistry, MIT, Cambridge, MA, USA.*

Core-nonpenetrating Rydberg states of molecules are a relatively untapped resource in molecular physics. Due to the $\ell(\ell+1)/r^2$ centrifugal barrier, the Rydberg electron in high-ℓ states is essentially decoupled from the ion-core. This decoupling leads to the system becoming atom-like, with long lifetimes, an "almost good" ℓ quantum number, and "pure-electronic" transitions that follow ΔJ^+=0 and Δv^+=0 propensity rules. Access to these nonpenetrating states is generally blocked by the necessity that the multistep excitation scheme traverses a "zone of death" in which nonradiative decay mechanisms are prohibitively fast.

Coherent population transfer methods, such as STImulated Raman Adiabatic Passage (STIRAP), allow population of core-nonpenetrating states without even transiently populating states in the "zone of death." We demonstrate coherent two-photon population transfer to Rydberg states of barium atoms using a pulsed dye laser and a chirped-pulse millimeter-wave spectrometer. Numerical calculations, using a density matrix formalism, reproduce our experimental results and provide insights into the fractional population transferred, optimal experimental conditions, and possibilities for future improvements, in particular extension to full STIRAP.

Intermission

RI06 3:27 – 3:42

HIGH HARMONIC GENERATION XUV SPECTROSCOPY FOR STUDYING ULTRAFAST PHOTOPHYSICS OF CO-ORDINATION COMPLEXES

ELIZABETH S RYLAND, MING-FU LIN, KRISTIN BENKE, MAX A VERKAMP, KAILI ZHANG, JOSH VURA-WEIS, *Department of Chemistry, University of Illinois at Urbana-Champaign, Urbana, IL, USA.*

Extreme ultraviolet (XUV) spectroscopy is an inner shell technique that probes the $M_{2,3}$-edge excitation of atoms. Absorption of the XUV photon causes a $3p \rightarrow 3d$ transition, the energy and shape of which is directly related to the element and ligand environment. This technique is thus element-, oxidation state-, spin state-, and ligand field specific. A process called high-harmonic generation (HHG) enables the production of ultrashort (~20fs) pulses of collimated XUV photons in a table-top instrument. This allows transient XUV spectroscopy to be conducted as an in-lab experiment, where it was previously only possible at accelerator-based light sources. Additionally, ultrashort pulses provide the capability for unprecedented time resolution (~50fs IRF). This technique has the capacity to serve a pivotal role in the study of electron and energy transfer processes in materials and chemical biology. I will present the XUV transient absorption instrument we have built, along with ultrafast transient $M_{2,3}$-edge absorption data of a series of small inorganic molecules in order to demonstrate the high specificity and time resolution of this tabletop technique as well as how our group is applying it to the study of ultrafast electronic dynamics of coordination complexes.

RI07 3:44 – 3:59

EXTENDING TABLETOP XUV SPECTROSCOPY TO THE LIQUID PHASE TO EXAMINE TRANSITION METAL CATALYSTS

KRISTIN BENKE, ELIZABETH S RYLAND, JOSH VURA-WEIS, *Department of Chemistry, University of Illinois at Urbana-Champaign, Urbana, IL, USA.*

M-edge spectroscopy of first row transition metals (3p to 3d excitation) is the low energy analogue of more well-known K- and L-edge spectroscopy, but can be implemented without the use of a synchrotron. Instead, M-edge spectroscopy can be performed as a tabletop method, relying on high harmonic generation (HHG) to produce ultrashort (~ 20 fs) pulses of extreme ultraviolet (XUV) light in the range of 10-100s of eV. We have shown tabletop M-edge spectroscopy to be a valuable tool in determining the electronic structure of metal-centered coordination complexes and have demonstrated its capacity to yield element-specific information about a compound's oxidation state, spin state, and ligand field. The power of this technique to distinguish these features makes it a promising addition to the arsenal of methods used to study metal-centered catalysts. A catalytic reaction can be initiated photochemically and the XUV probe can be used to track oxidative and structural changes to identify the key intermediates. Until recently tabletop XUV spectroscopy has been performed on thin film samples, but in order to examine homogeneous catalysis, the technique must be adapted to look at samples in the liquid phase. The challenges of adapting tabletop XUV spectroscopy to the liquid phase lie in the lower attenuation length of XUV light compared to soft and hard x-rays and the lower flux compared to synchrotron methods. As a result, the sample must be limited to a sub-micron thickness as well as isolated from the vacuum environment required for x-ray spectroscopy. I am developing a liquid flow cell that relies on confining the sample between two x-ray transmissive SiN membranes, as has been demonstrated for use at synchrotrons, but adapted to the unique difficulties encountered in tabletop XUV spectroscopy.

RI08 4:01 – 4:16

ULTRAFAST EXTREME ULTRAVIOLET SPECTROSCOPY OF METHYLAMMONIUM LEAD IODIDE PEROVSKITE FOR CARRIER SPECIFIC PHOTOPHYSICS

MAX A VERKAMP, MING-FU LIN, ELIZABETH S RYLAND, KRISTIN BENKE, JOSH VURA-WEIS, *Department of Chemistry, University of Illinois at Urbana-Champaign, Urbana, IL, USA.*

Methyl ammonium lead iodide (perovskite) is a leading candidate for next-generation solar cell devices. However, the fundamental photophysics responsible for its strong photovoltaic qualities are not fully understood. Ultrafast extreme ultraviolet (XUV) spectroscopy was used to investigate relaxation dynamics in perovskite with carrier specific signals arising from transitions from the common inner-shell level (I 4d) to the valence and conduction bands. Ultrashort (30 fs) pulses of XUV radiation in a broad spectrum (40-70 eV) were obtained using high-harmonic generation in a tabletop instrument. Transient absorption measurements with visible pump and XUV probe directly observed the dynamics of charge carriers after above-band and band-edge excitation.

RI09 4:18–4:33

LIQUID PHASE SUPERCONTINUUM FIBER-LOOP CAVITY ENHANCED ABSORPTION SPECTROSCOPY FOR H_2O IN ORGANICS

<u>MINGYUN LI</u>, KEVIN LEHMANN, *Departments of Chemistry and Physics, University of Virginia, Charlottesville, VA, USA.*

Last year we presented a way of liquid phase sensing for H_2O and D_2O samples using a side-polished-fiber (SPF) sensor. It is a setup to combine the advantages of Supercontinuum light source with fiber-loop sensing method to make liquid phase CEAS sensing easier and more reliable. After some calculation we found out that with a SPF sensor we could only make use of less than 0.2% of the light from Supercontinuum source, so we decided to make changes on sensors in order to make more light usable. Instead of a SPF or similar evanescent wave sensors, if the light can be guided through a sample directly in free space, we can get almost 100% of the light to be used. So we replaced our sensor by using a mirror and two fibers placed vertical to it side-by-side. The mirror reflects light from one fiber to the other. The free space coupling can make the most of our Supercontinuum source, and a much stronger signal is observed so far. We are now able to use our setup to monitor very low H_2O concentrations such as saturated H_2O solution in organics like CCl_4. Hopefully we can make our system more reliable in the future to make it use in more samples and lower concentrations.

RI10 4:35–4:50

MULTIPLEXED SATURATION SPECTROSCOPY WITH ELECTRO-OPTIC FREQUENCY COMBS

<u>DAVID A. LONG</u>, ADAM J. FLEISHER, *Chemical Sciences Division, National Institute of Standards and Technology, Gaithersburg, MD, USA;* DAVID F. PLUSQUELLIC, *Physical Measurement Laboratory, National Institute of Standards and Technology, Boulder, CO, USA;* JOSEPH T. HODGES, *Chemical Sciences Division, National Institute of Standards and Technology, Gaithersburg, MD, USA.*

Electro-optic frequency combs recently have been applied to a wide range of physical and spectroscopic measurements because of attributes including, simplicity, robustness, flexibility, phase coherence, and high spectral power density. As an illustrative example, I will focus upon multiplexed saturation spectroscopy of atomic potassium (^{39}K) using ultra-high resolution frequency combs which contain up to a million individual teeth with spacings between 2 kHz and 2 MHz. Through the use of a self-heterodyne detection method, we have been able to simultaneously observe phenomena such as hole burning, hyperfine pumping, and electromagnetically induced transparency. I will discuss these measurements as well as future applications in molecular and atomic spectroscopy.

RI11 4:52–5:07

DIRECT ABSORPTION SPECTROSCOPY WITH ELECTRO-OPTIC FREQUENCY COMBS

<u>ADAM J. FLEISHER</u>, DAVID A. LONG, *Chemical Sciences Division, National Institute of Standards and Technology, Gaithersburg, MD, USA;* DAVID F. PLUSQUELLIC, *Physical Measurement Laboratory, National Institute of Standards and Technology, Boulder, CO, USA;* JOSEPH T. HODGES, *Chemical Sciences Division, National Institute of Standards and Technology, Gaithersburg, MD, USA.*

The application of electro-optic frequency combs to direct absorption spectroscopy[a] has increased research interest in high-agility, modulator-based comb generation. This talk will review common architectures for electro-optic frequency comb generators as well as describe common self-heterodyne and multi-heterodyne (i.e., dual-comb) detection approaches. In order to achieve a sufficient signal-to-noise ratio on the recorded interferogram while allowing for manageable data volumes, broadband electro-optic frequency combs require deep coherent averaging,[b] preferably in real-time. Applications such as cavity-enhanced spectroscopy, precision atomic and molecular spectroscopy, as well as time-resolved spectroscopy will be introduced.

---

[a] D.A. Long et al., *Opt. Lett.* **39**, 2688 (2014)
[b] A.J. Fleisher et al., *Opt. Express* **24**, 10424 (2016)

RJ. Radicals
Thursday, June 22, 2017 – 1:45 PM
Room: 161 Noyes Laboratory

Chair: Neil J Reilly, University of Massachusetts Boston, Boston, MA, USA

RJ01   1:45 – 2:00

INFRARED SPECTRUM OF THE CYCLOBUTYL RADICAL IN He DROPLETS

ALAINA R. BROWN, PETER R. FRANKE, GARY E. DOUBERLY, *Department of Chemistry, University of Georgia, Athens, GA, USA.*

Gas phase cyclobutyl radical (C_4H_7) is produced via pyrolysis of cyclobutyl methyl nitrite ($C_4H_7(CH_2)ONO$). The nascent radicals are promptly solvated in liquid He droplets, allowing for the acquisition of the infrared spectrum in the CH stretching region. Anharmonic frequencies are predicted by VPT2+K simulations based upon a CCSD(T)/ANO0 force field. Several resonance polyads emerge in the 2800-3000 cm^{-1} region as a result of anharmonic coupling between the CH stretching fundamentals and CH_2 bend overtones and combinations. Evidence of rotational fine structure is observed for two bands. The vibrationally averaged cyclobutyl radical geometry and the C_4H_7 potential energy surface will be discussed. In agreement with the findings by Schultz[1] and coworkers, 1,3-butadiene is formed from cyclobutyl ring opening and H atom loss, given a sufficiently high pyrolysis temperature. However, signatures of 1-methylallyl and allylcarbinyl radicals, proposed[1] as intermediates along the above mentioned reaction path, are yet to be seen.

1. Schultz, J.C., Houle, F.A., Beauchamp, J.L. J. Am. Chem. Soc. 1984, 106, 7336-7347.

RJ02   2:02 – 2:17

$O(^3P)$ DOPED HELIUM DROPLETS

JOSEPH T. BRICE, GARY E. DOUBERLY, *Department of Chemistry, University of Georgia, Athens, GA, USA.*

Atomic oxygen (^3P) is generated via thermolysis in a commerical thermal gas cracker (Mantis Ltd. MGC-75). Complexes with HCN were investigated to qualitatively assess the doping efficiency of $O(^3P)$ into a helium droplet. Theoretical calculations of a linear $O \cdots HCN$ ($^3\Sigma$) complex at the CCSD(T)/aug-cc-pVTZ level are consistent with the rotational constants extracted from the rotational substructure in the experimental spectra, and with dipole moments approximated from Stark spectra. The thermal source will be used to study reactions between $O(^3P)$ and hydrocarbons in helium droplets, and preliminary data on this topic will be presented.

RJ03   2:19 – 2:34

THE O_2 + ETHYL REACTION IN HELIUM NANODROPLETS: INFRARED SPECTROSCOPY OF THE ETHYLPEROXY RADICAL

PETER R. FRANKE, GARY E. DOUBERLY, *Department of Chemistry, University of Georgia, Athens, GA, USA.*

Helium-solvated ethylperoxy radicals ($CH_3CH_2OO\bullet$) are formed via the *in situ* reaction between ethyl radicals and $\tilde{X}^3\Sigma_g^-$ O_2. The reactants are captured sequentially through the droplet pick-up technique. Helium droplets are doped with ethyl radical via pyrolysis of di-*tert*-amyl peroxide or *n*-propylnitrite in an effusive, low-pressure source. A mid-infrared spectrum of ethylperoxy is recorded with species-selective droplet beam depletion spectroscopy. Spectral assignments in the CH stretching region are made via comparisons to second-order vibrational perturbation theory with resonances (VPT2+K) based on coupled-cluster quartic force fields. *Gauche* and *trans* conformers are predicted to be nearly isoenergetic; however, the spectrum indicates that one dominant conformer is present. Indeed, in several previous studies in our group, where chemical reactions were conducted inside droplets, only a single conformer of the product was observed. Exploration of the ethylperoxy potential energy surface, particularly along the CCOO torsional and CO stretching coordinates, motivates an explanation that is based upon an adiabatic funneling mechanism that leads to the exclusive production of one conformer. The slower torsional degree of freedom is cooled more rapidly than the higher frequency stretching and bending coordinates owing to the stronger coupling between the torsional modes and the collective modes of the helium droplet. The reactants are cooled into the torsional well that stabilizes first during their approach on the PES.

RJ04 2:36–2:51

INFRARED SPECTRA OF THE 1-CHLOROMETHYL-1-METHYLALLYL AND 1-CHLOROMETHYL-2-METHYLALLYL RADICALS ISOLATED IN SOLID *PARA*-HYDROGEN

JAY C. AMICANGELO, *School of Science (Chemistry), Penn State Erie, Erie, PA, USA*; YUAN-PERN LEE, *Applied Chemistry, National Chiao Tung University, Hsinchu, Taiwan, Institute of Atomic and Molecular Sciences, Academia Sinica, Taipei, Taiwan.*

The reaction of chlorine atoms (Cl) with isoprene (C_5H_8) in solid *para*-hydrogen (p-H_2) matrices at 3.2 K has been studied using infrared spectroscopy. Mixtures of C_5H_8 and Cl_2 were co-deposited in p-H_2 at 3.2 K, followed by irradiation at 365 nm to cause the photodissociation of Cl_2 and the subsequent reaction of Cl atoms with C_5H_8. Upon 365 nm photolysis, a series of new lines appeared in the infrared spectrum, with the strongest appearing at 807.8 and 796.7 cm^{-1}. To determine the grouping of lines to distinct chemical species, secondary photolysis was performed using a low-pressure Hg lamp in combination with various filters. Based on the secondary photolysis behavior, it was determined that the majority of the new lines belong to two distinct chemical species, designated as set A (3047.2, 1482.2, 1459.5, 1396.6, 1349.6, 1268.2, 1237.9, 1170.3, 1108.8, 807.8, 754.1, 605.6, 526.9, 472.7 cm^{-1}) and set B (3112.7, 1487.6, 1382.6, 1257.7, 1229.1, 1034.8, 975.8, 942.4, 796.7, 667.9, 569.7 cm^{-1}). The most likely reactions to occur between Cl and C_5H_8 under the low temperature conditions in solid p-H_2 are the addition of the Cl atom to the four distinct alkene carbon atoms to produce the corresponding chlorine atom addition radicals (ClC_5H_8). Quantum-chemical calculations were performed at the B3PW91/6-311++G(2d,2p) level of theory for the four possible ClC_5H_8 radicals in order to determine the relative energetics and the predicted harmonic vibrational spectra for each radical. The calculations predict that the addition of Cl to each of the four carbons is exothermic, with relative energies of 0.0, 74.5, 67.4, and 7.9 kJ/mol for the addition to carbons 1 – 4, respectively. When the lines of set A and B are compared to the scaled harmonic vibrational spectra for all four of the possible Cl addition radicals, it is found that the best agreement for set A is with the radical produced by the addition to carbon 4 (1-chloromethyl-2-methylallyl radical) and the best agreement for set B is with the radical produced by addition to carbon 1 (1-chloromethyl-1-methylallyl radical). Therefore, the lines of set A and B are assigned to these radicals, respectively.

RJ05 2:53–3:08

THERMAL DECOMPOSITION OF METHYL ACETATE (CH_3COOCH_3) IN A FLASH-PYROLYSIS MICRO-REACTOR

JESSICA P PORTERFIELD, *Department of Chemistry and Biochemistry, University of Colorado, Boulder, CO, USA*; DAVID H. BROSS, *Chemical Sciences and Engineering Division, Argonne National Laboratory, Argonne, IL, USA*; BRANKO RUSCIC, *Computation Institute, The University of Chicago, Chicago, IL, USA*; JAMES H. THORPE, THANH LAM NGUYEN, *Department of Chemistry, The University of Texas, Austin, TX, USA*; JOSHUA H BARABAN, *Department of Chemistry and Biochemistry, University of Colorado, Boulder, CO, USA*; JOHN F. STANTON, *Department of Chemistry, The University of Texas, Austin, TX, USA*; JOHN W DAILY, *Department of Mechanical Engineering, University of Colorado Boulder, Boulder, CO, USA*; BARNEY ELLISON, *Department of Chemistry and Biochemistry, University of Colorado, Boulder, CO, USA.*

The thermal decomposition of methyl acetate (CH_3COOCH_3) has been studied in a set of flash pyrolysis micro-reactors. Samples were diluted to (0.06 – 0.13%) in carrier gases (He, Ar) and subjected to temperatures of 300 - 1600 K at roughly 20 Torr. After residence times of approximately 25 – 150 μseconds, the unimolecular pyrolysis products were detected by vacuum ultraviolet photoionization mass spectrometry at 10.487 eV (118.2 nm). Complementary product identification was provided by matrix isolation infrared spectroscopy. Decomposition began at 1000 K with the observation of (CH_2=C=O, CH_3OH), products of a four centered rearrangement with a $\Delta_{rxn}H_{298} = 39.1 \pm 0.2$ kcal mol^{-1}. As the micro-reactor was heated to 1300 K, a mixture of (CH_2=C=O, CH_3OH, CH_3, CH_2=O, H, CO, CO_2) appeared. A new novel pathway is calculated in which both methyl groups leave behind CO_2 simultaneously, $\Delta_{rxn}H_{298} = 74.5 \pm 0.4$ kcal mol^{-1}. This pathway is in contrast to step-wise loss of methyl radical, which can go in two ways: $\Delta_{rxn}H_{298}$ ($CH_3COOCH_3 \rightarrow CH_3 + COOCH_3$) $= 95.4 \pm 0.4$ kcal mol^{-1}, $\Delta_{rxn}H_{298}$ ($CH_3COOCH_3 \rightarrow CH_3COO + CH_3$) $= 88.0 \pm 0.3$ kcal mol^{-1}.

RJ06   **3:10 – 3:25**

BROADBAND MICROWAVE STUDY OF REACTION INTERMEDIATES AND PRODUCTS THROUGH THE PYROLYSIS OF OXYGENATED BIOFUELS

__CHAMARA ABEYSEKERA__, ALICIA O. HERNANDEZ-CASTILLO, SEAN FRITZ, TIMOTHY S. ZWIER, *Department of Chemistry, Purdue University, West Lafayette, IN, USA.*

The rapidly growing list of potential plant-derived biofuels creates a challenge for the scientific community to provide a molecular-scale understanding of their combustion. Development of accurate combustion models rests on a foundation of experimental data on the kinetics and product branching ratios of their individual reaction steps. Therefore, new spectroscopic tools are necessary to selectively detect and characterize fuel components and reactive intermediates generated by pyrolysis and combustion. Substituted furans, including furanic ethers, are considered second-generation biofuel candidates. Following the work of the Ellison group, an 8-18 GHz microwave study was carried out on the unimolecular and bimolecular decomposition of the smallest furanic ether, 2-methoxy furan, and it's pyrolysis intermediate, the 2-furanyloxy radical, formed in a high-temperature pyrolysis source coupled to a supersonic expansion. Details of the experimental setup and analysis of the spectrum of the radical will be discussed.

Intermission

RJ07   **3:44 – 3:59**

HIGH-RESOLUTION THz MEASUREMENTS OF BrO GENERATED IN AN INDUCTIVELY COUPLED PLASMA

__DEACON J NEMCHICK__, BRIAN DROUIN, *Jet Propulsion Laboratory, California Institute of Technology, Pasadena, CA, USA.*

Building upon the foundation provided by previous work, the $X_1\,^2\Pi_{3/2}$ and $X_2\,^2\Pi_{1/2}$ states of the transient radical, BrO, were interrogated in previously unprobed spectral regions (0.5 to 1.7 THz) by employing JPL developed high-resolution cascaded frequency multiplier sources. Like other members of the halogen monoxides (XO), this species has been the target of several recent atmospheric remote sensing studies and is a known participant in a catalytic ozone degradation cycle. For the current work, BrO is generated in an inductively coupled plasma under dynamic flow conditions and rotational lines are observed directly at their Doppler-limited resolution. New spectral transitions including those owing to both the ground ($\nu=0$) and excited ($\nu=1$ and 2) vibrational states of isotopologues composed of permutations of natural abundance ^{16}O, ^{18}O, ^{79}Br, and ^{81}Br are fit to a global Hamiltonian containing both fine and hyperfine terms. In addition to further refining existing spectroscopic parameters, new observations will be made available to remote detection communities through addition to the JPL catalog. New findings will be discussed along with future plans to extend these studies to other halogen monoxides (X=Cl and I) and the more massive halogen dioxides (OXO & XOO).

RJ08   **4:01 – 4:16**

DETECTION AND CHARACTERIZATION OF THE STANNYLENE (SnH$_2$) RADICAL IN THE GAS PHASE

__TONY SMITH__, *Ideal Vacuum Products LLC, Ideal Vacuum Products LLC, Albuquerque, NM, USA*; DENNIS CLOUTHIER, *Department of Chemistry, University of Kentucky, Lexington, KY, USA.*

The electronic spectrum of the jet-cooled SnH$_2$ radical has been detected by LIF spectroscopy. The radical was produced in a pulsed electric discharge through a precursor mixture of SnH$_4$ in argon. Each band in the LIF spectrum consists of a small number of rovibronic transitions to the lowest energy (K$_a$ = 0, J = 0,1,2,3) rotational levels in the excited state. High resolution spectra of the $^pP_1(1)$ line of the 2^2_0 band show 7 components whose relative intensities are characteristic of the tin major isotopic abundances. The emission spectra are also consistent with assigning the spectrum as due to SnH$_2$. The fluorescence lifetimes of the upper state rotational levels decrease with increasing J', indicative of a rotationally dependent predissociation process in the excited state, similar to that previously observed in SiH$_2$ and GeH$_2$. Fluorescence hole burning experiments have located the upper state K$_a$ = 2 levels which allow a determination of the molecular structure.

RJ09 4:18 – 4:33

FOURIER TRANSFORM ABSORPTION SPECTROSCOPY OF C_3 IN THE ν_3 ANTISYMMETRIC STRETCH MODE REGION

MICHEL VERVLOET, *AILES Beamline, Synchrotron SOLEIL, Saint-Aubin, France*; MARIE-ALINE MARTIN-DRUMEL, *CNRS, Institut des Sciences Moleculaires d'Orsay, Orsay, France*; DENNIS W. TOKARYK, *Department of Physics, University of New Brunswick, Fredericton, NB, Canada*; OLIVIER PIRALI[a], *Institut des Sciences Moléculaires d'Orsay, Université Paris-Sud, Orsay, France.*

The C_3 molecule has been detected in a variety of astrophysical objects thanks to the well-known 4050 Å ($A^1\Pi_u$–$X^1\Sigma_g^+$) electronic transition as well as the two IR active modes of the electronic ground state: ν_2 (~ 63.42 cm^{-1}) and ν_3 (~ 2040.02 cm^{-1})[b]. Previous laboratory data in the ν_3 region, obtained using diode laser spectroscopy and the photolysis of allene to produce C_3, permitted measurement of the fundamental $(0,0,1)\Sigma$–$(0,0,0)\Sigma$ as well as the hot bands: $(0,1,1)\Pi$–$(0,1,0)\Pi$; $(0,2,1)\Sigma$–$(0,2,0)\Sigma$; $(0,2,1)\Delta$–$(0,2,0)\Delta$ and provided insights on the anharmonicity of the $(0,n\nu_2,1)$ vibrational pattern[c].

We have recorded the absorption spectrum of C_3 in the 1800–2100 cm^{-1} region (at a resolution of 0.003 cm^{-1}) using the Bruker IFS 125 Fourier Transform spectrometer at the AILES beamline of Synchrotron SOLEIL. C_3 was produced in a DC discharge of methane heavily diluted in helium. The rovibrational temperature of C_3 produced in our discharge is noticeably higher than in Ref. [4], which allowed us to extend measurements to higher J values. More interestingly, we assigned new hot bands involving higher quanta of the ν_2 bending states: $(0,n\nu_2,1)$ with n ranging from 0 to 5. Despite the absence of Q branches for these bands, which results in a possible ambiguous J-assignment of P and R lines, the large variety of data considered in this work, in addition to our experimental data and including observations of comet spectra, allows confident assignments.

[a]also at: AILES beamline, Synchrotron SOLEIL, 91192 Gif sur Yvette, France
[b]L. Gausset, G. Herzberg, A. Lagerqvist, B. Rosen, Astrophysical Journal, 45–81 (1965); T. F. Giesen et al., The Astrophysical Journal, 551, L181–L184 (2001); K. W. Hinkle, J. J. Keady, P. F. Bernath, Science, 241, 1319–1322 (1988)
[c]K. Kawaguchi et al., J. Chem. Phys., 91, 1953–1957 (1989)

RJ10 4:35 – 4:50

IDENTIFICATION OF A JAHN-TELLER ACTIVE GAS PHASE SILOXY FREE RADICAL (Cl_3SiO) BY LIF SPECTROSCOPY

TONY SMITH, *Ideal Vacuum Products LLC, Ideal Vacuum Products LLC, Albuquerque, NM, USA*; DENNIS CLOUTHIER, *Department of Chemistry, University of Kentucky, Lexington, KY, USA.*

A very strong LIF spectrum was observed in the 655 - 600 nm region from the products of an electric discharge through a dilute mixture of silicon tetrachloride and oxygen in argon. The same spectrum was obtained from a Cl_3Si-O-$SiCl_3$ precursor in argon. The LIF bands do not have resolved rotational structure, suggesting that the carrier of the spectrum is a heavy molecule. Emission spectra show substantial differences depending on which upper state vibronic level is probed, and these differences are readily understood if we assume that the spectrum is due to the Cl_3SiO free radical with a significant Jahn-Teller effect in the ground ^2E state. This conclusion is reinforced by our own ab intio calculations of the ground and excited state vibronic energy levels, band contours, chlorine isotope effects, and electronic excitation energies. Cl_3SiO is the first siloxy radical to be detected in the gas phase.

RJ11 4:52 – 5:07

PHOTOELECTRON IMAGING SPECTROSCOPY AS A WINDOW TO UNEXPECTED MOLECULES

CHRISTOPHER C BLACKSTONE, *Chemistry and Biochemistry, University of Arizona, Tucson, AZ, USA.*

Targeting an anion with the formula CH_3O_3 for exploration with photoelectron imaging spectroscopy, we determine its identity to be dihydroxymethanolate, an anion largely absent in the literature, and the conjugate base of the hypothetical species orthoformic acid. Comparing the observed photoelectron spectrum to CCSD-EOM-IP and CCSD-EOM-SF calculations completed in QChem and Franck-Condon overlap simulations in PESCAL, we are able to determine with confidence the connectivity of the atoms in this molecule.

LASER SPECTROSCOPY OF THE JET-COOLED SiCF FREE RADICAL

TONY SMITH, *Ideal Vacuum Products LLC, Ideal Vacuum Products LLC, Albuquerque, NM, USA*; DENNIS CLOUTHIER, *Department of Chemistry, University of Kentucky, Lexington, KY, USA*.

The SiCF free radical has been detected through the $A^2\Sigma^+$-$X^2\Pi$ band system in the 605 - 550 nm region. The radical was produced in an electric discharge through a dilute mixture of CF_3SiH_3 in high pressure argon and studied by laser induced fluorescence. The vibronic levels of the ground and excited states have been measured through LIF and emission spectroscopy and a Renner-Teller analysis has been undertaken for the ground $^2\Pi$ levels. The observed vibrational frequencies, partially resolved rotational band contours, Renner-Teller parameter, and electronic excitation energy are in accord with our predictions from high level ab initio (CCSD(T)/aug-cc-pVTZ) calculations. Theory shows that the radical has a silicon-carbon double bond in the ground state and a much shorter triple bond in the excited state. This is the third in the series of SiCX (X = H, Cl, and F) free radicals we have produced and studied in the gas phase.

FA. Planetary atmospheres
Friday, June 23, 2017 – 8:30 AM
Room: 274 Medical Sciences Building

Chair: James Neil Hodges, Old Dominion University, Norfolk, Virginia, United States

---

FA01   8:30–8:45

SPECTRAL LINE SHAPES IN THE ν_3 Q BRANCH OF $^{12}CH_4$ NEAR 3.3 μm

V. MALATHY DEVI, D. CHRIS BENNER, *Department of Physics, College of William and Mary, Williamsburg, VA, USA*; ROBERT R. GAMACHE, *Department of Environmental, Earth, and Atmospheric Sciences, University of Massachusetts Lowell, Lowell, MA, USA*; MARY ANN H. SMITH, *Science Directorate, NASA Langley Research Center, Hampton, VA, USA*; ROBERT L. SAMS, *Chemical Physics, Pacific Northwest National Laboratory, Richland, WA, USA.*

Detailed knowledge of spectroscopic parameters for prominent Q branches of methane is necessary for interpretation and modeling of high resolution infrared spectra of terrestrial and planetary atmospheres. We have measured air-broadened line shape parameters in the Q branch of $^{12}CH_4$ in the ν_3 fundamental band for a large number of transitions in the 3000 to 3023 cm^{-1} region by analyzing 13 room-temperature laboratory absorption spectra. Twelve of these spectra were recorded with 0.01 cm^{-1} resolution using the McMath-Pierce Fourier transform spectrometer (FTS) of the National Solar Observatory (NSO) on Kitt Peak, and one higher-resolution ($\sim$0.0011 cm^{-1}) low pressure ($\sim$1 Torr) spectrum of methane was obtained using the Bruker IFS 120HR FTS at the Pacific Northwest National Laboratory (PNNL) in Richland, WA. The air-broadened spectra were recorded using various absorption cells with path lengths of 5, 20, 25, and 150 cm, total sample pressures between 50 and 500 Torr, and CH_4 volume mixing ratios of 0.01 or less. All 13 spectra were fit simultaneously covering the 3000-3023 cm^{-1} spectral region using a multispectrum nonlinear least squares technique[a] to retrieve accurate line positions, absolute intensities, Lorentz air-broadened widths and pressure-shift coefficients. Line mixing using the off-diagonal relaxation matrix element formalism[b] was measured for a number of pairs of transitions for the CH_4-air collisional system. The results will be compared to values reported in the literature.

[a]D. C. Benner, C. P. Rinsland, V. Malathy Devi, M. A. H. Smith, D. Atkins, *JQSRT* **53** (1995) 705-721.

[b]A. Levy, N. Lacome, C. Chackerian, Collisional line mixing, in *Spectroscopy of the Earth's Atmosphere and Interstellar Medium*, Academic Press, Inc., Boston (1992) 261-337.

FA02   8:47–9:02

LINE POSITIONS OF CENTRIFUGAL DISTORSION INDUCED ROTATIONAL TRANSITIONS OF METHANE MEASURED UP TO 2.6 THZ AT SUB-MHZ ACCURACY WITH A CW-THZ PHOTOMIXING SPECTROMETER

CÉDRIC BRAY, ARNAUD CUISSET, FRANCIS HINDLE, GAËL MOURET, ROBIN BOCQUET, *Laboratoire de Physico-Chimie de l'Atmosphère, Université du Littoral Côte d'Opale, Dunkerque, France*; VINCENT BOUDON, *Laboratoire ICB, CNRS/Université de Bourgogne, DIJON, France.*

Several Doppler limited rotational transitions of methane induced by centrifugal distortion have been measured with an unprecedented frequency accuracy using the THz photomixing synthesizer based on a frequency comb. Compared to previous synchrotron based FT-Far-IR measurements of Boudon et al.[a], the accuracy of the line frequency measurements is improved by one order of magnitude, this yields a corresponding increase of two orders of magnitude to the weighting of these transitions in the global fit. The rotational transitions in the $\nu_4 \leftarrow \nu_4$ hot band are measured for the first time by the broad spectral coverage of the photomixing CW-THz spectrometer providing access up to $R(5)$ transitions at 2.6 THz. The new global fit including the present lines has been used to update the methane line list of the HITRAN database. Some small, but significant variations of the parameter values are observed and are accompanied by a reduction of the 1-σ uncertainties on the rotational (B_0) and centrifugal distortion (D_0) constants.

[a]V. Boudon, O. Pirali, P. Roy, J.-B. Brubach, L. Manceron, J. Vander Auwera, J. Quant. Spectrosc. Radiat. Transfer, **111**, 1117–1129 (2010).

INFRARED ABSORPTION CROSS SECTIONS OF COLD PROPANE IN THE LOW FREQUENCY REGION BETWEEN 600 - 1300 cm^{-1}

<u>ANDY WONG</u>, *Department of Chemistry and Biochemistry, Old Dominion University, Norfolk, VA, USA*; ROBERT J. HARGREAVES, *Atmospheric, Oceanic & Planetary Physics, Oxford University, Oxford, United Kingdom*; BRANT E. BILLINGHURST, *EFD, Canadian Light Source Inc., Saskatoon, Saskatchewan, Canada*; PETER F. BERNATH, *Department of Chemistry and Biochemistry, Old Dominion University, Norfolk, VA, USA*.

Propane is one of several hydrocarbons present in the atmospheres of the Giant Planets, Jupiter and Saturn. In order to characterize the atmospheres of the Giant Planets, it is necessary to provide absorption cross sections which can be used to determine abundances. Absorption cross sections have been obtained from high resolution transmission spectra recorded at the Canadian Light Source Far Infrared beamline. The experimental conditions used mimic those of the atmospheres belonging to the Giant Planets using He and H_2 as foreign broadeners.

FIRST HIGH RESOLUTION IR SPECTRA OF 2-^{13}C-PROPANE. THE ν_9 B-TYPE BAND NEAR 366.767 cm^{-1} AND THE ν_{26} C-TYPE BAND NEAR 746.615 cm^{-1}. DETERMINATION OF GROUND AND UPPER STATE CONSTANTS.

<u>S.J. DAUNT</u>, ROBERT GRZYWACZ, *Department of Physics & Astronomy, The University of Tennessee-Knoxville, Knoxville, TN, USA*; WALTER LAFFERTY, *Optical Technology Division, National Institute of Standards and Technology, Gaithersburg, MD, USA*; JEAN-MARIE FLAUD, *CNRS, Universités Paris Est Créteil et Paris Diderot, LISA, Créteil, France*; BRANT E. BILLINGHURST, *EFD, Canadian Light Source Inc., Saskatoon, Saskatchewan, Canada*.

This is the first report in a project to record high resolution IR data of the ^{13}C and D substituted isotopologues of propane.

In this talk we will give details on the first high resolution ($\Delta \nu = 0.0009$ cm^{-1}) IR investigation of 2-^{13}C-propane. Spectra of the CCC skeletal bending mode near 336.767 cm^{-1} (B-type) and the wagging mode near 746.615 cm^{-1} (C-type) were recorded using the FTS on the Far-IR beamline of the Canadian Light Source (CLS). The spectra were assigned both traditionally and with the aid of the PGOPHER program of Colin Western.[a] The only available MW data on this molecule are the six K =0 J lines from Lide.[b] We therefore had to use the present data to determine a new set of ground state constants that included centrifugal distortion terms for this molecule. We compare these experimentally determined values with the recent *ab initio* values of Villa, Senent & Carvajal.[c] Upper state constants for both bands have been found that provide a good simulation of the spectra. The hope is that this data will be useful in identifying isotopic propane lines in Titan and other astrophysical objects.

[a] C. Western, J. Quant. Spectrosc. & Rad. Transf. **186**, 221 ff. (2017).
[b] Lide, J.Chem. Phys. **33**, p.1514ff. (1960).
[c] Villa, Senent & Carvajal, PCCP **15**, 10258 (2013).

FA05

FIRST HIGH RESOLUTION IR SPECTRA OF 1-^{13}C-PROPANE. THE ν_9 B-TYPE BAND NEAR 366.404 cm^{-1} AND THE ν_{26} C-TYPE BAND NEAR 748.470 cm^{-1}. DETERMINATION OF GROUND AND UPPER STATE CONSTANTS.

S.J. DAUNT, ROBERT GRZYWACZ, *Department of Physics & Astronomy, The University of Tennessee-Knoxville, Knoxville, TN, USA*; WALTER LAFFERTY, *Optical Technology Division, National Institute of Standards and Technology, Gaithersburg, MD, USA*; JEAN-MARIE FLAUD, *CNRS, Universités Paris Est Créteil et Paris Diderot, LISA, Créteil, France*; BRANT E. BILLINGHURST, *EFD, Canadian Light Source Inc., Saskatoon, Saskatchewan, Canada.*

We report in this talk on the first high resolution IR spectra ($\Delta\nu$ = 0.0009 cm^{-1}) of the 1-^{13}C-Propane isotopologue. Spectra were taken on the Bruker FTS instrument on the Far-IR beamline at the Canadian National Synchrotron (CLS) located at the University of Saskatchewan. The ν_9 B-type band centered near 366.404 cm^{-1} appears unperturbed and lines were assigned up to K = 17 and J = 50. Since the 1960 MW study of Lide[a] only used 6 J lines of K = 0 we had to use GSCD analyses to determine a fuller set of molecular constants for this molecule. Since normal propane has been detected using the ν_{26} C-type band in Titan and other astrophysical objects our main focus was on the analagous bands for the both the 1-^{13}C and 2-^{13}C isotopologues. Assigned lines up to K = 17, J = 50 in ν_{26} were analyzed with GSCD to independently obtain ground state rotational constants. These were consistent with those obtained from the ν_9 analysis. Upper state constants were also determined that reproduce the vast majority of this band. As in the normal and 2-^{13}C species a Coriolis resonance with the 2ν_9 state causes lines of most K levels above 15 to be shifted.[b] We did not have enough sample available at the time of these experiments to be able to record the 2ν_9 - ν_9 hot band transitions in the low frequency study of ν_9.

[a]Lide, J. Chem. Phys. **33**, p. 1514 ff. (1960)
[b]Flaud, Kwabia Tchana, Lafferty & Nixon, Mol. Phys. **108**, p. 699 ff. (2010)

FA06

FIRST HIGH RESOLUTION IR STUDY OF THE ν_{14} (A') A-TYPE BAND NEAR 421.847 cm^{-1} OF 2-^{13}C-PROPENE

S.J. DAUNT, ROBERT GRZYWACZ, *Department of Physics & Astronomy, The University of Tennessee-Knoxville, Knoxville, TN, USA*; BRANT E. BILLINGHURST, *EFD, Canadian Light Source Inc., Saskatoon, Saskatchewan, Canada.*

This is is the first high resolution IR study of any band of the 2-^{13}C-propene species. There have been only two previous high resolution studies of vibration-rotation bands of the normal species.[a] The band examined here is the ν_{14} (A') CCC skeletal bending near 421.847 cm^{-1} which has an A-Type asymmetric rotor structure. The spectra were recorded on the FTS at the Far-IR beamline of the Canadian Light Source with a resolution of $\Delta\nu$ = 0.0009 cm^{-1}. We have assigned and fitted around 2200 transitions and determined ground and upper state rotational constants. Lines with J up to 49 and K up to 12 were included. The subbands with K greater than 12 were perturbed and show torsional splittings that vary from small to extremely large. The fitting was done with the PGOPHER program of Colin Western.[b] The GS constants are in good agreement with the MW constants reported recently by Craig, Groner and co-workers.[c]

[a]Ainetschian, Fraser, Ortigoso & Pate, J. Chem. Phys. **100**, 729 ff. (1994); Lafferty, Flaud & Herman, J. Mol. Struct. **780-781**, 65 ff. (2006).
[b]Western, J. Quant. Spectrosc. Rad. Transf. **186**, 221 ff. (2017).
[c]Paper M109, 71st ISMS Symposium (2016); J. Mol. Spectrosc. **328**, 1-6 (2016).

Intermission

FA07

HIGH RESOLUTION INFRARED CAVITY ENHANCED ABSORPTION OF PROPYNE

PARASHU R NYAUPANE, MARLON DIAZ, ANN BARTON, CARLOS MANZANARES, *Department of Chemistry and Biochemistry, Baylor University, Waco, TX, USA.*

High resolution cavity enhanced absorption spectroscopy (CEAS) of the first overtone transition of the acetylenic C-H stretch of propyne will be presented. The spectrum has been obtained with a tunable diode laser with resolution 0.0003 cm^{-1} at room temperature. The output of the laser was modulated and coupled to an optical cavity. The experiments are performed at pressures from 70 mTorr to 1 Torr. The optical cavity is attached to a low temperature cryostat. We are planning to study the near-infrared spectroscopy of propyne and other molecules in a static cell at low temperature. The experimental set up can be used to study kinetics and spectroscopy of molecules in the atmosphere of planets and satellites of the outer solar system and molecular complexes in the gas phase at low temperatures using liquid He or liquid N2 as cryogens.

FA08

AB INITIO CHARACTERIZATION OF SULFUR COMPOUNDS AND THEIR CHEMISTRY FOR VENUS AND THE INTERSTELLAR MEDIUM

DAVID E. WOON, *Department of Chemistry, University of Illinois at Urbana-Champaign, Urbana, IL, USA.*

The atmosphere of Venus is known to contain trace amounts of SO, SO_2, OCS, H_2SO_4, and possibly H_2S and elemental sulfur oligomers, S_n. Modeling studies indicate that many more compounds containing sulfur and both sulfur and chlorine may also be present, given that the known compounds are photolyzed by solar radiation in the upper atmosphere of Venus and yield reactive radical species. A large number of exotic compounds containing S, O, H, C, and/or Cl of suspected or plausible significance for Venus chemistry have been characterized at the RCCSD(T)/aug-cc-pVTZ level, yielding structures, dipole moments, and dipole polarizabilities. Representative compounds and associated chemical reactions will be discussed. Both abstraction and addition-elimination reactions have been characterized.

FA09

INFRARED SPECTROSCOPIC AND THEORETICAL STUDY OF THE HC_nO^+ (N=5-12) CATIONS

WEI LI, JIAYE JIN, GUANJUN WANG, MINGFEI ZHOU, *Fudan University, Department of Chemistry, Shanghai, China.*

Carbon chains and derivatives are highly active species, which are widely existed as reactive intermediates in many chemical processes including atmospheric chemistry, hydrocarbon combustion, as well as interstellar chemistry. The carbon chain cations, HC_nO^+ (n = 5-12) are produced via pulsed laser vaporization of a graphite target in supersonic expansions containing carbon monoxide and hydrogen. The infrared spectra are measured via mass-selected infrared photodissociation spectroscopy of the CO "tagged" [$HC_nO\cdot CO$] cation complexes in the 1600-3500 cm^{-1} region. The geometries and electronic ground states of these cation complexes are determined by their infrared spectra in conjunction with theoretical calculations. All the HC_nO^+ (n = 5-12) core cations are characterized to be linear carbon chain derivatives terminated by hydrogen and oxygen. The HC_nO^+ cations with odd n have closed-shell singlet ground states with polyyne-like structures, while those with even n have triplet ground states with allene-like structures.

FA10

IMPACT OF INSERTION REACTION OF $O(^1D)$ INTO THE CARBONIC ACID MOLECULE IN THE ATMOSPHERE OF EARTH AND MARS

SOURAV GHOSHAL[a], MONTU K. HAZRA[b], *Chemical Sciences Division, Saha Institute of Nuclear Physics, Kolkata, West Bengal, India.*

In this talk, we present the energetics and kinetics of the insertion reaction of the $O(^1D)$ into the H_2CO_3 molecule that finally produces the percarbonic acid $[H_2C(O)O_3]$ molecule $(H_2CO_3 + O(^1D) \rightarrow H_2C(O)O_3)$. The rate constants have been calculated by the Variable-Reaction-Coordinate Variational Transition State Theory $(VRC - VTST)$. From our results, we show that the rate constants of the insertion reaction are significantly higher than the rate constants associated with the H_2O-assisted H_2CO_3 decomposition $(H_2CO_3 + H_2O \rightarrow CO_2 + 2H_2O)$, acetic acid (AA)-assisted H_2CO_3 decomposition $(H_2CO_3 + AA \rightarrow CO_2 + H_2O + AA)$ and OH radical-initiated H_2CO_3 degradation reaction $(H_2CO_3 + OH^. \rightarrow HCO_3^. + H_2O)$ —which are currently assumed to be the potentially important reaction channels to interpret the atmospheric loss of the H_2CO_3 molecule in the Earth. Finally, we also discuss the potential impact of the H$_2$O-assisted H_2CO_3 decomposition reaction, OH radical-initiated H_2CO_3 degradation reaction and the above-mentioned insertion reaction on equal footing toward the loss of H_2CO_3 molecule, especially, in the surface of Mars.

[a]Private Communication

[b]Private Communication. (Financial support from the BARD project, Department of Atomic Energy, Government of India, is gratefully acknowledged.)

FA11

PHOTOCHEMICAL FORMATION OF SULFUR-CONTAINING AEROSOLS

JAY A KROLL, VERONICA VAIDA, *Department of Chemistry and Biochemistry, University of Colorado, Boulder, CO, USA.*

In order to understand planetary climate systems, modeling the properties of atmospheric aerosols is vital. Aerosol formation plays an important role in planetary climates and is tied to feedback loops that can either warm or cool a planet. Sulfur compounds are known to play an important role in new particle aerosol formation and have been observed in a number of planetary atmospheres throughout our solar system. Our current understanding of sulfur chemistry explains much of what we observe in Earth's atmosphere; however, several discrepancies arise when comparing observations of the Venusian atmosphere with model predictions. This suggests that there are still problems in our fundamental understanding of sulfur chemistry. This is concerning given recent renewed interest in sulfate injections in the stratosphere for solar radiation management geo-engineering schemes. We investigate the role of sunlight as a potential driver of the formation of sulfur-containing aerosols. I will present recent work investigating the generation of large quantities of aerosol from the irradiation of mixtures of SO$_2$ with water and organic species, using a solar simulator that mimics the light that is available in the Earth's troposphere and the Venusian middle atmosphere. I will present on recent work done in our lab suggesting the formation of sulfurous acid, H$_2$SO$_3$, and describe experimental work that supports this proposed mechanism. Additionally I will present on new work showing the highly reactive nature of electronically excited SO$_2$ with saturated alkane species. The implications of this photochemically induced sulfur aerosol formation in the atmosphere of Earth and other planetary atmospheres will be discussed.

FB. (Hyper)fine structure, tunneling

Friday, June 23, 2017 – 8:30 AM

Room: 116 Roger Adams Lab

Chair: Isabelle Kleiner, CNRS et Universités Paris-Est et Paris Diderot, Créteil, France

---

FB01   8:30 – 8:45

MICROWAVE SPECTRUM OF 1-SILA-1-ISOCYANOCYCLOPENT-3-ENE

FRANK E MARSHALL, *Department of Chemistry, Missouri University of Science and Technology, Rolla, MO, USA*; DANIEL V. HICKMAN, GAMIL A GUIRGIS, *Chemistry, College of Charleston, Charleston, SC, USA*; MICIIAEL H. PALMER, *School of Chemistry, University of Edinburgh, Edinburgh, United Kingdom*; CHARLES J. WURREY, *Department of Chemistry, University of Missouri - Kansas City, Kansas City, MO, USA*; NICOLE MOON, THOMAS D. PERSINGER, <u>G. S. GRUBBS II</u>, *Department of Chemistry, Missouri University of Science and Technology, Rolla, MO, USA*.

The microwave spectrum of synthesized molecule, 1-sila-1-isocyanocyclopent-3-ene has been studied using chirped pulse and cavity Fourier transform microwave (CP-FTMW and FTMW) techniques. The rotational spectrum has been assigned along with hyperfine splitting due to the ^{14}N nucleus. Very limited sample could be synthesized at a time with this molecule, demonstrating the power of the CP-FTMW technique. Analysis of the molecule, along with structural possibilities will be discussed.

FB02   8:47 – 9:02

NUCLEAR QUADRUPOLE COUPLING IN SiH_2I_2 DUE TO THE PRESENCE OF TWO IODINE NUCLEI

ERIC A. ARSENAULT, *Department of Chemistry, Wesleyan University, Middletown, CT, USA*; DANIEL A. OBENCHAIN, *Institut für Physikalische Chemie und Elektrochemie, Gottfried-Wilhelm-Leibniz-Universität, Hannover, Germany*; <u>W. ORELLANA</u>, STEWART E. NOVICK, *Department of Chemistry, Wesleyan University, Middletown, CT, USA*.

The rotational spectrum of diiodosilane was measured with a jet-pulsed, cavity Fourier transform microwave spectrometer over the frequency range 8.8 GHz to 15 GHz and assigned for the first time. The complete nuclear quadrupole coupling (NQC) tensors for both iodine nuclei were obtained for the ^{28}Si, ^{29}Si, and ^{30}Si isotopologues of diiodosilane. In addition to the nuclear quadrupole coupling constants (NQCCs), rotational constants, centrifugal distortion constants, and nuclear-spin rotation constants were determined for each silicon isotopologue. Subtle, yet unmistakable, changes in the NQCCs of iodine upon isotopic substitution will be examined. A r_0 structure of diiodosilane was also fit via isotopic substitution, leading to the determination of bond lengths and angles: Si–I = 2.4236(19) Å, Si–H = 1.475(21) Å, $\angle$(I–Si–I) = 111.27(13)$^\circ$, and $\angle$(I–Si–H) = 105.9(19)$^\circ$. These results will be compared to the results of a previous gas electron diffraction study.

FB03

MICROWAVE SPECTRUM OF THE H_2S DIMER: OBSERVATION OF K_a=1 LINES

ARIJIT DAS, *Department of Inorganic and Physical Chemistry, Indian Institute of Science, Bangalore, India*; PANKAJ MANDAL, *Department of Chemistry, Indian Institute of Science Education and Research, Pune, Maharshtra, India*; FRANK J LOVAS, *Sensor Science Division, National Institute of Standards and Technology, Gaithersburg, MD, USA*; CHRIS MEDCRAFT, *School of Chemistry, Newcastle University, Newcastle-upon-Tyne, United Kingdom*; ELANGANNAN ARUNAN, *Department of Inorganic and Physical Chemistry, Indian Institute of Science, Bangalore, India*.

Large amplitude tunneling motions in $(H_2S)_2$ complicate the analysis of its microwave spectrum. The previous rotational spectrum of $(H_2S)_2$ was observed using the Balle-Flygare pulsed nozzle FT microwave spectrometers at NIST and IISc. For most isotopomers of $(H_2S)_2$ a two state pattern of a-type K_a=0 transitions had been observed and were interpreted to arise from $E_1{}^{+/-}$ and $E_2{}^{+/-}$ states of the six tunneling states expected for $(H_2S)_2$. K_a=0 lines gave us only the distance between the acceptor and donor S atoms.[a][b][c] The (B+C)/2 for E_1 and E_2 states were found to be 1749.3091(8) MHz and 1748.1090(8) MHz respectively. In this work, we have observed the K_a=1 microwave transitions which enable us to determine finer structural details of the dimer. The observation of the K_a=1 lines indicate that $(H_2S)_2$ is not spherical in nature, their interactions do have some anisotropy. Preliminary assignment of K_a=1 lines for the E_1 state results in B=1752.859 MHz and C=1745.780 MHz. We also report a new progression of lines which probably belongs to the parent isotopomers.

[a]F. J. Lovas, P. K. Mandal and E. Arunan, unpublished work

[b]P. K. Mandal Ph.D. Dissertation, Indian Institute of Science, (2005)

[c]F. J. Lovas, R. D. Suenram, and L. H. Coudert. 43rd Int.Symp. on Molecular Spectroscopy. (1988)

FB04

ROTATIONAL SPECTRA AND NUCLEAR QUADRUPOLE COUPLING CONSTANTS OF 4-HALOPYRAZOLES $C_3N_2H_3X$ (X = Br, I)

GRAHAM A. COOPER, CHRIS MEDCRAFT, *School of Chemistry, Newcastle University, Newcastle-upon-Tyne, United Kingdom*; ANTHONY LEGON, *School of Chemistry, University of Bristol, Bristol, United Kingdom*; NICK WALKER, *School of Chemistry, Newcastle University, Newcastle-upon-Tyne, United Kingdom*.

The microwave spectra of the heteroaromatic molecules 4-bromopyrazole and 4-iodopyrazole have been recorded for the first time, along with their *N*-deuterated isotopologues. These species have recently been found to be useful in structural determination of proteins due to their ability to attach at a variety of binding sites.[a] The nuclear quadrupole coupling constants have been fitted, and these have been used to determine the nature of the C-X bond, and related to the strength of the halogen bonds formed by the molecules.

[a]J. D. Bauman, J. J. E. K. Harrison, and E. Arnold, *IUCrJ* 2016, 3, 51–60

FB05

AN INVESTIGATION OF THE DIPOLE FORBIDDEN TRANSITION EFFECTS IN BROMOFLUOROCARBONS AS IT PERTAINS TO 3-BROMO-1,1,1,2,2-PENTAFLUOROPROPANE USING CP-FTMW SPECTROSCOPY

FRANK E MARSHALL, <u>NICOLE MOON</u>, THOMAS D. PERSINGER, DAVID JOSEPH GILLCRIST, N. E. SHREVE, *Department of Chemistry, Missouri University of Science and Technology, Rolla, MO, USA*; WILLIAM C. BAILEY, *Department of Chemistry-Physics, Kean University (Retired), Union, NJ, USA*; G. S. GRUBBS II, *Department of Chemistry, Missouri University of Science and Technology, Rolla, MO, USA*.

As part of a series of bromofluorocarbon species and analogues, the microwave spectrum of the molecule 3-bromo-1,1,1,2,2-pentafluoropropane has been measured on a CP-FTMW spectrometer located at Missouri S&T. The resultant spectrum is dense with transitions occurring at a rate of ≈1 transition/MHz! Within the spectrum, ^{79}Br and ^{81}Br isotopologues of multiple conformers of 3-bromo-1,1,1,2,2-pentafluoropropane have been identified. Rotational constants, centrifugal distortion parameters, nuclear quadrupole coupling constants and how each compare with theory for each conformer will be discussed.

Due to the large quadrupolar moment of bromine, heavy, brominated molecules are good candidates for dipole-forbidden transitions. Previous studies with bromoperfluoroacetone[a] provided a rich spectrum full of dipole forbidden transitions that 3-bromo-1,1,1,2,2-pentafluoropropane does not share. This difference will be explained using structural considerations along with the matrix elements needed to enact these transitions.

[a]F. E. Marshall, D. J. Gillcrist, T. D. Persinger, S. Jaeger, C. C. Hurley, N. E. Shreve, N. Moon, and G. S. Grubbs II, *J. Mol. Spectrosc.* **328** (2016) 59.

Intermission

FB06

A REINVESTIGATION OF THE ELECTRONIC PROPERTIES OF 2-BROMOPYRIDINE WITH HIGH-RESOLUTION MICROWAVE SPECTROSCOPY

<u>ANGELA Y. CHUNG</u>, ERIC A. ARSENAULT, STEWART E. NOVICK, *Department of Chemistry, Wesleyan University, Middletown, CT, USA*.

The rotational spectrum of 2-bromopyridine (C_5H_4BrN) was reinvestigated in the frequency range of 10-15.5 GHz by high-resolution Fourier transform microwave (FTMW) spectroscopy. The new observations of ^{14}N hyperfine splittings in previously studied transitions[a] belonging to both bromine isotopologues ($C_5H_4{}^{79}BrN$ and $C_5H_4{}^{81}BrN$) led to improved measurements of the rotational constants and bromine nuclear quadrupole coupling constants. The full nuclear quadrupole coupling (NQC) tensor of ^{14}N was resolved for the first time, in addition to five centrifugal distortion constants. A comparison of the two ^{14}N NQC tensors of $C_5H_4{}^{79}BrN$ and $C_5H_4{}^{81}BrN$ will be presented.

[a]Caminati, W.; Forti, P. *Chemical Physics Letters* **1972**, *15*(3), 343–349.

FB07

USING HYPERFINE STRUCTURE TO QUANTIFY THE EFFECTS OF SUBSTITUTION ON THE ELECTRON DISTRIBUTION WITHIN A PYRIDINE RING: A STUDY OF 2-, 3-, AND 4-PICOLYLAMINE

LINDSEY M McDIVITT, KORRINA M HIMES, JOSIAH R BAILEY, TIMOTHY J McMAHON, <u>RYAN G BIRD</u>, *Chemistry, University of Pittsburgh Johnstown, Johnstown, PA, USA*.

The ground state rotational spectra of the three methylamine substituted pyridines, 2-, 3-, and 4-picolylamine, were collected and analyzed over the frequency range of 7-17.5 GHz using chirped-pulsed Fourier transform microwave spectroscopy. All three molecules show a distinctive quadrupole splitting, which is representative of the local electronic environment around the two different ^{14}N nuclei, with the pyridine nitrogen being particularly sensitive to the pi-electron distribution within the ring. The role that the position of the methylamine group plays on the quadrupole coupling constants on both nitrogens will be discussed and compared to other substituted pyridines.

FB08   10:46 – 11:01

PURE ROTATIONAL SPECTRUM OF THE "NON-POLAR" DIMER OF FORMIC ACID

LUCA EVANGELISTI, WEIXING LI, *Dipartimento di Chimica G. Ciamician, Università di Bologna, Bologna, Italy*; QIAN GOU, *School of Chemistry and Chemical Engineering, Chongqing University, Chongqing, China*; ROLF MEYER, *Laboratorium für Physikalische Chemie, ETH Zurich, Zurich, Switzerland*; WALTHER CAMINATI, *Dipartimento di Chimica G. Ciamician, Università di Bologna, Bologna, Italy.*

The rotational spectra of three deuterated isotopologues of the dimer of formic acid have been measured, thank to the small dipole moment induced by asymmetric H –>D substitution(s). For the HCOOH-DCOOH species the concerted double proton transfer of the two hydroxyl hydrogens takes place between two equivalent minima and generates a tunneling splitting of 331.2(6) MHz. From this splitting a barrier to proton tunneling of about 30 kJ/mol has been estimated.

FB09   11:03 – 11:18

CROSS-CONTAMINATION OF FITTING PARAMETERS IN MULTIDIMENSIONAL TUNNELING TREATMENTS

NOBUKIMI OHASHI, , *Kanazawa University, Kanazawa, Japan*; JON T. HOUGEN, *Sensor Science Division, National Institute of Standards and Technology, Gaithersburg, MD, USA.*

In this talk we examine the two-dimensional tunneling formalism used previously to fit the hydrogen-transfer and internal-rotation splittings in the microwave spectrum of 2-methylmalonaldehyde in an effort to determine the origin of various counterintuitive results concerning the isotopic dependence of the internal-rotation splittings in that molecule. We find that the cause of the problem lies in a "parameter contamination" phenomenon, where some of the numerical magnitude of splitting parameters from modes with large tunneling splittings "leaks into" the parameters of modes with smaller tunneling splittings. We further find that such parameter contamination, which greatly complicates the determination of barrier heights from the least-squares-fitted splitting parameters, will be a general problem in spectral fits using the multi-dimensional tunneling formalism, since it arises from subtle mathematical features of the non-orthogonal framework functions used to set up the tunneling Hamiltonian. Transforming to a physically less intuitive orthonormal set of basis functions allows us to give an approximate numerical estimate of the contamination of tunneling parameters for 2-methylmalonaldehyde by combining a dominant tunneling path hypothesis with results recently given for the hydrogen-transfer–internal-rotation potential function for this molecule.

FB10   11:20 – 11:35

SPIN-SPIN AND SPIN-ROTATION FINE STRUCTURE OF THE METASTABLE $a\,^3\Sigma_u^+$ STATES OF MOLECULAR HELIUM

PAUL JANSEN, LUCA SEMERIA, FREDERIC MERKT, *Laboratorium für Physikalische Chemie, ETH Zurich, Zurich, Switzerland.*

In a recent series of experiments[a,b], we have determined term values of all rotational levels of the $X^+\,^2\Sigma_u^+$ ($\nu^+ = 0$) ground vibronic state of $^4He_2^+$ with rotational quantum number $N^+ \leq 19$ at an accuracy of 25 MHz using MQDT-assisted Rydberg-series extrapolation of metastable helium molecules in the $a\,^3\Sigma_u^+$ state. The precison of these experiments was limited by the 150 MHz linewidth of the pulsed laser system employed. In order to improve our resolution and possibly observe the spin-rotation splitting in the He_2^+ ion, we have replaced the pulsed laser by a CW laser system with a bandwidth of 1.5 MHz. This system was used to measure the spin-spin and spin-rotation fine structure of metastable He_2 in the $a\,^3\Sigma_u^+$ ($\nu'' = 0$) state. Metastable helium molecules were produced by striking a discharge in an expansion of neat helium gas. By cooling the source to a temperature of 10 K, the velocity of the molecular beam was reduced to 500 m/s and an experimental Doppler-limited linewidth of 25 MHz was observed. Fine-structure splittings for all rotational levels with $N'' \leq 27$ have been measured at an accuracy of 5 MHz and, when possible, have been compared to the values reported in earlier investigations.[c,d,e,f] This comparison revealed a discrepancy that increased with increasing values of N''. To verify our results, we have recently constructed a variaton of a classical molecular-beam magnetic-resonance setup that uses a multistage Zeeman decelerator and a RF stripline for de- and repopulation of the F_2 spin-rotational components with $J'' = N''$, respectively.

[a]P. Jansen, L. Semeria, L. Esteban Hofer, S. Scheidegger, J. A. Agner, H. Schmutz, and F. Merkt, *Phys. Rev. Lett.* **114**, 133202 (2015).
[b]L. Semeria, P. Jansen, and F. Merkt, *J. Chem. Phys.* **145**, 204301 (2016).
[c]W. Lichten, M. V. McCusker, and T. L. Vierima, *J. Chem. Phys.* **61**, 2200 (1974).
[d]W. Lichten and T. Wik, *J. Chem. Phys.* **69**, 98 (1978).
[e]M. Kristensen and N. Bjerre, *J. Chem. Phys.* **93**, 983 (1990).
[f]I Hazell, A. Norregaard, and N. Bjerre, *J. Mol. Spectrosc.* **172**, 135 (1995).

FB11   *Post-Deadline Abstract*   **11:37 – 11:52**

ROTATIONAL SPECTRA AND NUCLEAR QUADRUPOLE COUPLING CONSTANTS OF IODOIMIDAZOLES

<u>GRAHAM A. COOPER</u>, CARA J ANDERSON, CHRIS MEDCRAFT, *School of Chemistry, Newcastle University, Newcastle-upon-Tyne, United Kingdom*; ANTHONY LEGON, *School of Chemistry, University of Bristol, Bristol, United Kingdom*; NICK WALKER, *School of Chemistry, Newcastle University, Newcastle-upon-Tyne, United Kingdom*.

The microwave spectra of two isomers of iodoimidazole have been recorded and assigned with resolution of their nuclear quadrupole coupling constants. These constants have been analysed in terms of the conjugation between the lone pairs on the iodine atom and the aromatic π-bonding system, and the effect of this conjugation on the distribution of π-electron density in the ring. A comparison of these properties has been made between iodoimidazole and other 5- and 6-membered aromatic rings bonded to halogen atoms.

FC. Vibrational structure/frequencies

Friday, June 23, 2017 – 8:30 AM

Room: 1024 Chemistry Annex

Chair: Melanie A.R. Reber, University of Georgia, Athens, GA, USA

FC01   8:30 – 8:45

LASER SPECTROSCOPY OF VINYL ALCOHOL EMBEDDED IN HELIUM DROPLETS

HAYLEY BUNN, *School of Chemistry and Physics, The University of Adelaide, Adelaide, South Australia, Australia*; PAUL RASTON, *Department of Chemistry and Biochemistry, James Madison University, Harrisonburg, VA, USA*; GARY E. DOUBERLY, *Department of Chemistry, University of Georgia, Athens, GA, USA.*

Vinyl alcohol has two rotameric forms, known as *syn-* and *anti-*vinyl alcohol, where *syn* is the most stable. While both have been investigated by microwave[a] and far-infrared[b] spectroscopy, only the *syn* rotamer has been investigated by mid-infrared spectroscopy[c]. This is due to the low *anti* rotamer population (15%) at room temperature, in addition to the closeness in proximity of the mid-infrared bands between the rotamers; this results in overlapping bands that are dominated by *syn*-vinyl alcohol absorptions. In this investigation we increase the *anti*-vinyl alcohol population to 40% by using a high temperature "pyrolysis" source, and eliminate the spectral overlap by recording the spectra at low temperature in helium nanodroplets. We observe a number of bands of both rotamers in the OH, CH, and CO stretching regions that display rotational substructure. A highlight of this work is the observation of a Fermi dyad in the OH stretching region of *anti*-vinyl alcohol. Anharmonic frequency calculations suggest that this is due to a near degeneracy of the OH stretching state (ν_1) with a triple combination involving ν_7, ν_8, and ν_9.

[a]M. Rodler, J. Mol. Spec. 114, 23 (1985);S. Saito, Chem. Phys. Lett. 42, 3 (1976)

[b]H. Bunn, R. Hudson, A. S. Gentleman, and P. L. Raston, ACS Earth Space Chem. DOI: 10.1021/acsearthspacechem.6b00008 (2017)

[c]D-L Joo, A. J. Merer, D. J. Clouthier, J. Mol. Spec. 197, 68 (1999)

FC02   8:47 – 9:02

INFRARED SPECTRA OF THE *n*-PROPYL AND *i*-PROPYL RADICALS IN SOLID PARA-HYDROGEN

GREGORY T. PULLEN, PETER R. FRANKE, GARY E. DOUBERLY, *Department of Chemistry, University of Georgia, Athens, GA, USA*; YUAN-PERN LEE, *Applied Chemistry, National Chiao Tung University, Hsinchu, Taiwan, Institute of Atomic and Molecular Sciences, Academia Sinica, Taipei, Taiwan.*

We report the infrared spectra of the *n*-propyl and *i*-propyl radicals measured in solid para-hydrogen (p-H_2) matrices at 3.2 K. *n*-Propyl and *i*-propyl radicals were produced via the 248 nm irradiation of matrices formed by co-depositing p-H_2 and either 1-Iodopropane (*n*-propyl) or 2-Iodopropane (*i*-propyl). Secondary photolysis was used to group spectral lines all due to the same species. Lines in the C-H stretching region were compared to previous work using the Helium Nanodroplet Isolation (HENDI) technique,[a] and are in excellent agreement. In addition to a few lines previously measured in Ar matrices, we observe many previously unreported bands below 2000 cm^{-1}, which we attribute to the *n*-propyl and *i*-propyl radicals. The assignment of features below 2000 cm^{-1} are made via comparisons to anharmonic VPT2+K frequency computations.

[a]Peter R. Franke, Daniel P. Tabor, Christopher P. Moradi, Gary E. Douberly, Jay Agarwal, Henry F. Schaefer III, and Edwin L. Sibert III, *Journal of Chemical Physics* 145, 224304 (2016).

FC03   9:04 – 9:19

THE INFLUENCE OF ANHARMONIC, DISPERSION AND SOLVATION EFFECTS IN THE IR SPECTRA OF PROTO-NATED NEUROTRANSMITTERS SEROTONIN AND DOPAMINE

VIPIN BAHADUR SINGH, *Department of Physics, Udai Pratap Autonomous College, Varanasi, India.*

The nerve cells in our brain communicate with each other by specific molecules known as neurotransmitters to regulate mood, voluntary movement, learning and long-term memory. As a consequence, the knowledge of their flexible conformation and IR spectra provides an understanding of these highly specific and important biological phenomena at the molecular level. Here we characterize vibrational spectroscopic signatures of conformer-specific intramolecular interactions of the protonated dopamine and serotonin and related molecules (all have core ethylamine moiety in their structures) using the MP2 and dispersion corrected DFT calculations including the anharmonicity, dispersion and solvation (aqueous and alcohol solutions). The anharmonic calculations reveal rather large effects on the predicted IR frequencies and intensities of NH_3+ stretching and bending modes of folded gauche conformers. The best experimental frequency estimate for the bound NH^+-π stretch mode lies between the unscaled anharmonic and scaled harmonic values. The cation-π interaction in dopamine and serotonin (expected to play an important role in human aggression) is found stronger in alcohol solution in comparison to aqueous solution.

FC04   9:21 – 9:36

TRANSIENT RAMAN SPECTRA, STRUCTURE AND THERMOCHEMISTRY OF THE THIOCYANATE DIMER RADICAL ANION IN WATER

IRENEUSZ JANIK, G. N. R. TRIPATHI, IAN CARMICHAEL, *Radiation Laboratory, University of Notre Dame, Notre Dame, IN, USA.*

Time-resolved resonance-enhanced Stokes and anti-Stokes Raman spectra of the thiocyanate dimer radical anion, $(SCN)_2^-$, prepared by pulse radiolysis in water, have been obtained and interpreted in conjunction with theoretical calculations to provide detailed information on the molecular geometry and bond properties of the species. The structural properties of the radical are used to develop a molecular perspective on its thermochemistry in aqueous solution. Twenty-nine Stokes Raman bands of the radical observed in the 120-4200 cm^{-1} region are assigned in terms of the strongly enhanced 220 cm^{-1} fundamental, weakly enhanced 721 cm^{-1}, and moderately enhanced 2073 cm^{-1} fundamentals, their overtones and combinations. Calculations by range-separated hybrid (RSH) density functionals (ωB97x and LC-ωPBE) support the spectroscopic assignments of the 220 cm^{-1} vibration to a predominantly SS stretching mode and the features at 721 cm^{-1} and 2073 cm^{-1} to CS and CN stretching modes, respectively. The corresponding bond lengths are 2.705 (+0.036) Å, 1.663 (+0.001) Å and 1.158 (+0.002) Å. A first order anharmonicity of 1 cm^{-1} determined for the SS stretching mode suggests a convergence of vibrational states at an energy 1.5 eV, using the Birch-Sponer extrapolation. This value, estimated for the radical confined in solvent cage, compares well with the calculated gas-phase energy required for the radical (1.22 eV) to dissociate into SCN and SCN^- fragments. The enthalpy of dissociation drops to 0.63 eV in water when solvent dielectric effects on the radical and its dissociation products upon S-S bond scission are incorporated in the calculations. No frequency shift or spectral broadening was observed between light and heavy water solvents, indicating that the motion of solvent molecules in the hydration shell have no perceptible effect on the intramolecular dynamics of the radical. The Stokes and anti-Stokes Raman frequencies were found to be identical within the experimental uncertainty, suggesting that the frequency difference between the thermally relaxed and spontaneously created vibrational states of $(SCN)_2^-$ in water is too small to be observable.

FC05 9:38 – 9:53

INTRAMOLECULAR VIBRATIONAL ENERGY REDISTRIBUTION (IVR) IN SELECTED S_1 LEVELS ABOVE 1000 cm^{-1} IN PARA-FLUOROTOLUENE.

<u>LAURA E. WHALLEY</u>, ADRIAN M. GARDNER, WILLIAM DUNCAN TUTTLE, JULIA A DAVIES, KATHARINE L REID, TIMOTHY G. WRIGHT, *School of Chemistry, University of Nottingham, Nottingham, United Kingdom.*

With increasing vibrational wavenumber, the density of states of a molecule is expected to rise dramatically, especially so when low wavenumber torsions (internal rotations) are present, as in the case of *para*-fluorotoluene (*p*FT). This in turn is expected to lead to more opportunities for coupling between vibrational modes, which is the driving force for intramolecular vibrational energy redistribution (IVR). Previous studies[a,b] at higher energies have focussed on the two close lying vibrational levels at 1200 cm^{-1} in the S_1 electronic state of *p*FT which were assigned to two zero-order bright states (ZOBSs), whose characters predominantly involve C-CH_3 and C-F stretching modes. A surprising result of these studies was that the photoelectron spectra showed evidence that IVR is more extensive following excitation of the C-F mode than it is following excitation of the C-CH_3 mode, despite these levels being separated by only 35 cm^{-1}. This observation provides evidence that the IVR dynamics are mode-specific, which in turn may be a consequence of the IVR route being dependent on couplings to nearby states that are only available to the C-F mode.

In this work, in order to further investigate this behaviour, we have employed resonance-enhanced multiphoton ionisation (REMPI) spectroscopy and zero-kinetic-energy (ZEKE) spectroscopy to probe S_1 levels above 1000 cm^{-1} in *p*FT. Such ZEKE spectra have been recorded via a number of S_1 intermediate levels allowing the character and coupling between vibrations to be unravelled; the consequence of this coupling will be discussed with a view to understanding any IVR dynamics seen.

[a]C. J. Hammond, V. L. Ayles, D. E. Bergeron, K. L. Reid and T. G. Wright, *J. Chem. Phys.*, **125**, 124308 (2006)

[b]J. A. Davies, A. M. Green, A. M. Gardner, C. D. Withers, T. G. Wright and K. L. Reid, *Phys. Chem. Chem. Phys.*, **16**, 430 (2014)

Intermission

FC06 10:12 – 10:27

VIBRATION AND VIBRATION-TORSION LEVELS OF THE S_1 AND GROUND CATIONIC D_0^+ STATES OF PARA-FLUOROTOLUENE AND PARA-XYLENE BELOW 1000 cm^{-1}

<u>WILLIAM DUNCAN TUTTLE</u>, ADRIAN M. GARDNER, LAURA E. WHALLEY, TIMOTHY G. WRIGHT, *School of Chemistry, University of Nottingham, Nottingham, United Kingdom.*

We have employed resonance-enhanced multiphoton ionisation (REMPI) spectroscopy and zero-kinetic-energy (ZEKE) spectroscopy to investigate the first excited electronic singlet (S_1) state and the cationic ground state (D_0^+) of *para*-fluorotoluene (*p*FT) and *para*-xylene (*p*Xyl). Spectra have been recorded *via* a large number of selected intermediate levels, to support assignment of the vibration and vibration-torsion levels in these molecules and to investigate possible couplings.

The study of levels in this region builds upon previous work on the lower energy regions of *p*FT and *p*Xyl[a,b,c] and here we are interested in how vibration-torsion (vibtor) levels might combine and interact with vibrational ones, and so we consider the possible couplings which occur. Comparisons between the spectra of the two molecules show a close correspondence, and the influence of the second methyl rotor in *para*-xylene on the onset of intramolecular vibrational redistribution (IVR) in the S_1 state is a point of interest. This has bearing on future work which will need to consider the role of both more flexible side chains of substituted benzene molecules, and multiple side chains.

[a]A. M. Gardner, W. D. Tuttle, L. Whalley, A. Claydon, J. H. Carter and T. G. Wright, *J. Chem. Phys.*, **145**, 124307 (2016).

[b]A. M. Gardner, W. D. Tuttle, P. Groner and T. G. Wright, *J. Chem. Phys.*, (2017, in press).

[c]W. D. Tuttle, A. M. Gardner, K. O'Regan, W. Malewicz and T. G. Wright, *J. Chem. Phys.*, (2017, in press).

FC07

CO-ASSIGNMENTS OF THE CALCULATED QUANTUM-MECHANICAL MOLECULAR VIBRATIONAL FREQUENCIES OF *CIS*-ACROLEIN IN THE GROUND S_0 AND LOWEST EXCITED T_1 AND S_1 ELECTRONIC STATES

V.A. BATAEV, <u>YURII PANCHENKO</u>, ALEXANDER ABRAMENKOV, *Department of Chemistry, Lomonosov Moscow State University, Moscow, Russia.*

The optimization of geometrical parameters and calculation of the force fields of *cis*-acrolein (CH_2CHCHO) in the ground S_0 and lowest excited triplet (T_1) and singlet (S_1) electronic states are performed at the CASPT2/def2-TZVPP theoretical level. Co-assignments of the vibrational frequencies calculated from the corresponding force fields are validated by calculation of the mode-mixing matrices for the pairs of the (S_0, T_1) and (S_0, S_1) electronic states. The results obtained by comparing the vibrational frequencies calculated from the scaled force fields with the experimental vibrational frequencies made it possible to corroborate some details of the molecular structure in these electronic states.

FC08

AB INITIO CALCULATION OF THE INFRARED SPECTRUM FOR XeF6 MOLECULE

<u>LAN CHENG</u>, *Department of Chemistry, Johns Hopkins University, Baltimore, MD, USA.*

Scalar relativistic coupled cluster calculations are presented for the infrared spectrum of the xenon hexafluoride (XeF6) molecule in its Oh and C3v structures. Anharmonic contributions to vibrational frequencies and infrared intensities of the C3v structure are taken into account using second order vibrational perturbation theory (VPT2). The effect due to Fermi resonances in the VPT2 calculations is analyzed. A transition state linking the C3v and Oh structures has also been located in the potential energy surface. The fluxional character of the molecule is discussed.

FC09

MOLECULAR AND ELECTRONIC STRUCTURES OF CERIUM AND CERIUM SUBOXIDE CLUSTERS

<u>JARED O. KAFADER</u>, JOSEY E TOPOLSKI, CAROLINE CHICK JARROLD, *Department of Chemistry, Indiana University, Bloomington, IN, USA.*

Cerium-based materials have electronic properties optimal for their utilization in many applications including electronics and catalysis. Electronic spectroscopy of small cerium and cerium suboxide clusters helps us understand the role 6s, 5d, and 4f electrons play during the oxidation process. The spectroscopy of small cerium-containing molecules is relatively unexplored and to broaden this understanding we have completed the characterization of Ce_xO_y cluster anions and neutrals using photoelectron spectroscopy coupled with DFT calculations. The characterization of Ce_xO_y molecules have allowed for the determination of their electron affinities, bonding motifs, the assignment of numerous anion to neutral state transitions, modeling of anion/neutral structures, and electron orbital occupation.

FD. Clusters/Complexes

Friday, June 23, 2017 – 8:30 AM

Room: B102 Chemical and Life Sciences

Chair: Gerhard Schwaab, Ruhr University Bochum, Bochum, Germany

---

FD01   **8:30–8:45**

CHARACTERIZATION OF A CARBON DIOXIDE-HEXAFLOUROBENZENE COMPLEX USING MATRIX ISOLATION INFRARED SPECTROSCOPY

JAY C. AMICANGELO, BRADLEY K. GALL, MARYN N. HORN, *School of Science (Chemistry), Penn State Erie, Erie, PA, USA.*

Matrix isolation infrared spectroscopy was used to characterize a 1:1 complex of carbon dioxide (CO_2) with hexaflourobenzene (C_6F_6). Co-deposition experiments with CO_2 and C_6F_6 were performed at 20 K using argon as the matrix gas. New infrared peaks attributable to the CO_2-C_6F_6 complex were observed near the O-C-O antisymmetric stretching vibration of the CO_2 monomer and near the C-F stretching vibration of the C_6F_6 monomer. The initial identification of the newly observed infrared peaks to those of a CO_2-C_6F_6 complex was established by performing several concentration studies in which the sample-to-matrix ratios of the monomers were varied between 1:100 to 1:1600, by comparing the resulting co-deposition spectra with the spectra of the individual monomers, and by matrix annealing experiments (30 – 35 K). Co-deposition experiments were also performed using isotopically labeled carbon dioxide ($^{13}CO_2$) and the analogous peaks for the $^{13}CO_2$-C_6F_6 complex were observed. Quantum chemical calculations were performed for the CO_2-C_6F_6 complex at the MP2/aug-cc-pVDZ level of theory in order to explore the intermolecular potential energy surface of the complex and to obtain optimized complex geometries and predicted vibrational frequencies of the complex. The calculations for the exploration of the potential energy surface involved rigid scans along the intermolecular distance and various angle coordinates for several general orientations of the two monomers. Based on these calculations, full geometry optimizations were then performed and two stable complex minima were found: one in which the CO_2 is perpendicular and centered to the C_6F_6 ring (ΔE_{int} = -7.9 kJ/mol) and one in which the CO_2 is parallel to the C_6F_6 ring but displaced from the center (ΔE_{int} = -6.0 kJ/mol). Comparing the predicted vibrational spectra for both complexes to the observed experimental spectra, particularly for the O-C-O antisymmetric stretching region, it is concluded that both structures are present in the solid argon matrices.

FD02   **8:47–9:02**

VIBRATIONAL PREDISSOCIATION OF THE $\tilde{A}$ STATE OF THE C_3Ar COMPLEX IN THE EXCITATION ENERGY REGION OF 25410-25535 CM^{-1}

YI-JEN WANG, YEN-CHU HSU, *Institute of Atomic and Molecular Sciences, Academia Sinica, Taipei, Taiwan.*

About 11 C_3Ar bands near the 0 4^- 0-000 and 0 2^+ 0-000 transitions of the $\tilde{A}^1\Pi_u$–$\tilde{X}^1\Sigma^+_g$ system of C_3 have been studied by both laser-induced fluorescence and wavelength-resolved emission techniques. Two prominent pairs of C_3Ar features were observed to the red of each of these two C_3 transitions. Each pair consists of a type A band and a type C band, with the type C band lying about 3 cm^{-1} above the type A band. Rotational analysis showed that three of the bands are comparatively sharp, with line widths of 0.035 cm^{-1}, but the pair at 25504 and 25507 cm^{-1} shows clear evidence of diffuseness. The spectral widths of the rotational lines do not depend on the excitation energies in any simple way. Most of the features in the wavelength-resolved emission spectra can be assigned as emission from vibrationally excited levels of the $\tilde{A}$ state of the C_3 fragments down to the ground electronic state. Two different types of vibrational excitation of the C_3 fragments have been found: pure C_3-bending and antisymmetric C-C stretching. The branching ratios of the C_3 product states, the C_3-Ar vdW binding energy, and propensity rules for vibrational predissociation processes will be presented.

FD03   9:04–9:19

THE ν_3 FUNDAMENTAL VIBRATIONAL BAND OF SCCCS REVISITED

THOMAS SALOMON, *I. Physikalisches Institut, Universität zu Köln, Köln, Germany*; JOHN B DUDEK, *Department of Chemistry, Hartwick College, Oneonta, NY, USA*; SVEN THORWIRTH, *I. Physikalisches Institut, Universität zu Köln, Köln, Germany.*

The ν_3 fundamental vibrational band of carbon subsulfide, SCCCS, first studied by Holland and collaborators[a,b] has been reinvestigated using a combination of laser ablation production, free-jet expansion and quantum cascade laser spectroscopy. In addition to the fundamental band (located at $2100\,\mathrm{cm}^{-1}$) and associated hot bands originating from the lowest bending mode ν_7, the hot bands from the two energetically higher-lying bending modes ν_5 and ν_6 have been observed for the first time as has the $S^{13}CCCS$ isotopic species.

[a] F. Holland, M. Winnewisser, C. Jarman, H. W. Kroto, and K. M. T. Yamada 1988, *J. Mol. Spectrosc.* 130, 344
[b] F. Holland and M. Winnewisser 1991, *J. Mol. Spectrosc.* 147, 496

FD04   9:21–9:36

CORE ION STRUCTURES AND SOLVATION EFFECTS IN GAS PHASE $[Sn(CO_2)_n]^-$ CLUSTERS

MICHAEL C THOMPSON, J. MATHIAS WEBER, *JILA and the Department of Chemistry and Biochemistry, University of Colorado-Boulder, Boulder, CO, USA.*

We report infrared photodissociation spectra of $[Sn(CO_2)_n]$ (n=2-6) clusters. We explore core ion geometries through quantum chemical calculations and assign our experimental spectra through comparison with calculated vibrational frequencies. We discuss our results in the context of heterogeneous catalytic reduction of CO_2, and compare our results with previous work on other post-transition metal species.

Intermission

FD05   9:55–10:10

EXPERIMENTAL INSIGHT ON THE CONFORMATIONAL LANDSCAPE OF THE SF_6 DIMER: EVIDENCE FOR THREE CONFORMERS

PIERRE ASSELIN, *Department of Chemistry, MONARIS, CNRS, UMR 8233, Sorbonne Universités, UPMC Univ Paris 06, Paris, France*; ALEXEY POTAPOV[a], , *Laboratory Astrophysics Group of the Max Planck Institute for Astronomy at the Friedrich Schiller University, Jena, Germany*; VINCENT BOUDON, *Laboratoire ICB, CNRS/Université de Bourgogne, DIJON, France*; LAURENT BRUEL, *CEA Marcoule, DEN, Bagnols-sur-Cèze, FRANCE*; MARC-ANDRÉ GAVEAU, MICHEL MONS, *CEA Saclay, LIDYL, Gif-sur-Yvette, France.*

The rovibrational spectrum of both parallel and perpendicular bands of the SF_6 dimer near the ν_3 band of SF_6 monomer was reinvestigated[b] using high resolution jet-cooled infrared laser spectroscopy to provide deeper insight on its conformational landscape. Taking advantage of our versatile set-up[c], jet-cooled spectra were recorded by combining different geometries of supersonic expansions, SF_6 concentrations seeded in a carrier gas and axial distances. Relaxation effects could be evidenced at very low rotational temperature leading to different conformational populations. Three spectral features (noted #1, #2 and #3) belonging to three dimer conformers are unambiguously identified on the grounds of 3 distinct S-S distances derived from the rovibrational analysis of parallel band contours in the 932-935 cm^{-1} range. Symmetry assignment, a priori accessible from the perpendicular band structure of a spherical top dimer, could not be clearly proved. The dependence of such conformational infrared signatures as a function of expansion conditions provides additional information about population dependence and interconversion processes taking place between these three forms predicted to be nearly isoenergetic by theoretical calculations[d]. Based on experimental considerations, a qualitative picture of the nearly flat potential energy surface of the SF_6 dimer is proposed which could explain the dominant presence of #1 and #3 populations in fast/cold axisymmetric expansions and that of #1 and #2 populations in slow/hot planar ones.

[a] AP acknowledges COST Action CM1401 "Our Astro-Chemical History"
[b] R.-D. Urban and M. Takami, J. Chem. Phys. 103, 9132 (1995).
[c] P. Asselin, Y. Berger, T. R. Huet, R. Motiyenko, L. Margulès, R. J. Hendricks, M. R. Tarbutt, S. Tokunaga, B. Darquié, PCCP 19, 4576 (2017).
[d] T. Vazhappilly, A. Marjolin and K. J. Jordan, J. Phys. Chem. B 120, 1788 (2016).

FD06

SAMARIUM DOPED CERIUM OXIDE CLUSTERS: A STUDY ON THE MODULATION OF ELECTRONIC STRUCTURE

JOSEY E TOPOLSKI, JARED O. KAFADER, *Department of Chemistry, Indiana University, Bloomington, IN, USA*; VICMARIE MARRERO-COLON, *Chemistry, Universidad Metropolitana, San Juan, Puerto Rico*; CAROLINE CHICK JARROLD, *Department of Chemistry, Indiana University, Bloomington, IN, USA*.

Cerium oxide is known for its use in solid oxide fuel cells due to its high ionic conductivity. The doping of trivalent samarium atoms into cerium oxide is known to enhance the ionic conductivity through the generation of additional oxygen vacancies. This study probes the electronic structure of $Sm_xCe_yO_z$ ($x+y=3$, $z=2$-4) anion and neutral clusters. Anion photoelectron spectra of these mixed metal clusters exhibit additional spectral features not present in the previously studied cerium oxide clusters. Density functional theory calculations have been used to aid interpretation of collected spectra. The results of this work can be used to inform the design of materials used for solid oxide fuel cells.

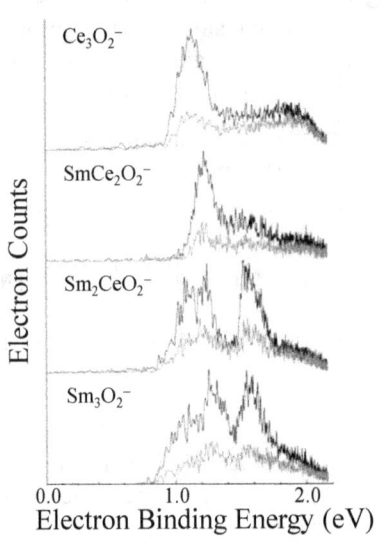

FD07

JET-COOLED INFRARED LASER SPECTROSCOPY IN THE UMBRELLA ν_2 VIBRATION REGION OF NH_3: IMPROVING THE POTENTIAL ENERGY SURFACE MODEL OF THE $NH_3 - Ar$ VAN DER WAALS COMPLEX

PIERRE ASSELIN, ATEF JABRI, *Department of Chemistry, MONARIS, CNRS, UMR 8233, Sorbonne Universités, UPMC Univ Paris 06, Paris, France*; ALEXEY POTAPOV[a], , *Laboratory Astrophysics Group of the Max Planck Institute for Astronomy at the Friedrich Schiller University, Jena, Germany*; JÉROME LOREAU, *Service de Chimie Quantique et Photophysique, Université Libre de Bruxelles, Brussels, Belgium*; AD VAN DER AVOIRD, *Institute for Molecules and Materials (IMM), Radboud University Nijmegen, Nijmegen, Netherlands*.

Taking advantage of our sensitive laser spectrometer coupled to a pulsed slit jet[b] we recorded near the ν_2 vibration a series of rovibrational transitions of the $NH_3 - Ar$ van der Waals (vdW) complex. These transitions involve in the ground vibrational state several internal rotor states corresponding to the orthoNH_3 and paraNH_3 spin modifications of the complex. They are labeled by $\Sigma_a(j,k)$, $\Sigma_s(j,k)$, $\Pi_a(j,k)$ and $\Pi_s(j,k)$ where $\Sigma(K=0)$ and $\Pi(K=1)$ indicate the projection K of the total rotational angular momentum J on the vdW axis, the superscripts s and a designate a symmetric or antisymmetric NH_3 inversion wave function, and j, k quantum numbers indicate the correlation between the internal-rotor state of the complex and the j, k rotational state of the free NH_3 monomer. Five bands have been identified, only one of which was partly observed before[c]. They include transitions starting from the $\Sigma_a(j=0$ or $j=1)$ state without any internal angular momentum, consequently they can be assigned from the band contour of a linear-molecule-like K=0, $\Delta J=1$ transition. The energies and splittings of the rovibrational levels of the $\nu_2 = 1 \leftarrow 0$ spectrum derived from the analysis of the Π_s, $\Sigma_s(j=1) \leftarrow \Sigma_a(j=0)$, k=0 bands and mostly of the Σ_s, Π_s and $\Sigma_a(j=1) \leftarrow \Sigma_a(j=1)$, k=1 bands bring relevant information about the ν_2 dependence of the $NH_3 - Ar$ interaction, the rovibrational dynamics of the $NH_3 - Ar$ complex and provide a sensitive test of a recently developed 4D potential energy surface that includes explicitly its dependence on the umbrella motion[d].

[a] AP acknowledges COST Action CM1401 "Our Astro-Chemical History"

[b] P. Asselin, Y. Berger, T. R. Huet, R. Motiyenko, L. Margulès, R. J. Hendricks, M. R. Tarbutt, S. Tokunaga, B. Darquié, PCCP 19, 4576 (2017),

[c] G. T. Fraser, A.S. Pine and W. A. Kreiner, J. Chem. Phys. 94, 7061 (1991).

[d] J. Loreau, J. Liévin, Y. Scribano and A. van der Avoird, J. Chem. Phys. 141, 224303 (2014).

FD08   *Post-Deadline Abstract*   **10:46 – 11:01**

VIBRATIONALLY EXCITED CARBON MONOXIDE PRODUCED VIA A CHEMICAL REACTION BETWEEN CARBON VAPOR AND OXYGEN

ELIJAH R JANS, *Department of Mechanical Engineering, The Ohio State University, Columbus, OH, USA*; ZAKARI ECKERT, KRAIG FREDERICKSON, BILL RICH, IGOR V. ADAMOVICH, *Department of Aerospace and Mechanical Engineering, The Ohio State University, Columbus, OH, USA.*

Measurements of the vibrational distribution function of carbon monoxide produced via a reaction between carbon vapor and molecular oxygen has shown a total population inversion on vibrational levels 4-7. Carbon vapor, produced using an arc discharge to sublimate graphite, is mixed with an argon oxygen flow. The excited carbon monoxide is vibrationally populated up to level v=14, at low temperatures, T=400-450 K, in a collision-dominated environment, 15-20 Torr, with total population inversions between v=4-7. The average vibrational energy per CO molecule formed by the reaction is 0.6-1.2 eV/molecule, which corresponds to 10-20% of the reaction enthalpy. Kinetic modeling of the flow reactor, including state specific vibrational processes, was performed to infer the vibrational distribution of the products of the reaction. The results show viability of developing of a new chemical CO laser from the reaction of carbon vapor and oxygen.

FE. Ions
Friday, June 23, 2017 – 8:30 AM
Room: 161 Noyes Laboratory

Chair: Timothy S. Zwier, Purdue University, West Lafayette, IN, USA

FE01   **8:30 – 8:45**

CHARGE OSCILLATION IN C−O STRETCHING VIBRATIONS: A COMPARISON OF CO_2^- ANION AND CARBOXYLATE FUNCTIONAL GROUPS

MICHAEL C THOMPSON, <u>J. MATHIAS WEBER</u>, *JILA and the Department of Chemistry and Biochemistry, University of Colorado-Boulder, Boulder, CO, USA.*

We compare the intensity ratio of symmetric to antisymmetric C−O stretching vibrational transitions in CO_2^- and $MCOO^-$ (M = H, Ag and Bi) using photodissociation spectroscopy. This ratio depends strongly on the bonding partner M, caused by a dynamic change in the molecular charge distribution during vibrational motion. Density functional theory calculations indicate that such charge oscillations can occur for both the symmetric and antisymmetric C−O stretching vibrations in these systems. In the symmetric C−O stretching modes, however, they are at play only if a bonding partner is present, which acts as a reservoir for charge during CO bond compression in the symmetric stretching vibration.

FE02   **8:47 – 9:02**

THRESHOLD IONIZATION SPECTROSCOPY OF $La(CH_3CN)$ AND $La(C_4H_9CN)$ RADICALS FORMED BY La REACTIONS WITH ALKANE NITRILES

<u>AHAMED ULLAH</u>, JONG HYUN KIM, WENJIN CAO, DONG-SHENG YANG, *Department of Chemistry, University of Kentucky, Lexington, KY, USA.*

La atom reactions with acetonitrile (CH_3CN) and pentanenitrile (C_4H_9CN) are carried out in a laser-vaporization supersonic molecular beam source. Metal-containing species are observed using time-of-flight mass spectrometry. In this talk, we report the mass-analyzed threshold ionization (MATI) spectroscopic characterization of two metal-containing radicals, $La(CH_3CN)$ and $La(C_4H_9CN)$, formed by La associations with acetonitrile and pentanenitrile, respectively. Adiabatic ionization energies of the two La-alkane nitrile species and their vibrational frequencies are measured from the MATI spectra. Metal-ligand binding modes and molecular structures are investigated by comparing the spectroscopic measurements with density functional theory calculations and spectral simulations. For both alkane nitriles, the preferred La binding site is identified to be the nitrile group with a π-bind mode, the resultant metal complexes are three-membered metallacycles. While a single isomer is observed for $La(CH_3CN)$, two rotational conformers are identified for $La(C_4H_9CN)$. The binding and structures of these metal-alkane nitrile radicals are different from those formed by metal ion reactions, where metal ions were reported to favor σ binding with the nitrogen atom.[a]

---

[a] K. Eller, W. Zummack, H. Schwarz, L. M. Roth, B. S. Freiser, *J. Am. Chem. Soc.*, **1991**, *113*, 833-839

FE03   **9:04 – 9:19**

MASS-ANALYZED THRESHOLD IONIZATION SPECTROSCOPY AND SPIN-ORBIT COUPLING OF CERIUM-HYDROCARBON COMPLEXES

<u>YUCHEN ZHANG</u>, SUDESH KUMARI, *Department of Chemistry, University of Kentucky, Lexington, KY, USA*; MICHAEL W SCHMIDT, MARK S GORDON, *Department of Chemistry, Iowa State University, Ames, IA, USA*; DONG-SHENG YANG, *Department of Chemistry, University of Kentucky, Lexington, KY, USA.*

$Ce(C_2H_2)$ and $Ce(C_4H_6)$ are produced by the Ce-mediated ethylene activation and investigated by mass-analyzed threshold ionization (MATI) spectroscopy, isotopic substitutions, and relativistic quantum chemical computations. The MATI spectrum of $Ce(C_2H_2)$ exhibits two nearly identical band systems separated by 128 cm^{-1}, and that of $Ce(C_4H_6)$ shows three similar band systems separated by 55 and 105 cm^{-1}. These separations are not affected by deuteration. The observed band systems for the two Ce-hydrocarbon species are attributed to the spin-orbit splitting arising from interactions of triplet and singlet states. $Ce(C_2H_2)$ is a metallacyclopropene in C_{2v} symmetry, and $Ce(C_4H_6)$ is a metallacyclopentene in C_s symmetry. The low-energy valence electron configurations of the neutral and ionic states of each species are Ce $4f^16s^1$ and Ce $4f^1$, respectively. The remaining two electrons that are associated with the isolated Ce atom or ion are spin paired in a molecular orbital that is a bonding combination between a Ce 5d orbital and a hydrocarbon π^* antibonding orbital.

FE04 9:21 – 9:36

CONFORMATION-SPECIFIC INFRARED AND ULTRAVIOLET SPECTROSCOPY OF COLD [YAPAA+H]$^+$ AND [YGPAA+H]$^+$ IONS: A STEREOCHEMICAL "TWIST" ON THE β-HAIRPIN TURN

ANDREW F DeBLASE, CHRISTOPHER P HARRILAL, JOHN T LAWLER, NICOLE L BURKE, SCOTT A McLUCKEY, TIMOTHY S. ZWIER, *Department of Chemistry, Purdue University, West Lafayette, IN, USA.*

Incorporation of the unnatural D-proline (^{D}P) stereoisomer into a polypeptide sequence is a typical strategy to encourage formation of β-hairpin loops because natural sequences are often unstructured in solution. Using conformation-specific IR and UV spectroscopy of cold (10 K) gas-phase ions, we probe the inherent conformational preferences of the ^{D}P and ^{L}P diastereomers in the protonated peptide [YAPAA+H]$^+$, where only intramolecular interactions are possible. Consistent with the solution phase studies, one of the conformers of [YADPAA+H]$^+$ is folded into a charge-stabilized β-hairpin turn. However, a second predominant conformer family containing two sequential γ-turns is also identified, with similar energetic stability. A single conformational isomer of the ^{L}P diastereomer, [YALPAA+H]$^+$, is found and assigned to a structure that is not the anticipated "mirror image" β-turn. Instead, the ^{L}P stereo center promotes a cis alanine-proline amide bond. The assigned structures contain clues that the preference of the ^{D}P diastereomer to support a trans-amide bond and the proclivity of ^{L}P for a cis-amide bond is sterically driven and can be reversed by substituting glycine for alanine in position 2, forming [YGLPAA+H]$^+$. These results provide a basis for understanding the residue-specific and stereo-specific alterations in the potential energy surface that underlie these changing preferences, providing insights to the origin of β-hairpin formation.

FE05 9:38 – 9:53

PROBING IR-INDUCED ISOMERIZATION OF A MODEL PENTAPEPTIDE IN A CRYO-COOLED ION TRAP USING IR-UV DOUBLE RESONANCE

CHRISTOPHER P HARRILAL, ANDREW F DeBLASE, JOSHUA L FISCHER, JOHN T LAWLER, SCOTT A McLUCKEY, TIMOTHY S. ZWIER, *Department of Chemistry, Purdue University, West Lafayette, IN, USA.*

In the past decade, infrared and ultraviolet spectroscopy in cryo-cooled ion traps have become workhorse techniques to characterize the gas-phase 3D structures of biological ions. Often, multiple conformers of a single molecular ion are observed. While slow collisional cooling should result in funneling of many structures into a single minimum, recent studies show evidence for the kinetic trapping of entropically-favored structures near room temperature when these species are cooled to about 10K. In order to elucidate how the initial population fractionates during the cooling process, we use a variety of conformer specific IR-UV double resonance techniques to measure population distributions of the peptide ion [YGPAA+H]$^+$ in the gas phase at 10K. Previous studies conducted in our lab show the YGPAA peptide adopts two spectroscopically distinct conformers which differ principally in the cis/trans configuration of the carboxylic acid group at the C-terminus. By using IR-UV hole filing spectroscopy (HFS) and population transfer spectroscopy (PTS) we demonstrate the ability to selectively excite and interconvert between conformations and to quantitatively measure the distribution of conformer populations within the ion trap. Experimentally, we find a 65:35 ratio for the trans:cis conformer population. These conformers are connected through a single calculated transition state, allowing intramolecular isomerization rates and equilibrium population distributions to be calculated by Rice-Ramsperger-Kassel-Marcus (RRKM) theory. The relationship between the observed population ratio and the temperature-dependent equilibrium constant will be discussed.

FE06

CHARACTERIZING PEPTIDE β-HAIRPIN LOOPS VIA COLD ION SPECTROSCOPY OF MODEL COMPOUNDS

JOHN T LAWLER, ANDREW F DeBLASE, CHRISTOPHER P HARRILAL, *Department of Chemistry, Purdue University, West Lafayette, IN, USA*; JOSHUA L FISCHER, *Chemistry, Purdue University, West Lafayette, IN, USA*; SCOTT A McLUCKEY, TIMOTHY S. ZWIER, *Department of Chemistry, Purdue University, West Lafayette, IN, USA.*

The introduction of non-native D-amino acids into peptides is known to reduce conformational entropy in peptides. D-proline has been shown to promote the formation of β-hairpin loops when paired with Gly, providing a framework for building these loops with different lengths of anti-parallel beta-sheet. This study seeks to characterize and compare the conformational preferences of a model protonated pentapeptide containing DPG, $[YAP^D GA+H]^+$, with its L-Pro counterpart via conformation specific cold ion spectroscopy as a foundation for future consideration of larger beta-hairpin models.

The UV spectrum of $YAP^D GA$ of the Tyr chromophore is beautifully sharp, but contains a complicated set of transitions that could arise from the presence of more than one conformer. To assess this possibility, we recorded non-conformation specific IR "gain" spectra in the hydride stretch region. The IR spectrum so obtained displays a set of five strong IR transitions that bear a close resemblance to those found in one of the conformers of its close analog, $[YAP^D AA+H]^+$, signaling that a single conformer dominates the population. Two transitions at 3392 and 3464 cm-1 are slightly shifted versions of the C10 and C14 hydrogen bonds found in one of the conformers of $[YAP^D AA+H]^+$, and are characteristic of formation of a β-hairpin loop. Notably, in $[YAP^D GA+H]^+$, there is at most a minor second conformer with a free carboxylic acid OH, appearing weakly in the IR "gain" spectrum. As expected, the UV spectrum of $YAP^L GA$ is more congested, which suggests the presence of multiple conformers. Further investigation into this peptide will reveal the conformational preferences of the L-pro containing molecule. Preliminary data affirms that D-proline containing peptides show reduced conformational states when compared to their natural counterparts.

Intermission

FE07

FELIX SPECTROSCOPY OF LIKELY ASTRONOMICAL MOLECULAR IONS: HC_3O^+, $C_2H_3CNH^+$, and $C_2H_5CNH^+$

SVEN THORWIRTH, OSKAR ASVANY, SANDRA BRÜNKEN, PAVOL JUSKO, STEPHAN SCHLEMMER, *I. Physikalisches Institut, Universität zu Köln, Köln, Germany*; MARIE-ALINE MARTIN-DRUMEL, *CNRS, Institut des Sciences Moléculaires d'Orsay, Orsay, France*; MICHAEL C McCARTHY, *Atomic and Molecular Physics, Harvard-Smithsonian Center for Astrophysics, Cambridge, MA, USA.*

Infrared signatures of three molecular ions of relevance to the interstellar medium and planetary atmospheres have been detected at the Free Electron Laser for Infrared eXperiments, FELIX, at Radboud University (Nijmegen, The Netherlands) in combination with the 4K FELion 22-pole ion trap facility. Mid-infrared vibrational modes of protonated tricarbon monoxide, HC_3O^+, protonated vinyl cyanide, $C_2H_3CNH^+$, and protonated ethyl cyanide, $C_2H_5CNH^+$, were detected using resonant photodissociation of the respective Ne-complexes by monitoring the depletion of their cluster mass signal as a function of wavenumber. The infrared fingerprints compare very favorably with results from high-level quantum-chemical calculations performed at the CCSD(T) level of theory.

FE08

SUB-DOPPLER ROVIBRATIONAL SPECTROSCOPY OF THE H_3^+ CATION AND ISOTOPOLOGUES

CHARLES R. MARKUS, JEFFERSON E. McCOLLUM, *Department of Chemistry, University of Illinois at Urbana-Champaign, Urbana, IL, USA*; THOMAS S DIETER, *Department of Physics, University of Illinois at Urbana-Champaign, Urbana, IL, USA*; PHILIP A KOCHERIL, *Department of Chemistry, University of Illinois at Urbana-Champaign, Urbana, IL, USA*; BENJAMIN J. McCALL, *Departments of Chemistry and Astronomy, University of Illinois at Urbana-Champaign, Urbana, IL, USA.*

Molecular ions play a central role in the chemistry of the interstellar medium (ISM) and act as benchmarks for state of the art *ab initio* theory. The molecular ion H_3^+ initiates a chain of ion-neutral reactions which drives chemistry in the ISM, and observing it either directly or indirectly through its isotopologues is valuable for understanding interstellar chemistry. Improving the accuracy of laboratory measurements will assist future astronomical observations. H_3^+ is also one of a few systems whose rovibrational transitions can be predicted to spectroscopic accuracy (<1 cm^{-1}), and with careful treatment of adiabatic, nonadiabatic, and quantum electrodynamic corrections to the potential energy surface, predictions of low lying rovibrational states can rival the uncertainty of experimental measurements[a]. New experimental data will be needed to benchmark future treatment of these corrections.

Previously we have reported 26 transitions within the fundamental band of H_3^+ with MHz-level uncertainties[bcd]. With recent improvements to our overall sensitivity[e], we have expanded this survey to include additional transitions within the fundamental band and the first hot band. These new data will ultimately be used to predict ground state rovibrational energy levels through combination differences which will act as benchmarks for *ab initio* theory and predict forbidden rotational transitions of H_3^+. We will also discuss progress in measuring rovibrational transitions of the isotopologues H_2D^+ and D_2H^+, which will be used to assist in future THz astronomical observations.

[a]L. G. Diniz, J. R. Mohallem, A. Alijah, M. Pavanello, L. Adamowicz, O. L. Polyansky and J. Tennyson, *Phys. Rev. A* (2013), **88**, 032506.

[b]J. N. Hodges, A. J. Perry, P. A. Jenkins II, B. M. Siller, and B. J. McCall, *J. Chem. Phys.* (2013), **139**, 164201.

[c]A. J. Perry, J. N. Hodges, C. R. Markus, G. S. Kocheril, and B. J. McCall, *J. Mol. Spectrosc.* (2015), **317**, 71–73.

[d]A. J. Perry, C. R. Markus, J. N. Hodges, G. S. Kocheril, and B. J. McCall, *71st International Symposium on Molecular Spectroscopy* (2016), **MH03**

[e]C. R. Markus, A. J. Perry, J. N. Hodges, and B. J. McCall, *Opt. Express* (2017), **25**, 3709–3721.

FE09

SPECTROSCOPY OF THE LOW LYING STATES OF CaO$^+$

ROBERT A. VANGUNDY, MICHAEL HEAVEN, *Department of Chemistry, Emory University, Atlanta, GA, USA.*

Diatomic molecular ions that contain alkaline earth atoms are of interest for experiments involving ultra-cold molecular ions. The alkaline earth atomic cations are well suited for laser cooling as they have transitions that are analogous to those of the alkali metals. Hence, Coulomb crystals are readily formed in rf traps. Reactions of these atomic ions yield diatomic products that are sympathetically cooled to low translational temperatures by the surrounding atomic ions. In principle, spectroscopic measurements may be used to probe the internal energies of the molecular ions. However, gas phase spectroscopic data for the ions of interest are lacking. In the present study we have investigated CaO$^+$ using pulsed field ionization-zero kinetic energy photoelectron spectroscopy (PFI-ZEKE). Molecular constants for low energy vibrational levels for the ground state ($^2\Pi_{3/2}$) and two electronic states ($^2\Pi_{1/2}$ and $^2\Sigma^+$) have been determined. These measurements also provide the first accurate value for the ionization energy of CaO. Comparisons with high-level theoretical calculations will be discussed.

SPECTROSCOPY OF HIGHLY CHARGED TIN IONS FOR AN EXTREME ULTRAVIOLET LIGHT SOURCE FOR LITHOGRAPHY

FRANCESCO TORRETTI, ALEXANDER WINDBERGER, <u>WIM UBACHS</u>, RONNIE HOEKSTRA, OSCAR VERSOLATO, *(ARCNL), Advanced Research Center for NanoLithography, Amsterdam, Netherlands*; ALEXANDER RYABTSEV, *Molecular Spectroscopy, Institute of Spectroscopy, Troitsk, Moscow, Russia*; ANASTASIA BORSCHEVSKY, *Van Swinderen Institute, Universiteit Groningen, Groningen, Netherlands*; JULIAN BERENGUT, *School of Physics, University of New South Wales, Sydney, Australia*; JOSE CRESPO LOPEZ-URRUTIA, *MPI-K, Max Planck Institute fur Kernphysik, Heidelberg, Germany.*

Laser-produced tin plasmas are the prime candidates for the generation of extreme ultraviolet (EUV) light around 13.5 nm in nanolithographic applications. This light is generated primarily by atomic transitions in highly charged tin ions: Sn^{8+}-Sn^{14+}. Due to the electronic configurations of these charge states, thousands of atomic lines emit around 13.5 nm, clustered in a so-called unresolved transition array. As a result, accurate line identification becomes difficult in this regime. Nevertheless, this issue can be circumvented if one turns to the optical: with far fewer atomic states, only tens of transitions take place and the spectra can be resolved with far more ease. We have investigated optical emission lines in an electron-beam-ion-trap (EBIT), where we managed to charge-state resolve the spectra. Based on this technique and on a number of different *ab initio* techniques for calculating the level structure, the optical spectra could be assigned [1,2]. As a conclusion the assignments of EUV transitions in the literature require corrections. The EUV and optical spectra are measured simultaneously in the controlled conditions of the EBIT as well as in a droplet-based laser-produced plasma source providing information on the contribution of Sn^{q+} charge states to the EUV emission.

[1] A. Windberger, F. Torretti, A. Borschevsky, A. Ryabtsev, S. Dobrodey, H. Bekker, E. Eliav, U. Kaldor, W. Ubachs, R. Hoekstra, J.R. Crespo Lopez-Urrutia, O.O. Versolato, Analysis of the fine structure of Sn^{11+} - Sn^{14+} ions by optical spectroscopy in an electron beam ion trap, Phys. Rev. A 94, 012506 (2016).

[2] F. Torretti, A. Windberger, A. Ryabtsev, S. Dobrodey, H. Bekker, W. Ubachs, R. Hoekstra, E.V. Kahl, J.C. Berengut, J.R. Crespo Lopez-Urrutia, O.O. Versolato, Optical spectroscopy of complex open 4d-shell ions Sn^{7+} - Sn^{10+}, arXiv:1612.00747

AUTHOR INDEX

Tripathi, G. N. R. – FC04
Tsang, L. F. – WK03
Tseng, Chih-Yu – WA12
Tsukiyama, Koichi – TA02, TA03, TA06
Tsunekawa, Shozo – TI05
Tubergen, Michael – TC01
Tuttle, William Duncan – TD08, TD09, TD10, RH02, FC05, FC06
Twagirayezu, Sylvestre – WJ06

U

Ubachs, Wim – MH04, RA01, FE10
Uhlemann, Thomas – WC10
Ullah, Ahamed – FE02
Urata, Yuki – WA02
Uriarte, Iciar – TH04, WC04

V

Vaccaro, Patrick – WD02, WG01, WH05, WI09, RG05, RG09
Vaida, Veronica – MJ11, FA11
Vallejo-López, Montserrat – WC04
van Bokhoven, Jeroen A. – WK04
van der Avoird, Ad – TF08, WJ12, FD07
van der Meer, Lex – WD05
van der Poel, Aernout P.P. – MH04
van Dishoeck, Ewine – MA01
van Langevelde, Huib Jan – TF08
Van Voorhis, Troy – WB08, WI11
van Wijngaarden, Jennifer – TC04, RF08
Vander Auwera, Jean – MH03
VanGundy, Robert A. – FE09
Vasilchenko, Semyon – MJ09
Vazart, Fanny – RH08
Vealey, Zachary – WD02, WH05
Verkamp, Max A – RI06, RI08
Versolato, Oscar – FE10
Vervloet, Michel – RJ09
Vicente, Timón – WI02
Viehland, Larry A. – RH02
Vieira, Joaquin – RF02, RF03
Vigorito, Annalisa – TD01, WC03
Viquez Rojas, Claudia I – MG10, TB10
Vishnevskiy, Yury V. – TE09
Vispoel, Bastien – MJ05
Visser, Bradley – WK04
Viswanathan, K S – WC06
Vlemmings, Wouter H.T. – TF08
Vorobyev, Vasily V. – TG01
Voros, Tamas – WA10
Voss, Jonathan – WC11

Vura-Weis, Josh – WB05, RI06, RI07, RI08

W

Wagner, Albert F. – WH04
Wagner, Georg – TJ04, TJ10, WJ03
Wagner, J. Philipp – TE08
Wagner, Roman J. V. – WH06
Walker, Nick – TC10, TI01, RH11, FB04, FB11
Wallberg, Jens – MI03
Wallington, Timothy – TJ10
Walton, Jay R. – WI01
Wang, Guanjun – TE01, FA09
Wang, Kevin – RG08
Wang, Lai-Sheng – TB06
Wang, Lichang – WI05
Wang, Peng – MK06
Wang, Xiao-Gang – MI07
Wang, Yi-Jen – FD02
Wang, Zhibo – WG03
Watanabe, Naoki – TF01
Webb, Morgan – TG04
Weber, J. Mathias – WD10, WD11, FD04, FE01
Wehres, Nadine – MF03, TH07, TH08
Welch, Bradley – WI08
Weng, Xuefei – MK09
West, Channing – WG02, WG07, WG09, RG03, RG10, RG12
Westerfield, J. H. – TH06, WF08
Whalley, Laura E. – TD08, FC05, FC06
Widicus Weaver, Susanna L. – WA05, WA06, RF11
Wiedner, Martina C. – TH09
Wilkins, Olivia H. – MF11
Wilson, Andrew J. – WD01
Wilzewski, Jonas – WJ03
Windberger, Alexander – FE10
Winters, Caroline – MK11
Witsch, Daniel – WA07
Wodtke, Alec – TK03, WH06, WH07
Wong, Andy – FA03
Woon, David E. – FA08
Wright, Timothy G. – TD08, TD09, TD10, RH02, FC05, FC06
Wronkovich, Miles A. – TC07
Wu, Lu – WK06
Wullenkord, Julia – WE10
Wurrey, Charles J. – TI09, FB01
Wyllie, Ella – MH11

X

Xie, Changjian – WF04

Xu, Hong – WJ07
Xu, Li-Hong – TI04, TI06
Xu, Yunjie – TH01, TK06, WG03, WG04, RG03, RG06
Xue, Ci – MF08

Y

Yachmenev, Andrey – MH01, WG06
Yamada, Koichi MT – MK01, TA09
Yamamoto, Satoshi – TA06
Yang, Dong-Sheng – WK06, FE02, FE03
Yang, Jinyu – WJ09
Yang, Yao-Lun – TF02
Yarnall, Yukiko – WA09
Ye, Jun – MH08, WH01
Yi, Hongming – TJ02
Yoon, Leonard H. – TC06
Yoon, Young – TB11
Yoshida, Satoru – RI02
Young, Justin W. – MH07
Yousefi, Mahdi – MJ07, TJ07
Yun, Sirkhoo – WK09
Yurchenko, Sergei N. – MH01, TJ06
Yusef-Zadeh, Farhad – MF06

Z

Zak, Emil J – MG03, TG06, TJ02, WG06
Zakharenko, Olena – TI04, WA04
Zaleski, Daniel P. – WE08, WH02
Zarringhalam, Hanif – MH09
Zhang, Di – WC01
Zhang, Kaili – RI06
Zhang, Xueqiang Alex – TE11
Zhang, Yuchen – WK06, FE03
Zheng, Yanping – MK09
Zhou, Mingfei – MA02, TE01, FA09
Zhou, Qunfei – TG03
Zhou, X – TG09
Zhou, Yan – MH08, RI05
Zindel, Daniel – TD06, WG11
Zinga, Samuel – WA06
Zingsheim, Oliver – MF03, TC08, RF07
Ziurys, Lucy M. – MH06, TA04, WK11, RF04
Zobov, Nikolay Fedorovich – TJ06
Zou, Luyao – RF11
Zou, Wenli – WK03
Zwier, Timothy S. – MG10, TB09, TB10, WC01, WC02, WC07, WC08, WC09, RJ06, FE04, FE05, FE06

BrightSpec

Molecular Rotational Resonance Spectroscopy

Extend your research...

in millimeter and microwave pure rotational spectroscopy with...

- Complete, optimized instrument designs
- Comprehensive software for experiment setup, control, analysis
- Method library and sampling interfaces for solids, liquids, gases
- Spectral library with >200 measured compounds

BrightSpec One with AutoSampler

Frequency Ranges Include:	Educational	R & D	Industrial
8-18 GHz		✓	✓
18-26 GHz (K Band)	✓	✓	✓
75-110 GHz (W Band)		✓	✓
260-290 GHz		✓	✓
530-580 GHz		✓	

NEW in 2017
K-Band Discovery Series FT-MRR

Bandwidth	18 - 26 GHz
Pulsed Excitation Power	Up to 1.6 W
Duty Cycle	0 - 98%
External Synchronization	4 marker channels
Measurement Modes	Segmented Chirp, Targeted, Nutation, Double Resonance

K-Band Discovery Spectrometer

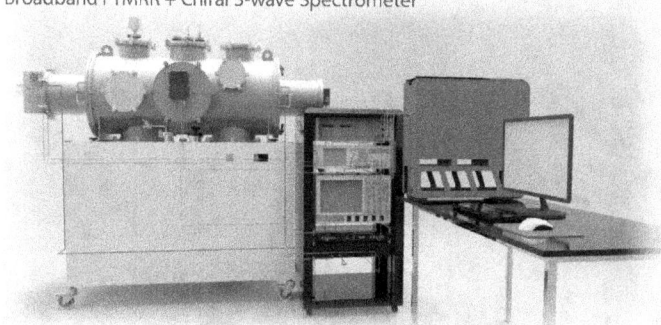
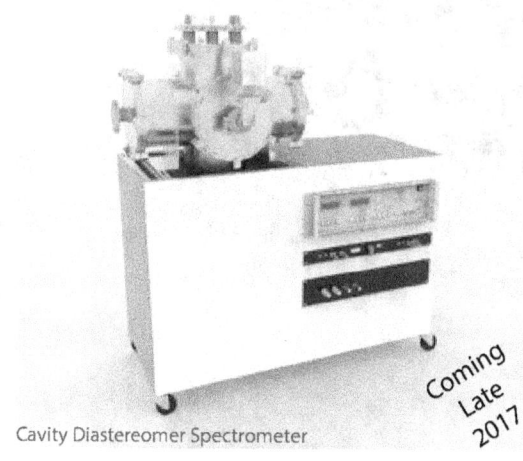
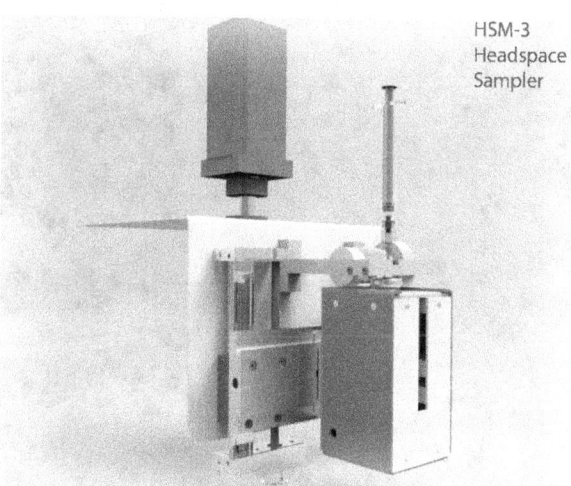

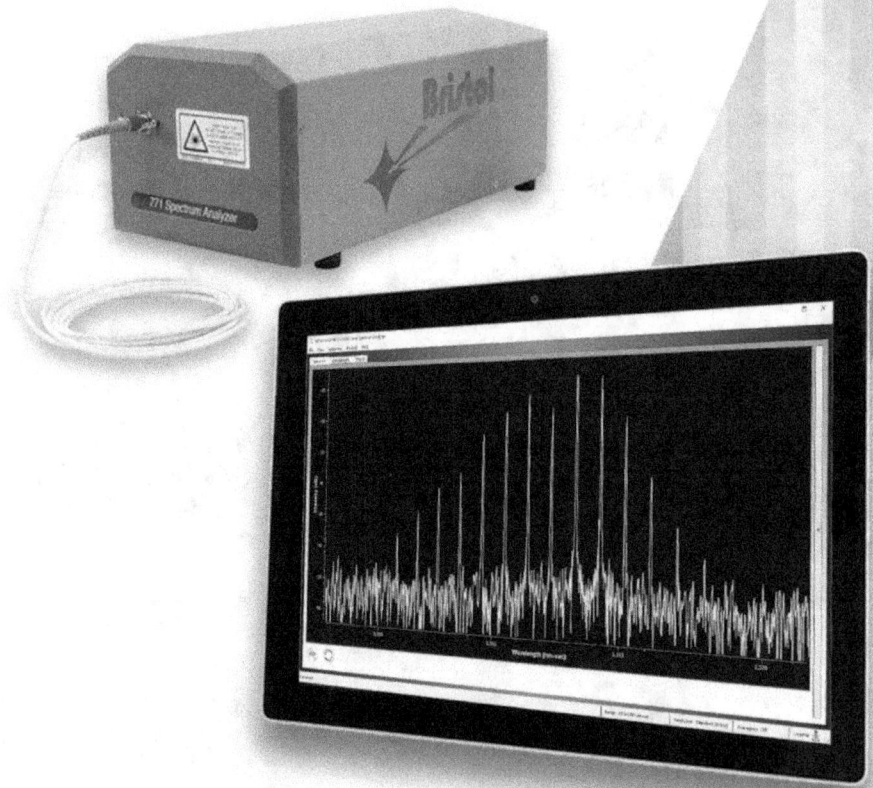

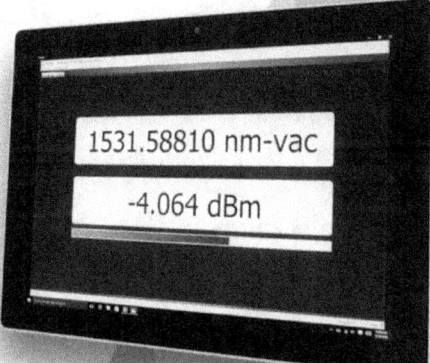
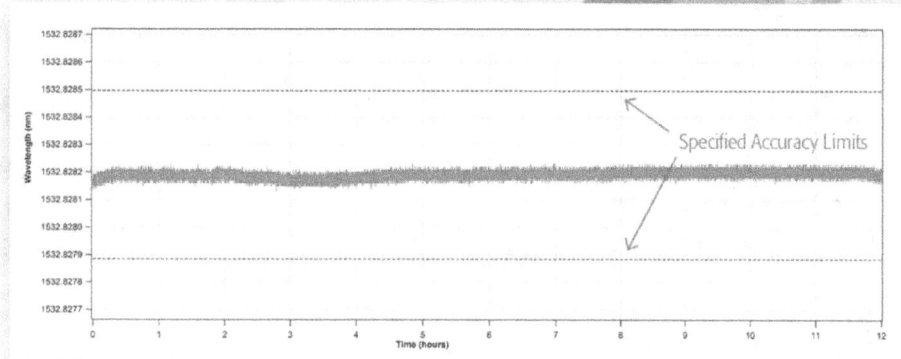

ELSEVIER

JOURNAL OF MOLECULAR SPECTROSCOPY

Recent special issues published in the Journal

NEW VISIONS OF SPECTROSCOPIC DATABASES
Edited by Nicole Jacquinet, Jean-Marie Flaud, Robert R. Gamache, Adriana Predoi-Cross & J. Vander Auwera - http://bit.ly/JMSdb1 & http://bit.ly/JMSdb2

POTENTIOLOGY AND SPECTROSCOPY IN HONOR OF ROBERT LE ROY
Edited by S. Yu, I. Gordon & P.-N. Roy - http://bit.ly/JMSRoy

MOLECULAR SPECTROSCOPY IN TRAPS
Edited by S. Schlemmer, S. Willitsch & T. Steimle - http://bit.ly/JMSTraps

SPECTROSCOPY AND INTER/INTRAMOLECULAR DYNAMICS IN HONOR OF W. CAMINATI
Edited by J.-U. Grabow, A. Lesarri & S. Melandri - http://bit.ly/JMSDyna

Special issues still open for submissions

MOLECULAR SPECTROSCOPY, ATMOSPHERIC COMPOSITION AND CLIMATE CHANGE
Edited by Vincent Boudon, Trevor Sears & Pierre-François Coheur - Call for papers: http://bit.ly/JMSClimate

Submit your manuscript: http://bit.ly/JMSSubmit

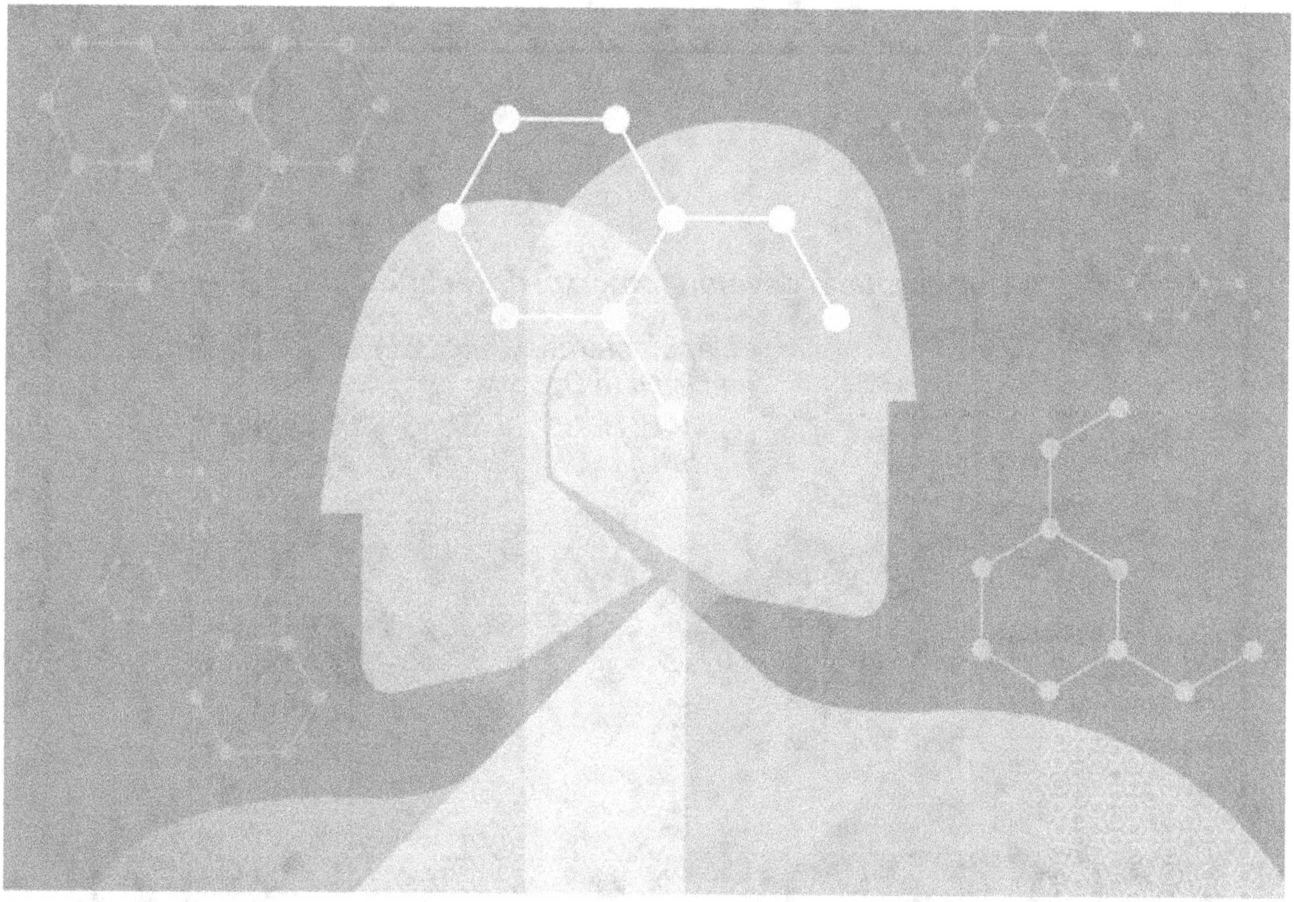

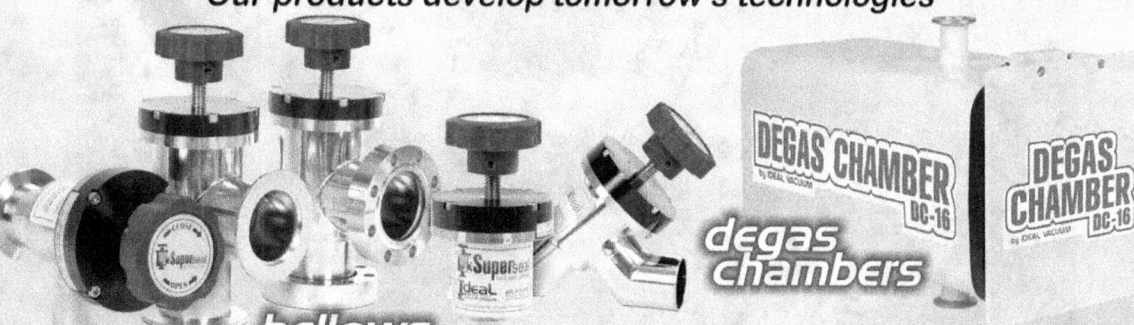

National Radio Astronomy Observatory (NRAO)

NRAO/AUI/NSF

Founded in 1956, the NRAO provides state-of-the-art radio telescope facilities for use by the international scientific community. NRAO telescopes are open to all astronomers regardless of institutional or national affiliation. Observing time on NRAO telescopes is available on a competitive basis to qualified scientists after evaluation of research proposals on the basis of scientific merit, the capability of the instruments to do the work, and the availability of the telescope during the requested time.

www.nrao.edu

NRAO is a facility of the National Science Foundation operated by Associated Universities, Inc.

North American ALMA Regional Center (NA-ARC)

The North American scientific community access to the **Atacama Large Millimeter/Submillimeter Array (ALMA)**

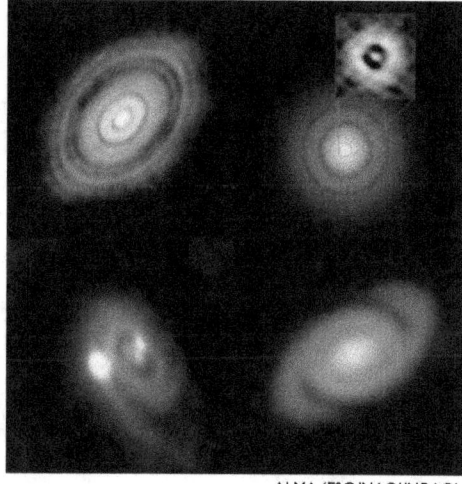

Headquartered at the **National Radio Astronomy Observatory (NRAO)** in Charlottesville, Virginia

ALMA (ESO/NAOJ/NRAO)

Outreach services to inquisitive parties interested in ALMA science:

- **Data Reduction and Expert Analysis Assistance:** Travel and lodging support during visits to NRAO from investigators of successful ALMA programs or archival researchers.

- **Data Reduction Parties:** 10-12 PIs and their students visit the NA ARC for expert training.

- **Summer Student Program:** Introducing undergraduate/graduate students to innovative research.

- **Student Observing Support:** Funds graduate students working on eligible ALMA proposals.

- **Graduate Pre-Doctoral Program:** Conduct thesis research under the supervision of an NRAO scientist.

- **ALMA Ambassador Postdoctoral Fellows Program:** Provides training and $10,000 research grant to postdoctoral researchers interested in expanding their ALMA/interferomerty expertise and sharing that knowledge with their home institutions through ALMA proposal writing workshops.

To learn more about the NA-ARC and ALMA, visit:
science.nrao.edu/facilities/alma

Quantel Welcomes You to the
International Symposium on Molecular Spectroscopy
72nd Meeting — June 19-23, 2017

Sponsor of the Women's Networking Reception —
Wednesday June 21st

Quantel
2 bis, avenue du Pacifique
Z.A. de Courtaboeuf — BP 23
91941 Les Ulis Cedex — France
33 (0)1 69 29 16 45

Quantel USA
49 Willow Peak Dr.
Bozeman, MT 59718
1-877-QUANTEL

www.quantel-laser.com

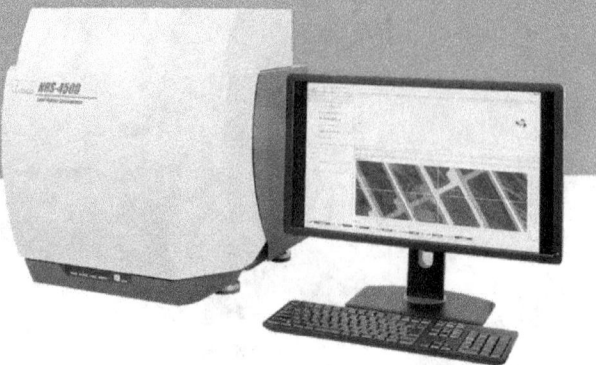
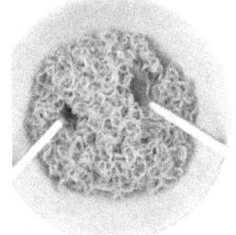

— The Coblentz Society – fostering understanding and application of vibrational spectroscopy —

Call for Coblentz Award
Nominations

The Coblentz Award is presented annually to an outstanding young molecular spectroscopist under the age of 40. This award is the Society's original award (first awarded in 1964), and is the complement of the 'Craver Award' that recognizes young spectroscopists for efforts in applied analytical vibrational spectroscopy. The candidate must be under the age of 40 on January 1 of the year of the award. The award comprises an honorarium, a plaque with a prism from the periscope of a World War II Navy submarine, and a travel allowance.

More information can be found at:
http://www.coblentz.org/awards/the-coblentz-award

Nominations for the 2018 Coblentz Award must include a detailed description of the nominee's accomplishments, a curriculum vitae or resume, and minimum of three supporting letters. Nominations for 2018 close on **July 15, 2017**. Files of candidates will be kept active for 3 years or until the age of eligibility is exceeded. Annual updates of candidate files are encouraged and will be solicited from the nomination source by the award's committee chair.

Please send nomination packages by email to nominations@coblentz.org

The Pacific Conference on Spectroscopy and Dynamics is an international forum to explore molecular spectroscopy in gas and condensed phases.

The Pacific Conference on Spectroscopy and Dynamics is one of the longest running scientific conferences in the United States. This meeting focuses on novel experimental techniques and theoretical methodologies, and their applications to emerging problems in reaction dynamics, biological systems, and environmental/materials/energy science.

In its 65th year, this 3-day, single session conference features invited and contributed talks, two poster sessions, and ample opportunities for discussion.

The 2018 Conference will be hosted at the Bahia Resort in San Diego, CA. Sunny skies, warm weather, and diverse outdoor activities make San Diego an ideal location to gather with fellow researchers in molecular spectroscopy and dynamics.

Invited Speakers (more to come!)

Thomas Allison, Stony Brook University
Christopher Bardeen, University of California at Riverside
Stephen Blanksby, Queensland University of Technology
Steven Boxer, Stanford University
Jan Cami, The University of Western Ontario
Robert Cave, Harvey Mudd College
Delphine Farmer, Colorado State University
Renee Frontiera, University of Minnesota
Ted Goodson III, University of Michigan
Eric Potma, University of California at Irvine
Garry Rumbles, National Renewal Energy Laboratory
Timothy Zwier, Purdue University

On behalf of the Executive Committee, we look forward to welcoming you to the Pacific Coast in January for great science.

Call for Abstracts will open in Fall, 2017. For more information: judyk@ucsd.edu
www.westernspectroscopy.org

ISMS MEETING VENUE INFORMATION

All contributed talks will be held in the Chemistry complex (and immediately adjoining buildings). The plenary talks will be held across the quad (about 600') in Foellinger Auditorium.

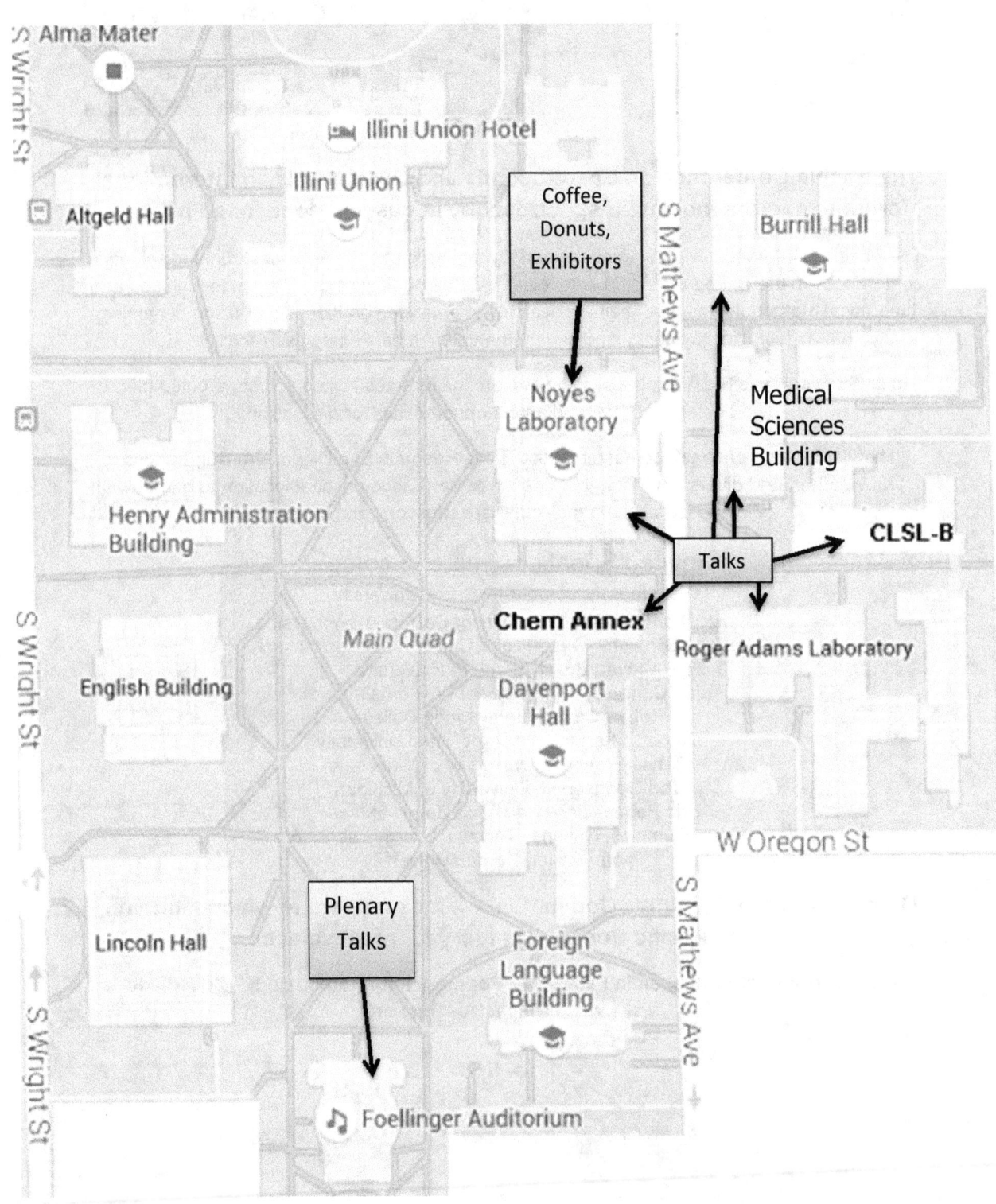

ACCESSIBLE ENTRANCES

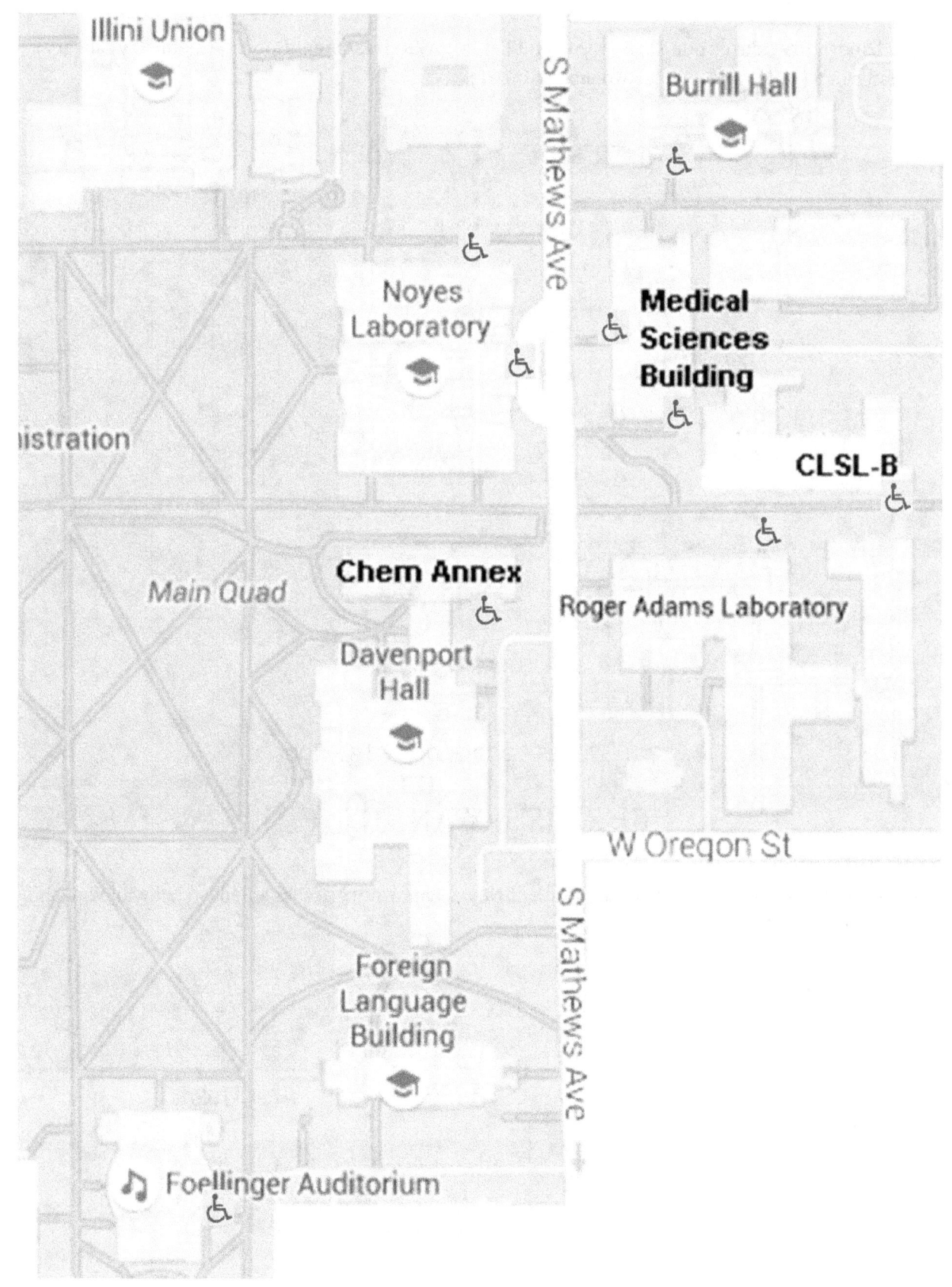

NOYES LABORATORY (NL)

Noyes Laboratory houses our Registration and Exhibitor/Refreshment Rooms (Chemistry Library), the Computer Lab (151), and one lecture hall (NL 161).

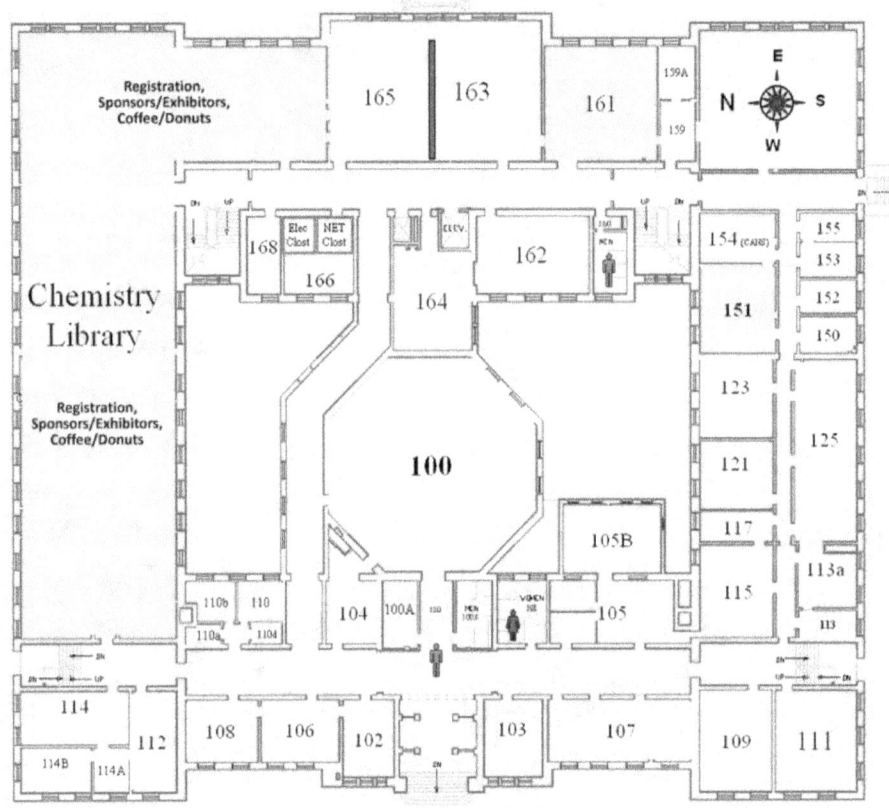

Noyes Laboratory - 1st Floor

CHEMISTRY ANNEX (CA)

Chemistry Annex is immediately to the south of Noyes Laboratory across a pedestrian walkway. It has one lecture hall (CA 1024)

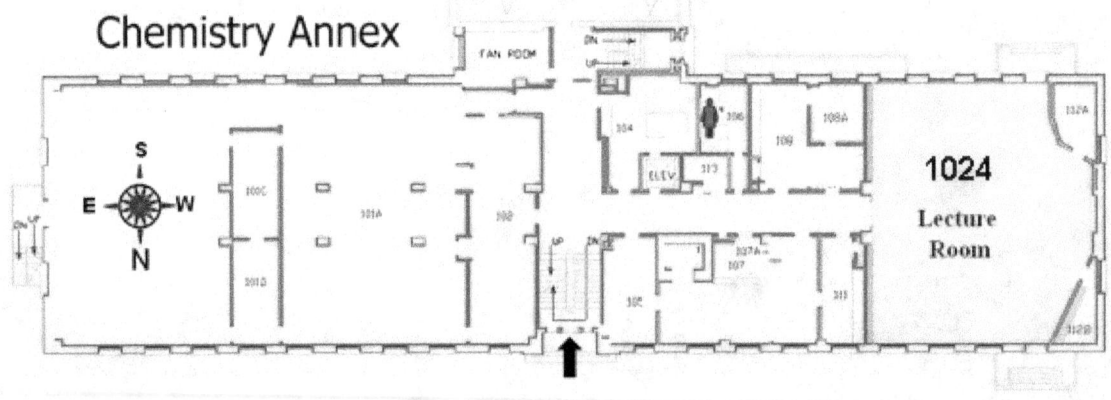

ROGER ADAMS LABORATORY (RAL)

Roger Adams Laboratory is across the street to the east of Chemistry Annex. It has one lecture hall (RAL 116). Please note that in Roger Adams Lab, the ground level is called "Ground" and the First Floor is equivalent to the Second Floor in the other buildings.

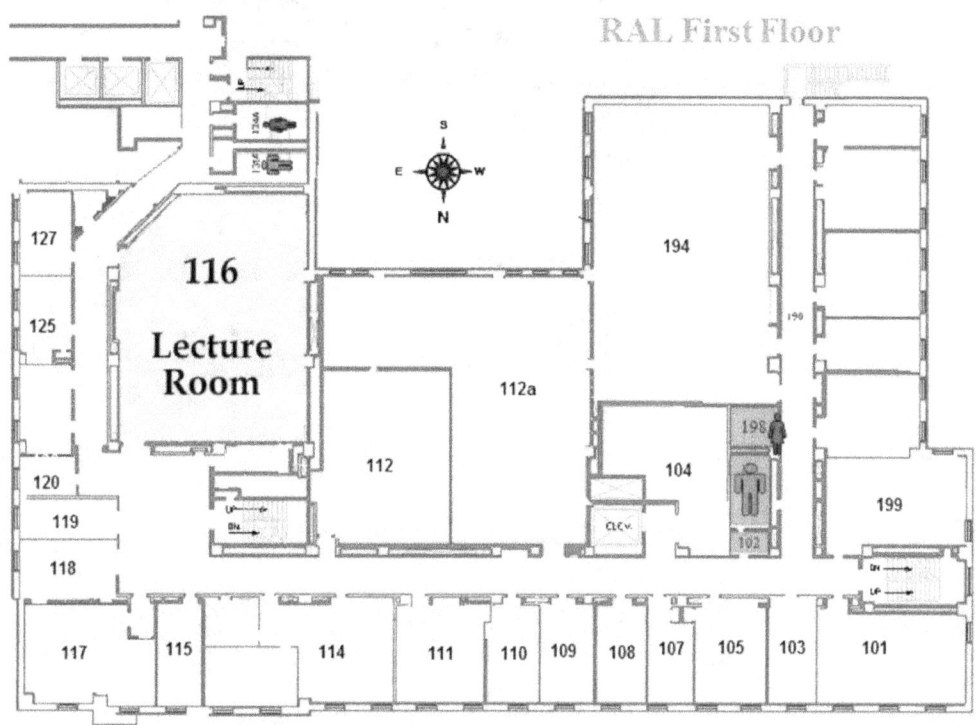

CHEMICAL AND LIFE SCIENCES (CLSL)

CLSL is a multi-wing building located across the street to the east of Noyes Laboratory. The lecture hall (CLSL B102) is in the B wing across the pedestrian walkway to the northeast of Roger Adams.

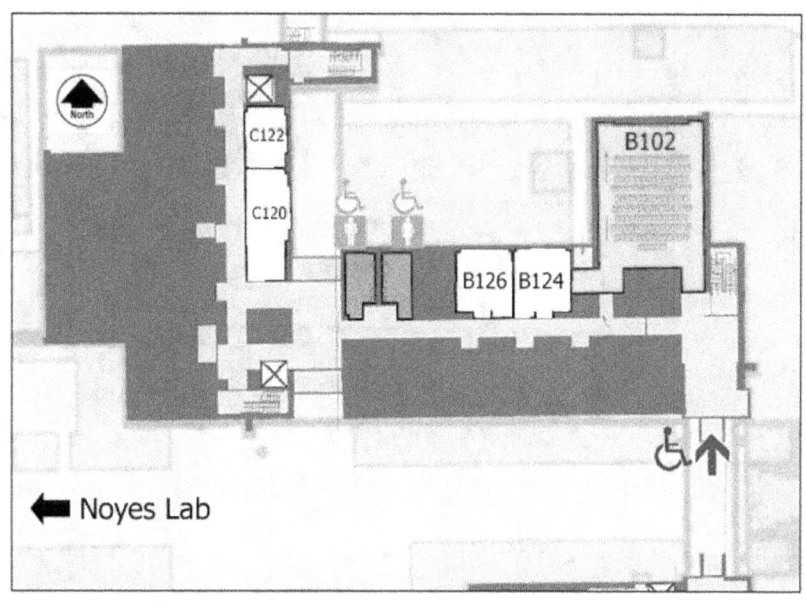

MEDICAL SCIENCES BUILDING (MSB)

Medical Sciences is across the pedestrian walkway to the north of RAL. It has one lecture hall (274).

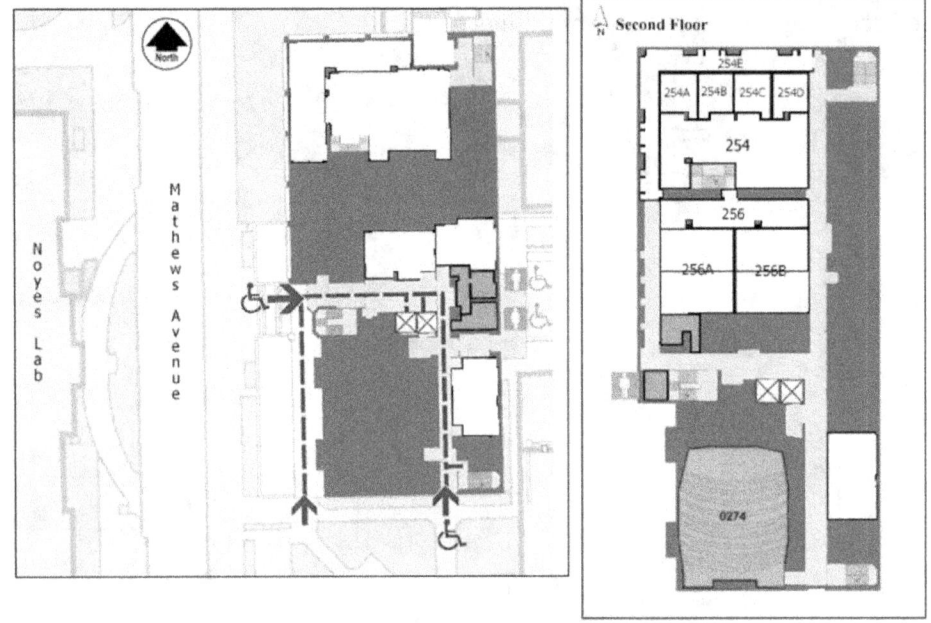

BURRILL HALL

Burrill Hall is due north of Medical Sciences. It has one lecture hall (140).

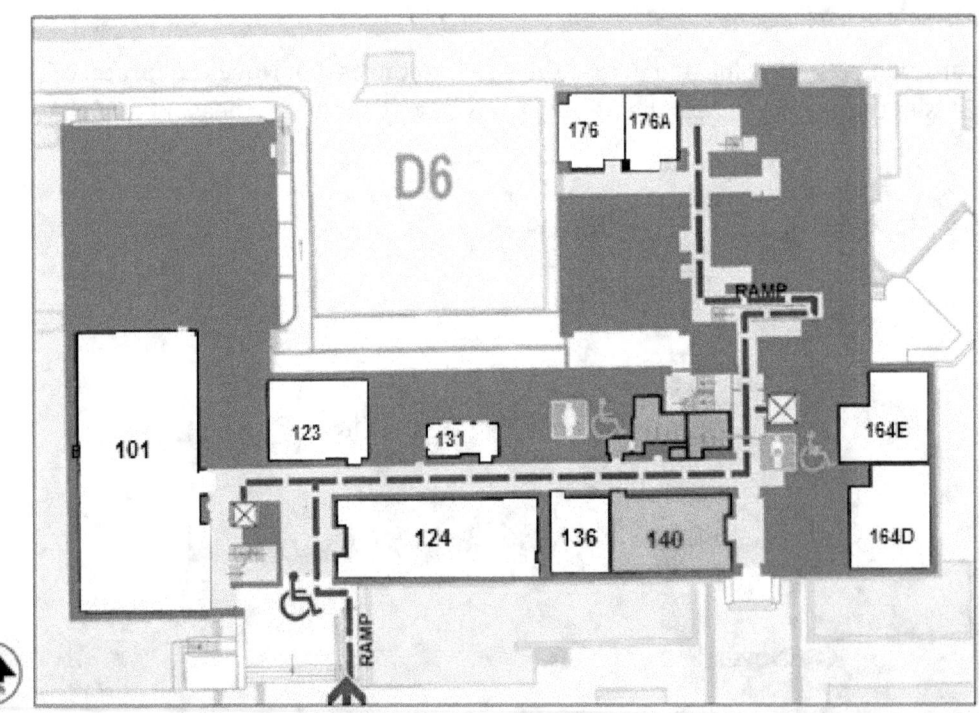

Foellinger Auditorium (Plenary and Intermission)

Foellinger Auditorium is located at the south end of the Quad. The main doors on the north (quad) side will open at 8:10 AM (the side ADA/wheelchair door will be open around 8:00 AM). There is seating on the main level and the upper balcony. There is no elevator in the building.

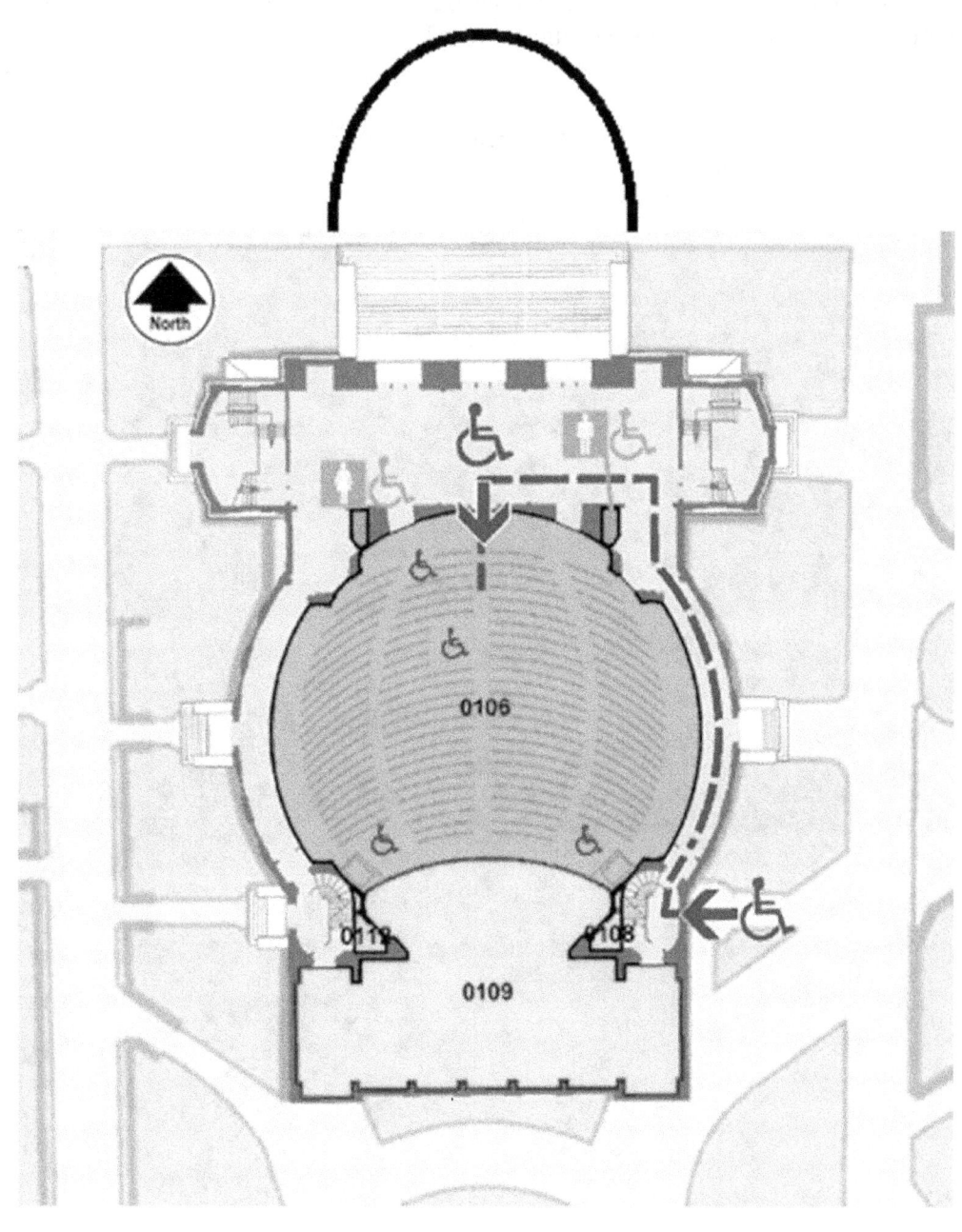

PARKING (E14) TO BOUSFIELD DORM

If you purchase a parking permit and are staying at the dorm, you will park in lot E14 (any spot). E14 is nearly due south of Bousfield Hall Dorm.

Parking enforcement begins at 6:00 AM on Monday, so you will need to have your car in lot E14 with your permit displayed before then. There are many parking meters on E. Peabody Drive (and in the lot across from Bousfield) if you wish to park closer for short periods (25 cents/15 minutes – generally between 6 AM and 6 PM, but check the meter because some go until 9 PM).

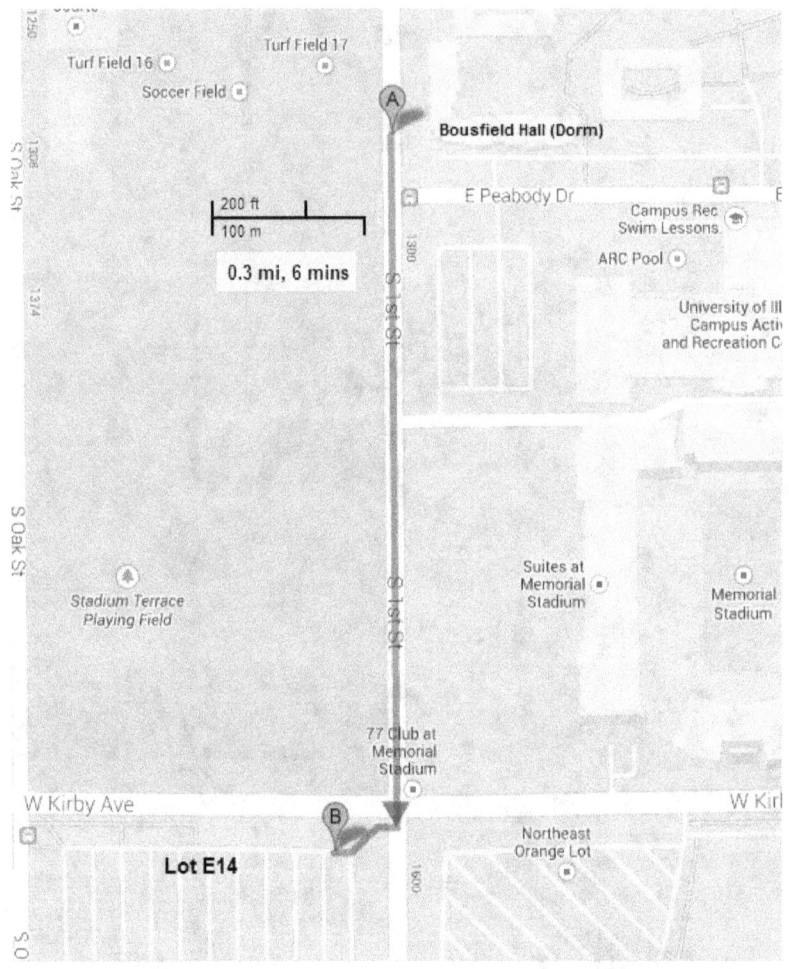

BOUSFIELD/WASSAJA DORM to MEETING VENUE (walking)

Bousfield & Wassaja Halls are just under a mile (15-20 minute walk) from the main symposium buildings

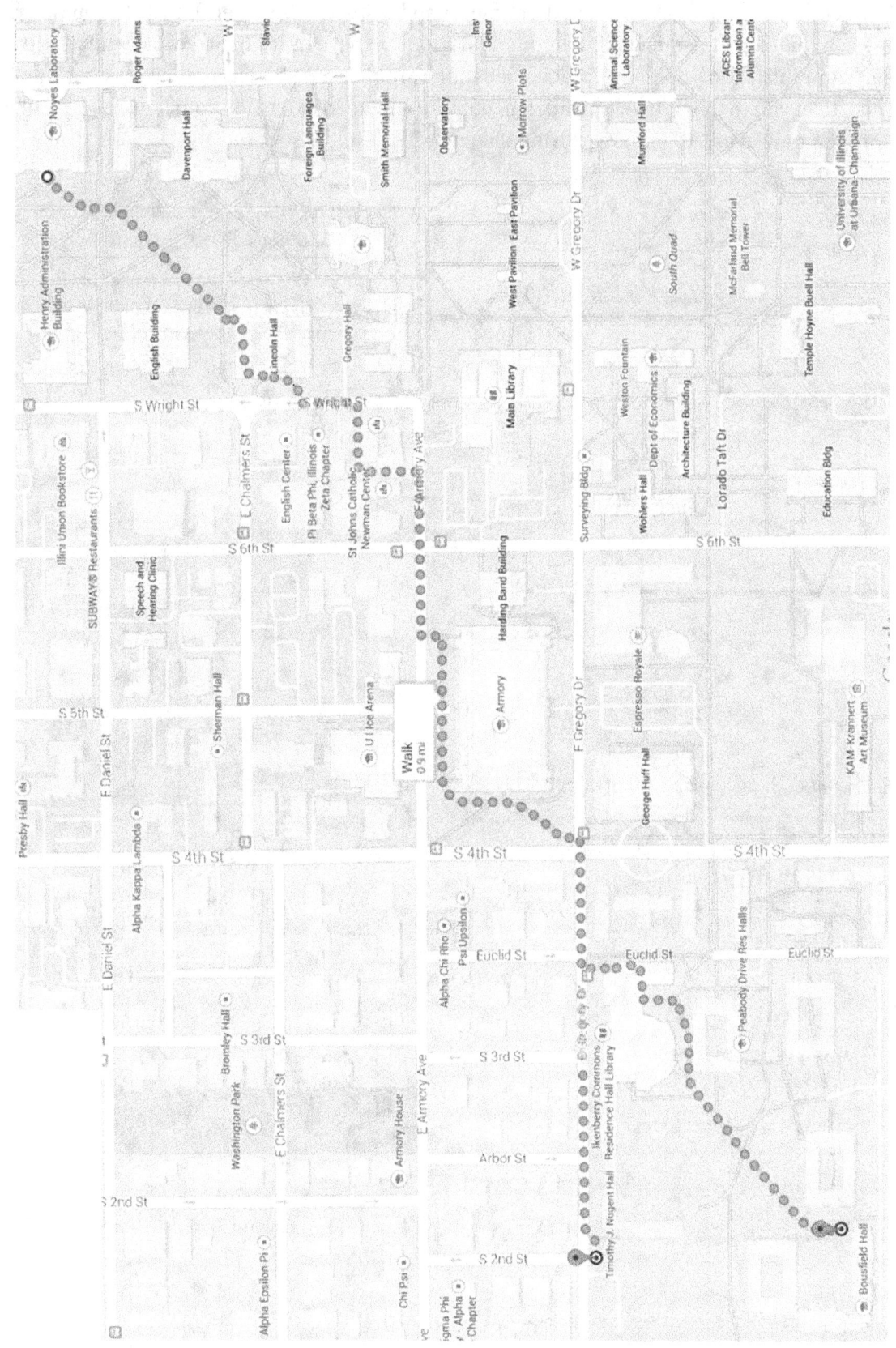

BOUSFIELD/WASSAJA DORM to MEETING VENUE (bus)

There is convenient and free bus service between Bousfield/Wassaja Dorms and 1 block from the meeting venue. The Yellow Line picks up on the corner of First and Peabody (Bousfield), and also on Gregory Drive (Wassaja) in front of Ikenberry Commons, and drops off at the Wright Street Terminal (just outside of the Henry Administration Building). Return locations are the same but across the street. The Yellow Line will also take you to downtown Champaign, but you will need to pay for your return (only iStops are free). Approximately every 10 minutes during the day.

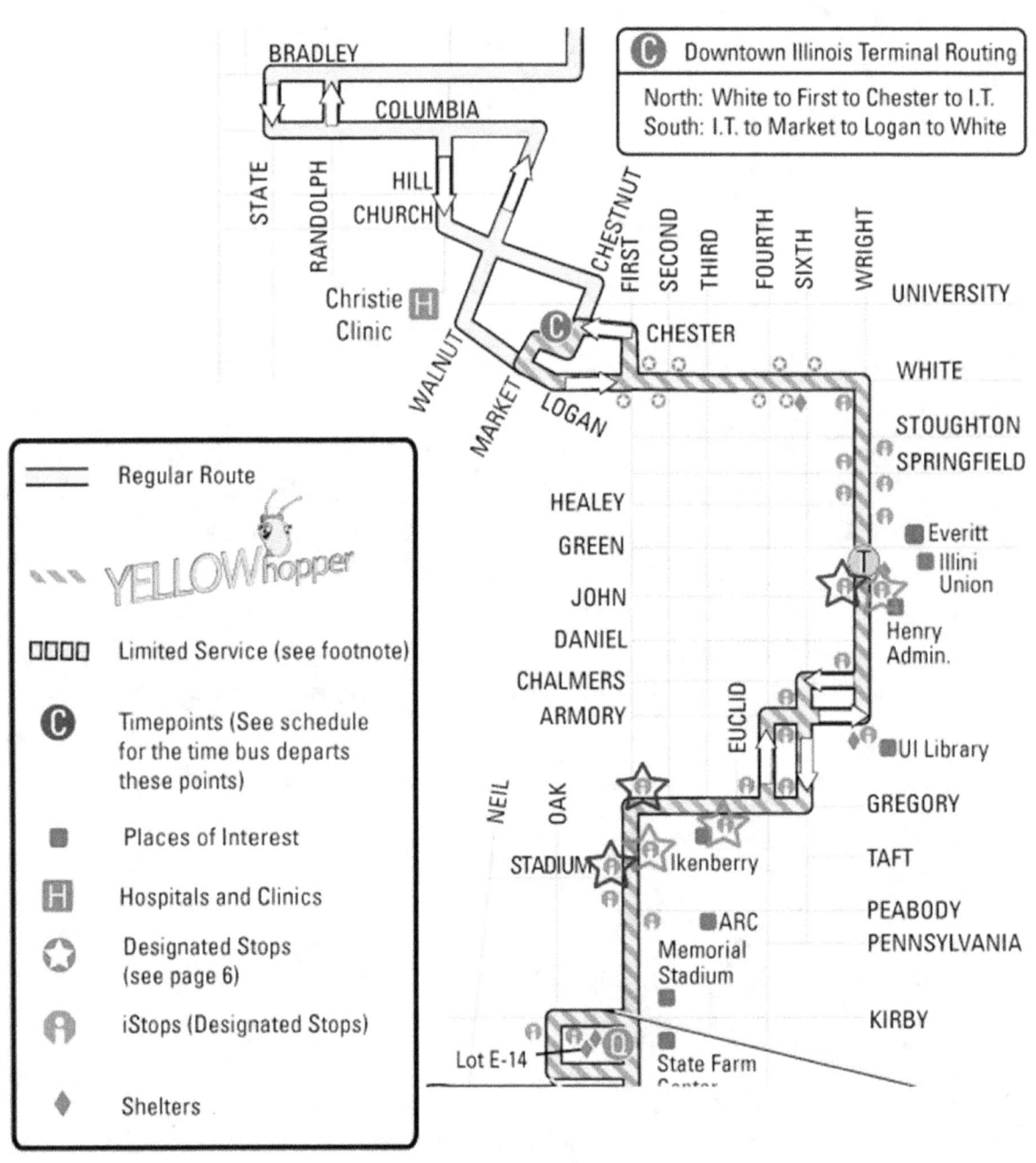

The Gold Line picks up on the corner of First and Peabody, and also on Gregory Drive in front of Ikenberry Commons and drops off at the Krannert Center (across the street from CLSL-B). Return locations are across the street. Runs every ~10 minutes during the day (offset from the Yellow Line by 5 minutes).

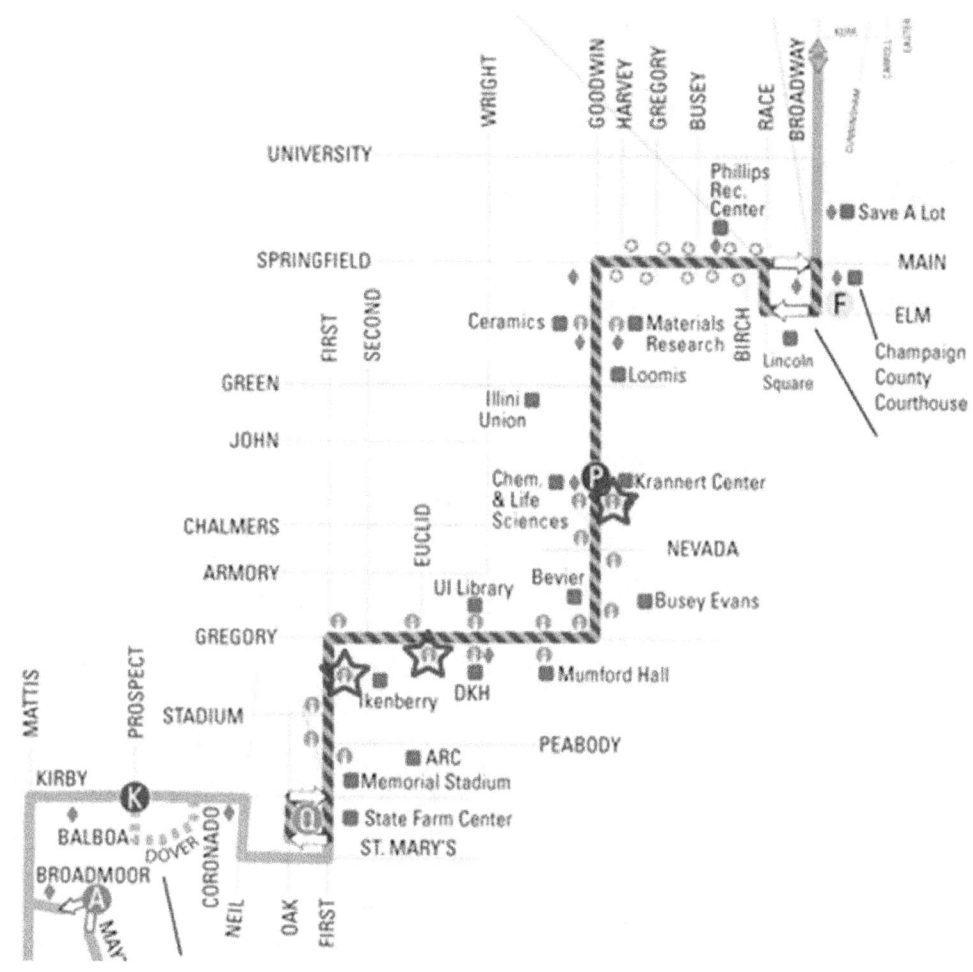

Bus Stops (Yellow Line = Left Arrow, Gold Line = Right Arrow, Foellinger Auditorium (Plenary) and Noyes Lab = Stars)

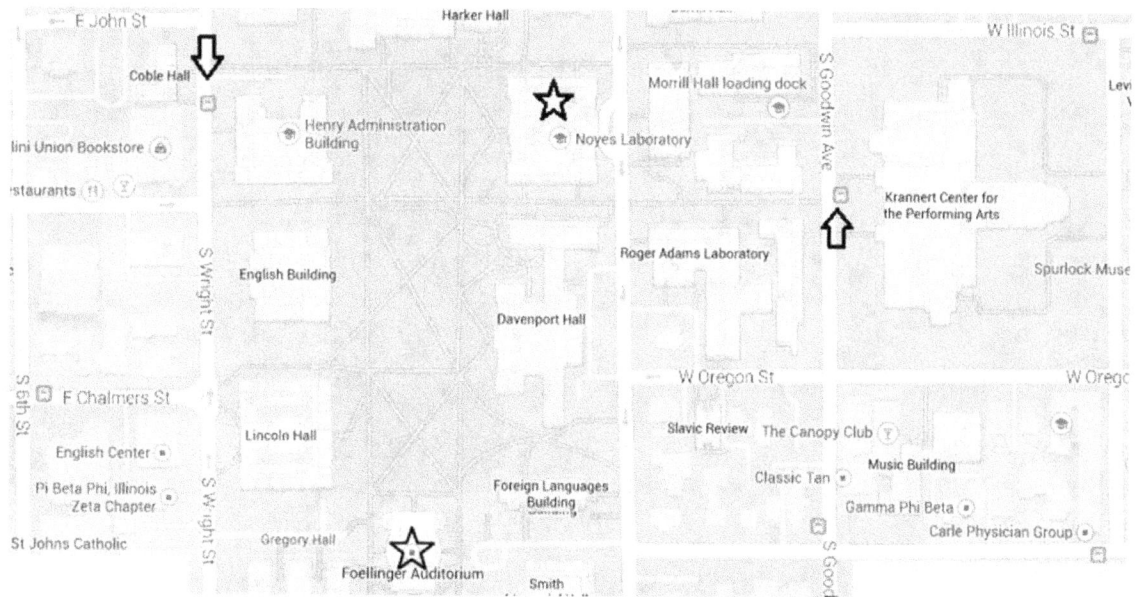

NOTES

NOTES

NOTES

NOTES

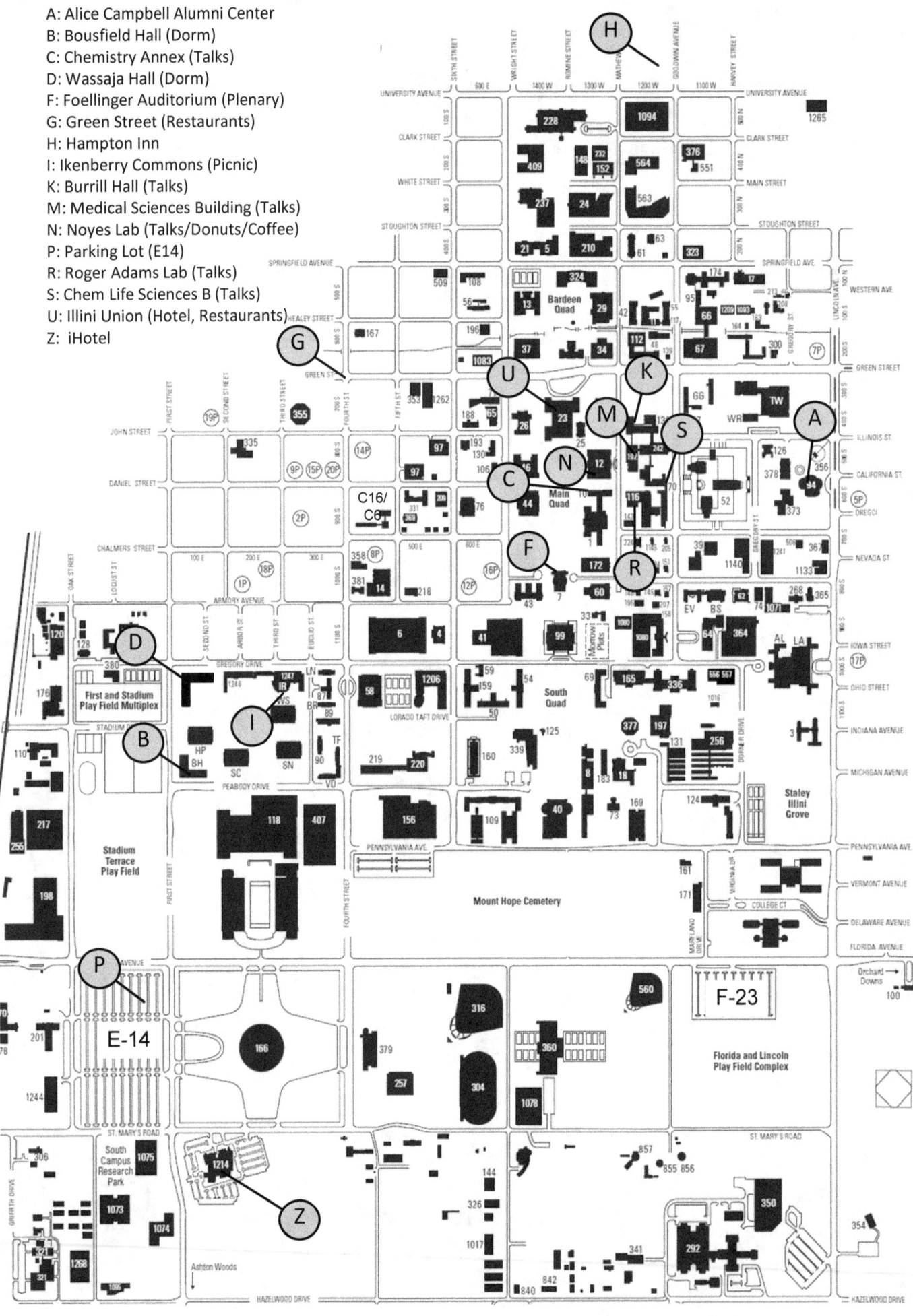

A: Alice Campbell Alumni Center
B: Bousfield Hall (Dorm)
C: Chemistry Annex (Talks)
D: Wassaja Hall (Dorm)
F: Foellinger Auditorium (Plenary)
G: Green Street (Restaurants)
H: Hampton Inn
I: Ikenberry Commons (Picnic)
K: Burrill Hall (Talks)
M: Medical Sciences Building (Talks)
N: Noyes Lab (Talks/Donuts/Coffee)
P: Parking Lot (E14)
R: Roger Adams Lab (Talks)
S: Chem Life Sciences B (Talks)
U: Illini Union (Hotel, Restaurants)
Z: iHotel